3rd International Conference on Measurement Instrumentation and Electronics (ICMIE 2018)

MATEC Web of Conferences Volume 208 (2018)

Lille, France
4 – 6 July 2018

Editors:

H. Bevrani
W. Shuhui

ISBN: 978-1-5108-7250-9

Printed from e-media with permission by:

Curran Associates, Inc.
57 Morehouse Lane
Red Hook, NY 12571

Some format issues inherent in the e-media version may also appear in this print version.

Printed by Curran Associates, Inc. (2018)

For additional information, please contact EDP Sciences – Web of Conferences at the address below.

EDP Sciences – Web of Conferences
17, Avenue du Hoggar
Parc d'Activité de Courtabœuf
BP 112
F-91944 Les Ulis Cedex A
France

Phone: +33 (0) 1 69 18 75 75
Fax: +33 (0) 1 69 28 84 91

contact-edps@webofconferences.org

Additional copies of this publication are available from:

Curran Associates, Inc.
57 Morehouse Lane
Red Hook, NY 12571 USA
Phone: 845-758-0400
Fax: 845-758-2633
Email: curran@proceedings.com
Web: www.proceedings.com

TABLE OF CONTENTS

Preface

ICMIE 2018

This issue of Proceedings gathers the papers presented at 2018 the 3rd International Conference on Measurement Instrumentation and Electronics (ICMIE 2018) held on July 4-6, 2018 in Lille, France. ICMIE was initiated in 2016, which is an international conference covering research and development in the field of measurement instrumentation, electronics and allow the participation from all over the world.

60 papers were submitted on the conference, and ICMIE finally accepted 32 papers after a double blinded peer review process by international reviewers and technical program committee members. Divided into 5 chapters, the papers provide a wide spectrum of researches on wide range of measurement instrumentation and electronics. The chapters are devoted to the framework of signal analysis and processing, electronics and communication engineering, modern electronic system & measurement and control technology, power engineering and system modeling, as well as computer science and intelligent technology. Specific research results by conference participants were presented and examined in the light of the frameworks outlined above, which is of interest to academics, researchers and professionals in this field.

Three keynote speeches were presented from Prof. Dennis Leung (Fellow of the IMechE and Energy Institute), The University of Hong Kong, Hong Kong; Prof. Bruno François, The Ecole Centrale de Lille, France; Prof. Hassan Bevrani, University of Kurdistan, Iran separately. All the talks were very impressive for the high level of professionalism, and in many cases original ideas and activities have been accomplished or proposed.

The credit for the success of the conference is to be shared with many colleagues. First and foremost, the advisory chair, conference chairs, program chairs, technical program committee members gave precious inputs and were always side by side with the organizers. We are also indebted to session chairs, international reviewers, conference secretariat who dedicated to make the conference run smoothly and properly, and ensure the proceedings quality. Last but not the least, we should express our thanks to all delegates, who showing the high level of international interest in the subject. It is exactly your participation that make the conference to its success.

The Proceedings provide the permanent record of what were presented. It indicated the state of development at the time of writing of all aspects of this important topic and will be invaluable to all researchers in the field for that reason. We truly believe the participants will find the discussion fruitful, and we hope you enjoy and find your engagement with their ideas valuable in sustaining your own professional development in the field of measurement instrumentation and electronics.

Yours Sincerely
Conference Organizing Committee
June 18, 2018

Statement of Peer review

In submitting conference proceedings to *Web of Conferences*, the editors of the proceedings certify to the Publisher that

1. They adhere to its **Policy on Publishing Integrity** in order to safeguard good scientific practice in publishing.
2. All articles have been subjected to peer review administered by the proceedings editors.
3. Reviews have been conducted by expert referees, who have been requested to provide unbiased and constructive comments aimed, whenever possible, at improving the work.
4. Proceedings editors have taken all reasonable steps to ensure the quality of the materials they publish and their decision to accept or reject a paper for publication has been based only on the merits of the work and the relevance to the journal.

Title, date and place of the conference

2018 3rd International Conference on Measurement Instrumentation and Electronics (ICMIE)
Lille, France
July 4-6, 2018

Proceedings editor(s):

Prof. Hassan Bevrani, University of Kurdistan, Iran

Prof. Wang Shuhui, Tongji University, China

Date and editor's signature

MATEC Web of Conferences **208**, 01001 (2018) https://doi.org/10.1051/matecconf/201820801001
ICMIE 2018

Doppler shift extraction of wideband signal using spectrum scaling matching

ZHANG Fengzhen, *LI* Guijuan, *ZHANG* Zhaohui and *HU* Chen

Dalian Scientific Test and Control Technology Institute, Science and Technology on Underwater Test and Control Laboratory, Dalian City, China

Abstract. Doppler shift is an important feature of moving targets. It can be used to extract target velocity, distance, track and other movement parameters. According to the problem of extracting Doppler shift for wideband signals with unstable line spectrum or no line spectrum, we proposed a Doppler shift extraction method for wideband signals based on spectral scaling matching. Firstly, a spectrum reference matrix corresponding to different relative Doppler shift is generated. Then, the matching degree of Doppler signal spectrum and reference matrix is measured by linear correlation coefficient. Finally, the Doppler shift of wideband signals is extracted through matching degree optimization. Simulation results show that the proposed method can extract the Doppler shift characteristics of wideband signals effectively.

1 Introduction

Doppler shift refers to the frequency difference between the target transmitting signal and the receiving signal of the observation point caused by the Doppler effect. Doppler shift involves target parameters about velocity, distance, trajectory, etc. It has an important application in many engineering problems. The line spectral Doppler shift of target radiated noise was used to estimate the distance and velocity parameters of underwater targets in [1, 2, 3]. Passive location was achieved by using the Doppler frequency shift characteristic of the aircraft's noise signal in [4, 5].

The key and prerequisite for estimation of target movement parameters were accurate and effectiveness of Doppler shift extraction. There are many methods for extracting Doppler frequency shift from line spectrum signal. Short Time Fourier Transform (STFT) can be directly used to extract Doppler shift by tracking the frequency variation of a line spectrum. There are other improved methods, such as a combined method of STFT and high resolution frequency estimation based on subspace rotation invariance technology, to improve the accuracy and resolution of frequency estimation [6].

However, for wideband signals with unstable line spectrum or no line spectrum, the methods based on line spectrum detection are ineffective to extract the Doppler shift. This causes that the above methods [1,2,3,4] can't be applied to estimate the movement parameters of wideband signal target.

According to the above problem, the paper proposed a Doppler shift extraction method for wideband signals. This method takes advantage of the stretching effect principle of signal spectrum with Doppler shift. A reference matrix corresponding to a variety of relative Doppler shift coefficients is established. Then the matching degree of Doppler frequency shift spectrum and reference matrix is measured by linear correlation coefficient to extract the Doppler frequency shift feature. The method solves the problem of extracting Doppler shift from wideband signals with unstable line spectrum or no line spectrum. So the target scope will be expanded about the movement parameters estimation based on Doppler shift.

2 Principle and method

The Doppler shift will cause the expansion of the signal spectrum. Similarly, the spectral sequences obtained from different Discrete Fourier Transform (DFT) point numbers correspond to different frequency resolutions. The Doppler shift of wideband signals can be extracted by establishing the relationship between the two.

2.1 Signal spectrum expansion caused by Doppler shift

According to the Doppler effect, the moving target signal received by the observation point will produce Doppler frequency shift. For frequency component f_i of the target signal, the frequency offset is $\Delta f_i(t)$:

$$\Delta f_i(t) = \frac{v_r(t)}{c} f_i \tag{1}$$

Where, $v_r(t)$ denotes the real-time velocity of the target relative to the observation point, c denotes the signal propagation velocity in medium. For broadband signals, the Doppler shift of different frequency components is proportional to its frequency. The Doppler frequency shift is large for the high-frequency component and small for the low frequency component.

The ratio of the frequency shift to the corresponding frequency can be defined as the relative Doppler coefficient without considering Doppler frequency shift at the specific frequency of the signal. It can be expressed as $a(t)$:

$$a(t)=\frac{\Delta f_i(t)}{f_i}=\frac{v_r(t)}{c} \tag{2}$$

The $a(t)$ is the ratio between the velocity $v_r(t)$ of the target relative to the observation point and the signal propagation velocity c in the medium actually. It reflects the frequency shift ratio of broadband signal at any frequency. For $a>0$, the target is moving closer to the observation point, and the signal frequency is shifted up. Conversely, $a<0$ indicates that the target is away from the observation point, and the signal frequency is shifted downward.

Fig. 1 is the spectrum of a wideband signal after Doppler shift with a relative Doppler shift coefficient $a=0.02$. It is shown that the spectrum of Doppler shift signal has a significant extension effect on the structure compared with the original.

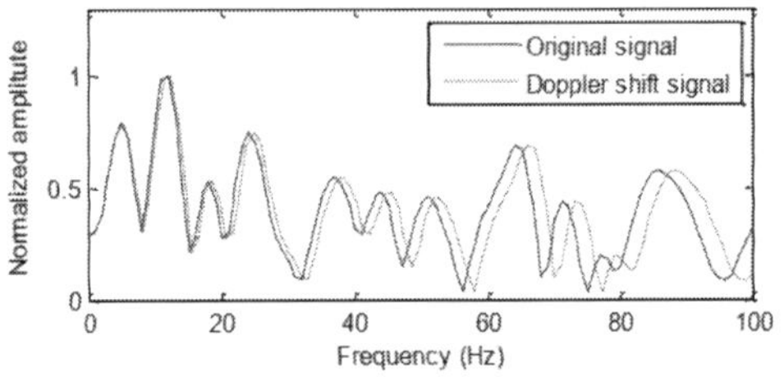

Fig 1. Spectrum expansion of Doppler shift signal.

2.2 Spectrum expansion caused by different DFT points

The DFT series of time domain signals $s(n)$ is defined [7] as:

$$S(k)=\sum_{m=1}^{N-1}s(n)e^{-j\frac{2\pi}{N}km},\ (m=0,1,...,N-1) \tag{3}$$

For the time series of signal with sampling rate F_s, the frequency resolution of M point DFT spectrum sequence is F_s/M. So, the position of line spectrum f_i in the frequency sequence is Mf_i/F_s.

Therefore, for discrete time series, there is a kind of "expand or contract" relationship between the frequency sequences with different DFT points. By increasing the point of DFT, the frequency spectrum of the signal has a kind of stretching effect compared with the original. The frequency spectrum of the signal has a kind of compression effect compared with the original spectrum with DFT points reduced.

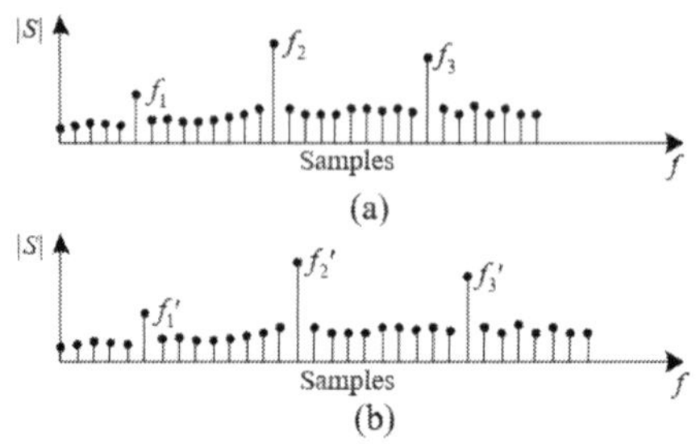

Fig 2. Spectrum expansion diagram of different DFT points.

Assume the spectrum sequence of signal is shown in Fig. 2(a). When the sequence is extended to the situation shown in Fig. 2(b), its frequency increases with directly proportional to its frequency. Without loss of generality, the frequency variation f_i' of spectral line f_i is as follows:

$$f_i'=f_i+\Delta f_i=f_i(1+\frac{\Delta f_i}{f_i}) \tag{4}$$

Therefore, in order to obtain the frequency expansion of Fig. 2(b) relative to Fig. 2(a), the method of reducing the DFT points can be used to compress the extended spectrum, and match with the original spectrum sequence.

It's easy to know, when the coordinate of the extended spectrum line f_i' is compressed to the frequency f_i, the best match between the spectrum and the original spectrum can be made. In this case,

$$\frac{Nf_i}{F_s}=\frac{(N+\Delta N)f_i'}{F_s} \tag{5}$$

Combined (4) and (5), it can be obtained:

$$\frac{-\Delta N}{N+\Delta N}=\frac{\Delta f_i}{f_i} \tag{6}$$

Therefore, if the appropriate number of DFT points is chosen to match the signal spectrum scalability of the signal frequency shift, the signal Doppler shift coefficient can be obtained by matching the signal spectrum expansion of the signal frequency shift.

$$a=-\frac{\Delta N}{N+\Delta N} \tag{7}$$

For the signal with the same sampling rate F_s and signal analysis window length N, the resolution r_1 of Doppler shift coefficient of the method is

$$r_1\le\frac{1}{N+1} \tag{8}$$

When using the conventional line spectrum analysis and tracking method to extract the Doppler shift coefficient, the accuracy is related to the frequency of the line spectrum. The resolution is the best when the line spectrum is located at the upper limit $F_s/2$ of effective frequency. Therefore, the Doppler frequency shift coefficient extracting resolution r_2 of the method is

$$r_2 \geq \frac{F_s}{N} \Big/ \frac{F_s}{2} = \frac{2}{N} \tag{9}$$

Consequently,

$$r_2 > 2r_1 \tag{10}$$

It can be seen that the resolution of the Doppler shift coefficient extracted by the proposed method is at least 2 times higher than that by the traditional line spectrum analysis and tracking method.

2.3 Extraction method and processes

According to the above principle of Doppler shift coefficient extraction, the steps of extracting the Doppler frequency shift characteristics of a portion of signal $s(n)$ are as follows:

(1) Firstly, select a segment of data in the signal as a reference signal. Fast Fourier transform (FFT) with $N, N\pm1, N\pm2, \ldots N\pm\Delta N$ points are conducted to the reference signal respectively. And the DFT series are combined into the spectrum expansion reference matrix $\mathbf{F_0}$ in which different spectrum sequences have different scaling and corresponding different Doppler frequency shift coefficients. The selection of ΔN should ensure that the maximum Doppler shift of the signal can be analyzed. And the reference signal segment usually selects the front of the signal, especially when it is used in real time.

(2) Divide the signal into segments, process them by N point FFT and obtained the spectrum sequence F. Then, use linear correlation coefficient to measure the matching degree between F and each spectral sequence F_{0i} of the spectrum scaling reference matrix $\mathbf{F_0}$.

$$R_i = \frac{\mathrm{cov}(F, F_{0i})}{\sqrt{\mathrm{var}(F)\cdot \mathrm{var}(F_{0i})}} \tag{11}$$

Where, $\mathrm{cov}(F, F_{0i})$ is the covariance of F and F_{0i}, $\mathrm{var}(F)$ and $\mathrm{var}(F_{0i})$ are the variance of F and F_{0i}, respectively.

(3) Find the FFT points M corresponding to the maximum value of the linear correlation coefficient sequence R. Then, the Doppler shift coefficient of the signal relative to the reference signal is

$$a = \frac{N}{M} - 1 \tag{12}$$

According to the above steps, the Doppler frequency shift coefficients of signals can be extracted by processing the segmented signals successively.

3 Simulation signal processing

The parameters of the signal simulation are set as follows. Suppose that in a two-dimensional coordinate system, a broadband signal target moves at a speed of v along the straight line parallel to the x-axis. The nearest vertical distance between the observation point R and the sound source trajectory is d. The propagation velocity of the signal in the medium is c. The location of the target and observation point is shown in Fig. 3.

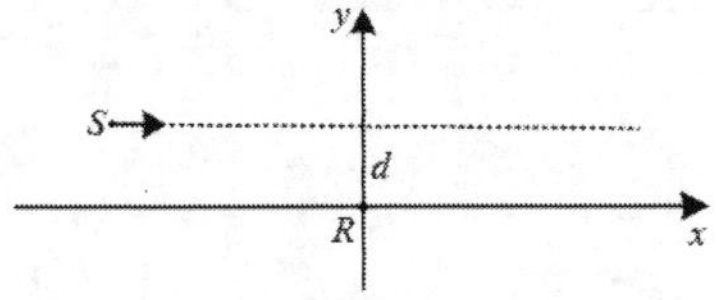

Fig 3. Diagram of the moving target and the observation position.

Suppose that the broadband signal source radiate steady broadband random noise into the medium. The noise signal is obtained by bandpass filtering of white Gauss noise. The frequency band of the signal is 1kHz~3kHz. The sampling frequency is 12kHz. The spectrum of the simulation signal is shown in Fig. 4. There is no line spectrum component in the signal.

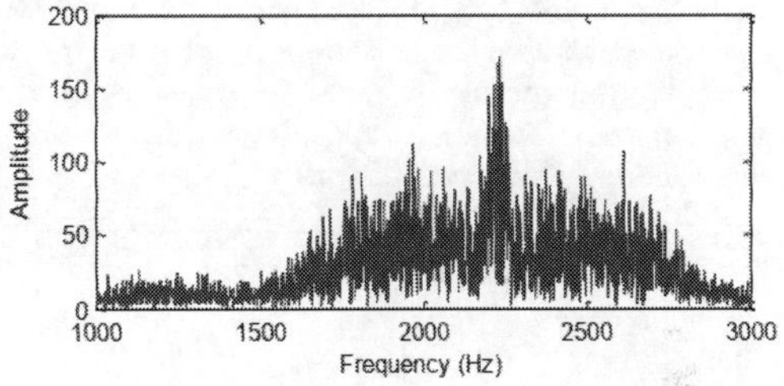

Fig 4. Signal spectrum diagram.

The signal received by the observation point R is the Doppler shifted signal. In the 20s time before and after the closest source location, the time-frequency spectrum of the simulation signal is shown in Fig. 5. It can be seen that it has obvious spectrum scaling effect at different time. The signal spectrum moves up when the source is closing to the observation point. And the signal spectrum moves down when the sound source away from the observation point.

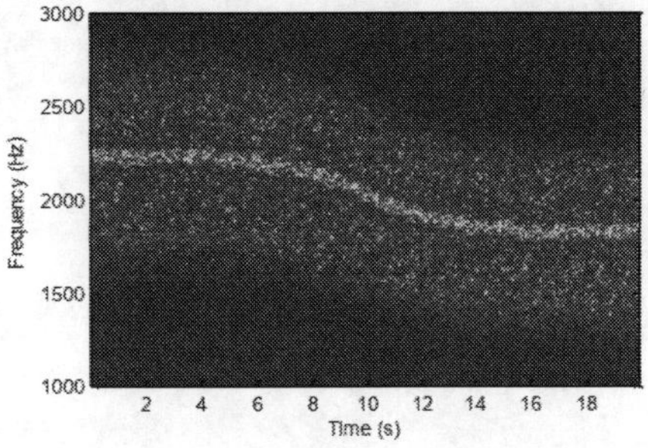

Fig5. Time-frequency spectrum of the signal.

According to the method proposed in this paper, the wideband signal with Doppler shift is processed. Firstly, the reference matrix should be set up, and then the scaling and matching of the frequency spectrum of the segmented signal and the spectrum of the reference matrix should be measured to calculate the Doppler shift coefficient of the signals of each segment.

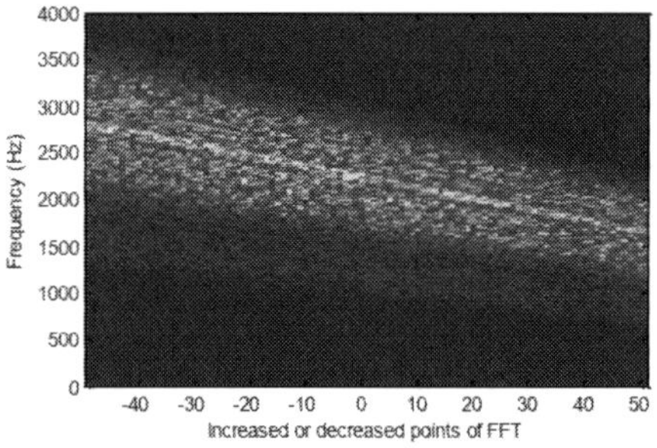

Fig 6. Image of spectrum matching reference matrix.

For band limited signals, when constructing reference matrix, only the spectrum sequence values in the effective frequency band should be reserved, but enough margin should be left out. The simulation signal is processed with the window width of 1 second. The spectrum sequence values below 4kHz is retained to establish the reference matrix. The reference matrix is represented as the image shown in Fig. 6. For the spectrum of segmented signals, the same number of points in the reference matrix is retained, in order to match the frequency spectrum easily.

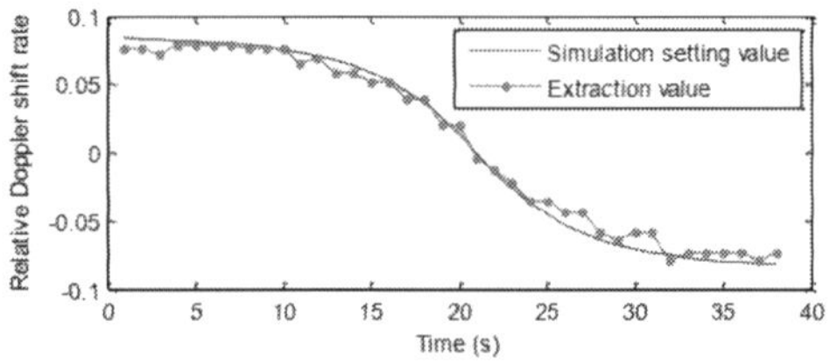

Fig 7. Relative Doppler shift coefficient extraction results.

According to the proposed method, the Doppler frequency shift coefficient extracted from the simulated signal is shown in Fig. 7. It is shown that the extraction results of the signal Doppler coefficient have good accuracy, the maximum absolute error is 0.015, the minimum absolute error is 0.01. The error is very small compared with the simulation set value.

4 Conclusion

In this paper, an extraction method of signal Doppler shift is proposed for the broadband signal with unstable line spectrum or no line spectrum. The method uses the consistency of signal spectrum expansion caused by Doppler shift and spectrum expansion caused by different DFT points. The relationship between the spectral analysis points and the Doppler shift coefficient is established. By means of the reference matrix corresponding to different Doppler frequency shift coefficients, the matching degree between the signal spectrum and the reference matrix can be measured by linear correlation coefficient. Then, the characteristics of the relative Doppler frequency shift coefficient are extracted by optimization. The proposed method has higher accuracy of Doppler shift analysis than the traditional method based on short time Fourier transform. Doppler frequency shift signal of moving target was simulated and processed. Results show that the method can extract the Doppler shift characteristics of the broadband signal without line spectrum with good precision. This method will expand the target application scope of target movement parameters estimation based on Doppler shift and has practical application value.

References

1. WU Guoqing, MA Li, ACTA ACUSTICA, **31**, 14 (2006)
2. XU Lingji, YANG Yixin, Journal of Harbin Engineering University, **34**, 1090 (2013)
3. GAO Wei, CHEN Chuan, Technical Acoustics, **29**, 374 (2010).
4. David C. Reid, Abdelhak M. Zoubir, Boualem Boashash. J. Acoust. Soc. Am., **102**, 207 (1997)
5. David R. Dall'Osto, Peter H. Dahl, J. Acoust. Soc. Am., **127**, 1771 (2010)
6. XU Ling-ji, YANG Yi-xin, TIAN Feng, Technical Acoustics, **30**, 523 (2011)

Experimental Research on Relationship between UHF Partial Discharge Signals and Discharge Quantity in GIS

Tao Yang[1], *Jianrong* Wu[1], *Guozhi* Zhang[2], *Liang* Huang[2], *Qiansu* Lv[2], *Lin* Zhang[3], *Li* Huang[4], *Siyang* He[5], *Shuai* Wang[6], *Xiaoxing* Zhang[2], *Jian* Zhang[2]

[1]Electric Power Science Research Institute, Guizhou Power Grid Co., Ltd, Guiyang 550000, China
[2]School of Electrical Engineering, Wuhan University, Wuhan 430000, China
[3]Guiyang Power Supply Burea, Guizhou Power Grid Co., Ltd, Guiyang 550000, China
[4]Liupanshui Power Supply Burea, Guizhou Power Grid Co., Ltd, Liupanshui 553000, China
[5]Duyun Power Supply Burea, Guizhou Power Grid Co., Ltd, Duyun 558000, China
[6]Tongren Power Supply Burea, Guizhou Power Grid Co., Ltd, Tongren 554300, China

Abstract. Apparent discharge quantity is usually recognized as judgement standard for severity of partial discharge in the field of electrical equipment insulation diagnosis. In recent years, the ultra-high-frequency (UHF) method has been widely used due to its unique advantages on on-line monitoring and positioning, but the relationship between UHF partial discharge signals and discharge quantity is still a difficult problem of UHF method. In order to research relationship between UHF partial discharge signals and discharge quantity in GIS, a large number of experiments, based on four kinds of typical insulation defects in the GIS, are carried out. Degree of correlation, between parameters including UHF energy, quadratic integral, peak voltage, peak-to-peak value, wave area and discharge quantity, the square of discharge quantity, is studied based on coefficient of determination. The experimental results show there is linear relation between UHF signal energy and the square of discharge quantity with high related degree.

1 Introduction

GIS, employing SF6 as insulation media, has received extensive application in electric network transformation and construction due to its many advantages such as small land occupancy, stable and reliable operation, suffering little by outer environment as well as less maintenance work[1]. Operating experiences indicate that the inner inevitable defects of GIS would still cause troubles and gradually deteriorate the devices although their high reliability, leading to the occurrences of major accidents. As a main factor of the induction of these accidents, the Partial discharge (PD) has aroused considerable researches by scholars and institutions. Currently, the relatively successful detection method for PD is pulse current method and ultra-high frequency (UHF) method [2-7].

The pulse current method is the only quantitative detection method that meets the international criterion [8]. Although it has some advantages of high sensitivity and quantitative detection for discharge capacity, its limitations of low detecting frequency and susceptibility to electromagnetic interference result in its inapplicability for on-site test. In terms of UHF method, it utilizes the antenna sensors installed inside or outside of GIS to receive UHF signals motivated by PD with frequency ranging from 300 to 3000 MHz and to proceed further detection and analyses. Such means possess the merits of high ant jamming capability, adoptability for fault localization and fault type recognition. However, this technology at present fails to detect the discharge capacity of inner defects quantitatively.

In recent years, a small number of researches have theoretically analyzed and experimentally investigated the dependence between UHF signals and discharge capacity, with related experiments limited in the focus of needle plate (metal protrusions) defects [10-11]. This paper performed a large number of experiments about four kinds of typical defects including GIS metal protrusions, free metal particles, surface pollutants of the insulators and air gap between high voltage conductor and insulators, putting forward the relationship between the discharge capacity and UHF signals. This progress would provide important reference value and guiding role for UHF method to realize the quantitative detection of partial discharge in power equipment.

2 Experiment system and experimental method

2.1 Experimental equipment

In this experiment, pulse current method is adopted based on IEC 60270 standard; In UHF test, the measured signal is UHF electromagnetic signal, which is the amplitude of electromagnetic signal with mV as a unit. Experimental circuit is shown in figure 1

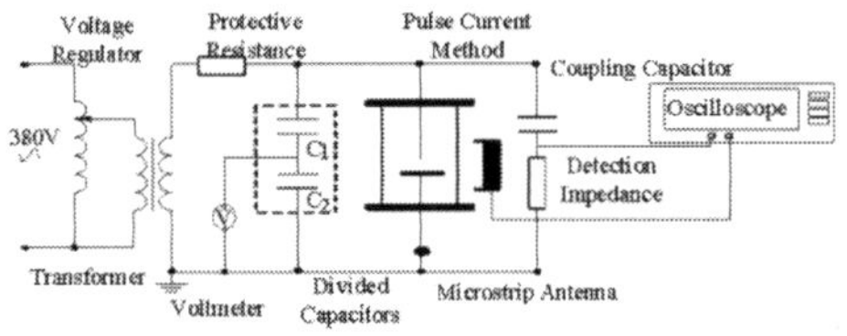

Fig 1. Experimental circuit

2.2 Analysis method on experiment data

In order to study the relationship between UHF signal and its apparent discharge, UHF signal is processed via a variety of mathematical methods, which are respectively expressed as UHF signal peak voltage Umax, peak-to-peak value Upp, wave area Sw, energy Es and quadratic integral I_{II}. Discharge quantity Q and discharge quantity square Q^2 of the pulse current signal are simultaneously calculated. The UHF signal energy Es can be calculated by:

$$E_s = \frac{\Delta t}{R}\sum U_i^2 \quad (1)$$

Where, U_i is the voltage amplitude of NO.i sampling point of UHF signals. R is the load impedance, and R is 50Ω. Δt is the time difference between adjacent sampling points, and Δt is 0.1ns.

Regression analysis are applied on all the experimental data analyzed regression. The concept of coefficient of determination f_d is introduced to describe the degree of regression. The so-called coefficient of determination is a parameter which is used to determine the degree of closeness of the sample observations around the sample regression line. Ranging between 0 and 1, the larger the value of the coefficient of determination is, the higher the degree of correlation among variables, and conversely, the lower the correlation between variables. The formula is following

$$f_d = \frac{S_R}{S_T} = \frac{\sum_{i=1}^{n} W_i(\hat{y}_i - \bar{Y})^2}{\sum_{i=1}^{n} W_i(Y_i - \bar{Y})^2} \quad (2)$$

Where, Y_i is the NO.i observation value. W_i is the weight corresponding to NO.i observations value, generally is set to 1. $\bar{Y}$ is the average value of observed values. $\hat{y}_i$ is the fitted value of NO.i observations value. S_R is the regression sum of squares, reflecting the influence of the change of independent variables on the change of dependent variables. S_T is total sum of square, or sum of squares of the difference between observed and fitted values, reflecting the total deviation of the N observations and their mean values of the dependent variable.

3 Experimental result

3.1 Metal protrusion defect

The partial discharge caused by the metal protrusion defect belongs to the typical corona discharge. In this paper, needle-plate electrode is used to simulate metal protrusion defect. The discharge is generated by adjusting the distance between needle and plate with the needle electrode connected to the high voltage side, and the plate electrode grounded.

Under the same metal protrusion defect, the partial discharge signals at different charge quantity are obtained by applying different voltages. The f_d value of each parameter of the UHF signals is shown in table I. Compared the f_d value of each parameter at different voltages in the table, it can be found that the UHF signals energy always has a high linearity with the square of charge quantity. Moreover, when the voltage is increased, the dispersion of metal protrusion defect discharge does not change, and the stability of the discharge remains unchanged.

Table 1. Setting Word's margins.

voltage/ kV	U_{max}	U_{pp}	S_w	E_s linear	E_s quadratic	Q^2	I_{II}
7	0.947	0.922	0.976	0.976	0.977	0.978	0.974
9	0.955	0.928	0.979	0.973	0.979	0.977	0.979
10	0.949	0.921	0.975	0.958	0.976	0.976	0.975

Taking the discharge at 9kV voltage as an example, the relationship between the UHF signal and the charge quantity is obtained, as shown in figure 2. It can be seen from the figure that the effect applying parabolic fitting is better than that of linear fitting about the relationship between the UHF signal energy and the charge quantity, and there is a linear relationship between the UHF signal energy and the square of charge quantity.

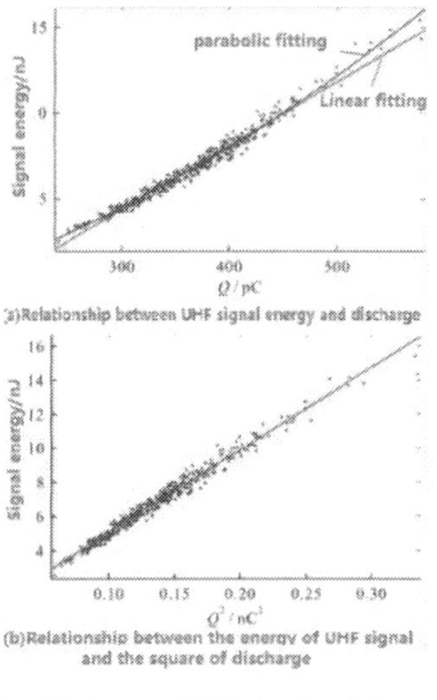

Fig 2. Relationship between UHF signal and charge quantity 9kV volatage

It is shown in figure 3 that the relationship between the signal energy and the square of charge quantity under the same partial discharge source at 7.0, 9.0 and 10.0kV. With the increase of the voltage, the UHF signal energy has an increasing tendency, but the experimental data can be regressed to the same line under different voltages. Therefore, this relationship can be used to quantify the charge quantity for metal protrusion defect.

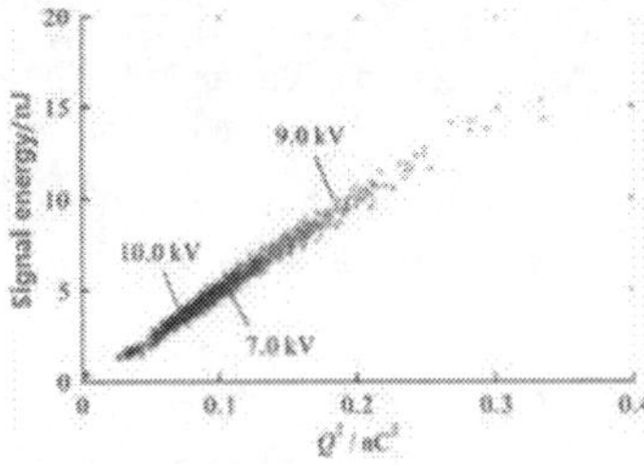

Fig 3. Relationship between UHF signal and charge quantity at different volatage

3.2 Free metal particles defect

Under the action of the external electric field, the free metal particles can induce charges and will jump or shift, which will greatly reduce the breakdown voltage of the gas insulation. In this paper, a series of rectangular thin aluminum sheets, which sizes are 2mm×2mm, are placed between the high and low voltage electrodes of the experimental device to simulate this discharge type.

Under the same free metal particles defect, the partial discharge signal at different charge quantity is obtained by applying different voltages. The f_d value of each parameter of the UHF signals is shown in table 2. It can be seen that with the increase of the voltage, the higher the coefficient of determination of the data, which shows that the discharge of free metal particles tends to be stable at higher voltage. But in general, the dispersion of the datum is large. Meanwhile, no matter which kind of voltages, it is the best that the linear relationship between the UHF signal energy and the square of charge quantity under the free metal particles defect. Taken the discharge caused by free metal particles at 4kV voltage as an example, the relationship between the UHF signal energy and the charge quantity is obtained, as shown in figure 4.

Figure 5 shows the relationship between the energy of the UHF signal and the square of discharge under 3.6, 4.0 and 6.0kV.

As can be seen from the figure 5, although the low voltage data linearity is slightly lower, but the experimental data at different voltages can be roughly returned to the same straight line.

Table 2. f_d of UHF signals on Free metal particles defect.

voltage/ kV	U_{max}	U_{pp}	S_w	E_s		Q^2	I_{II}
				linear	quadratic		
3.6	0.773	0.747	0.439	0.767	0.794	0.794	0.520
4.0	0.824	0.672	0.609	0.842	0.849	0.848	0.674
6.0	0.761	0.652	0.950	0.916	0.920	0.919	0.914

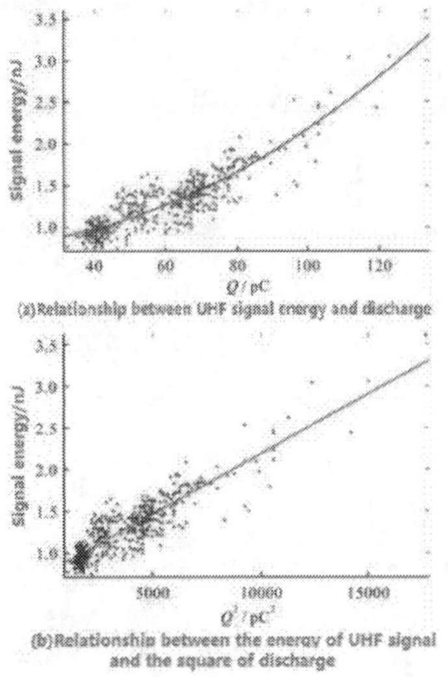

Fig 4. Relationship between UHF signal and charge quantity 4kV volatage

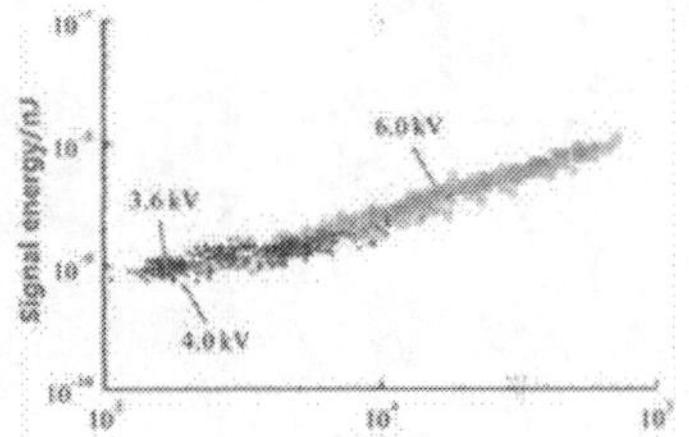

Fig 5. Relationship between UHF signal and charge quantity at different volatage

3.3 Insulator surface contamination defect

The discharge of metal contamination on insulator surface is between metal particle and insulator surface, therefore, the charge accumulation on insulator surface has an important influence on the discharge process. In this paper, the insulators with metal particles are placed between the high and low voltage electrodes to simulate the defects. The particle is aluminum sheet, the length is 10mm, the distance between the particles and the high voltage is 5mm.

By applying different voltages, the partial discharge signal under different discharge power is obtained, and the f_d value of each parameter of the UHF signal is shown in Table 3 Contrary to free metal particle defect, the linearity of the data is getting worse with the increase of voltage. The results show that the PD dispersion of metal contaminants on insulator surface increases at higher voltage, and the discharge is more and more unstable. However, it is still the best linear relationship between the UHF signal energy and the square of the discharge energy, the same as the first two defects.

The relationship between UHF signal and discharge under 7kV voltage is shown in Figure 6. Figure 7 shows the relationship between the discharge energy and the square of the discharge power at the same local discharge power at 5.2, 7 and 9.0kV. As can be seen from the diagram, the quantitative analysis of the linear relationship between the UHF signal energy and the square of the discharge will have a certain error in the case of high voltage.

Table 3. f_d of UHF signals on insulator surface contamination defect.

voltage/ kV	U_{max}	U_{pp}	S_w	E_s		Q^2	I_{II}
				linear	quadratic		
5.2	0.817	0.866	0.971	0.936	0.971	0.970	0.961
7.0	0.690	0.690	0.882	0.875	0.901	0.897	0.852
9.0	0.651	0.689	0.842	0.854	0.868	0.868	0.799

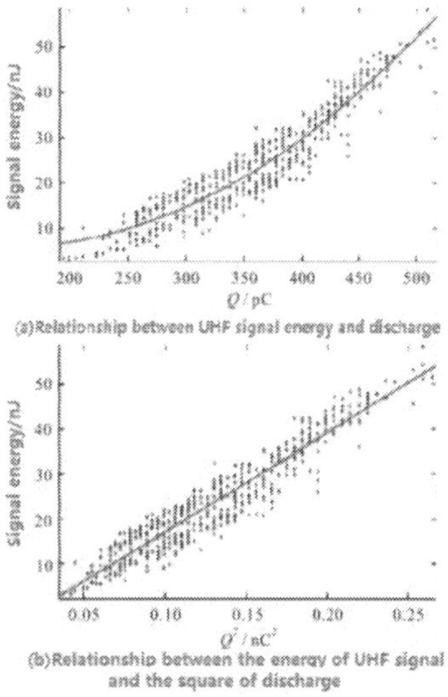

Fig 6. Relationship between UHF signal and charge quantity 7kV volatage

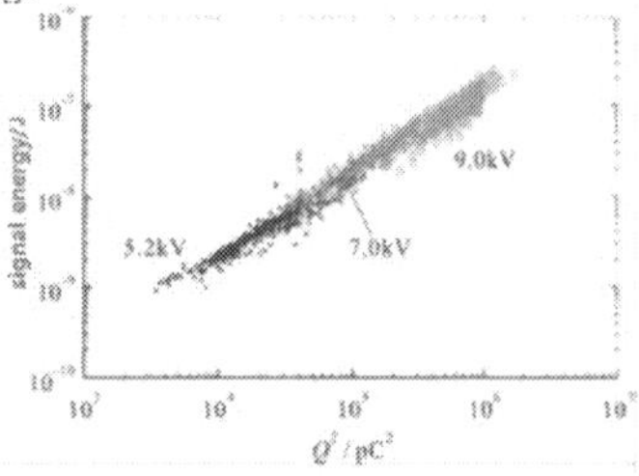

Fig 7. Relationship between UHF signal and charge quantity at different volatage

3.4 Air gap defect between insulator and high voltage conductor

There is an air gap between the medium and the electrode, which leads to internal partial discharge, which is related to the characteristics of the medium and the shape, size and position of the gap and the nature of the gas in the air gap. The insulator is placed between the high and low voltage electrodes, and the 3mm gap between the insulator and the high voltage electrode is used to simulate the defects.

The f_d value of each parameter of the UHF signal is shown in Table 4. As can be seen from the table, the UHF signal energy and the square of discharge has always been a high degree of linearity. At the same time, when the voltage rises, the dispersion of the air gap defect between the insulator and the high voltage conductor decreases, and the discharge is more stable.

Table 4. *f_d of UHF* signals *on Electrode Gap defect.*

voltage/ kV	U_{max}	U_{pp}	S_w	E_s		Q^2	I_{II}
				linear	quadratic		
2.6	0.868	0.901	0.891	0.856	0.918	0.912	0.877
6.0	0.911	0.931	0.825	0.915	0.945	0.947	0.818
9.0	0.901	0.904	0.924	0.967	0.977	0.977	0.931

The relationship between UHF signal and discharge under 9.0kV voltage is shown in Figure 8. Figure 9 shows the relationship between the discharge energy and the square of the discharge at the same local discharge power at 2.6, 6 and 9.0kV. As can be seen from the figure, the experimental data at different voltages can be returned to the same straight line, so the UHF signal can be used to quantify the gap defects.

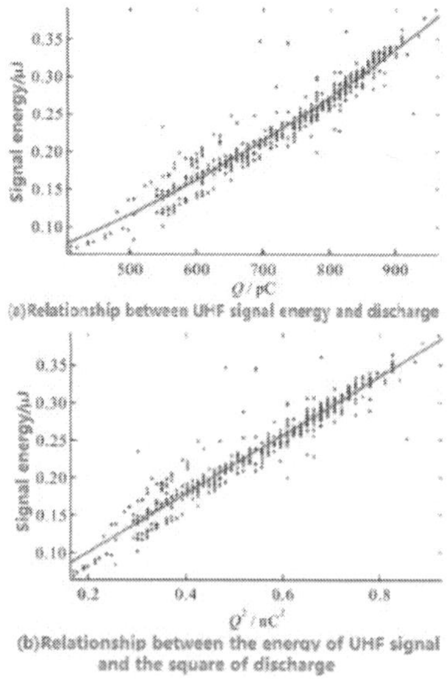

Fig 8. Relationship between UHF signal and charge quantity 9kV volatage

 https://doi.org/10.1051/matecconf/201820801002

4 Conclusion

Based on a large number of experiments, the relationship between the UHF signal parameters and the discharge of 4 typical defects of GIS partial discharge is compared.

1) The linear relationship between the UHF signal energy and the square of the discharge of 4 kinds of defects in GIS has the highest coefficient of determination. That is to say, the linear relationship between the UHF energy and the square of the discharge (or the quadratic relationship between UHF energy and discharge) is the best regression curve.

2) UHF signal can be used to quantify the discharge of 4 kinds of typical defects, in which the error of metal defect is the lowest, which can reach 3%. The free metal particles have the highest defects, which are basically more than 10%. The error is: metal protrusion< air gap<metal contamination on insulator surface<free metal particle.

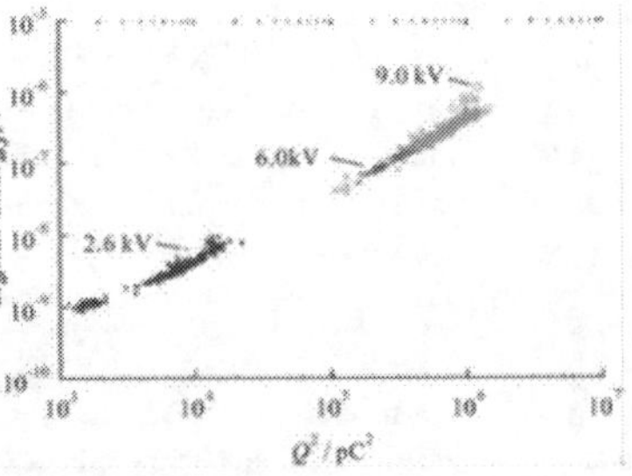

Fig 9. Relationship between UHF signal and charge quantity at different volatage.

References

1. K. D. Srivastava and M. M. Morcos, "A review of some critical aspects of insulation design of GIS/GIL systems," *2001 IEEE/PES Transmission and Distribution Conference and Exposition. Developing New Perspectives (Cat. No.01CH37294)*, Atlanta, GA, pp. 787-792 **vol.2**(2001)
2. M. D. Judd, O. Farish and B. F. Hampton, "The excitation of UHF signals by partial discharges in GIS," in *IEEE T DIELECT EL IN* **vol. 3**, no. 2, pp. 213-228 (1996)
3. J. S. Pearson *et al.*, "Partial discharge diagnostics for gas insulated substations," in *IEEE T DIELECT EL IN*, **vol. 2**, no. 5, pp. 893-905, Oct (1995)
4. M. D. Judd, O. Farish and B. F. Hampton, "Broadband couplers for UHF detection of partial discharge in gas-insulated substations," in *IEE Proceedings - Science, Measurement and Technology*, **vol. 142**, no. 3, pp. 237-243 (1995)
5. M. D. Judd, B. F. Hampton and O. Farish, "Modelling partial discharge excitation of UHF signals in waveguide structures using Green's functions," in *IEE Proceedings - Science, Measurement and Technology*, **vol. 143**, no. 1, pp. 63-70(1996)
6. Zhou Qian, Tang Ju, Xu Zhongrong, Meng Yanhui and Zhang Xiaoxing, "Study on partial discharge of metallic outshoot insulation defect in GIS," *Proceedings of 2005 International Symposium on Electrical Insulating Materials, (ISEIM 2005)*. pp. 431-434 **Vol. 2** (2005).
7. Tang, Ju, et al. "Investigation of Partial Discharge on Typical Defects with UHF Detection Method for GIS." *Przegląd Elektrotechniczny* **88**.12:351-355(2012).
8. Standard, I. E. C. "High-voltage test techniques: partial discharge measurements." IEC-60270 (2000).
9. X. Zhang, S. Xiao, N. Shu, J. Tang and W. Li, "GIS partial discharge pattern recognition based on the chaos theory," in *IEEE T DIELECT EL IN*, **vol. 21**, no. 2, pp. 783-7 (2014)
10. G. P. Cleary and M. D. Judd, "UHF and current pulse measurements of partial discharge activity in mineral oil," in *IEE Proceedings - Science, Measurement and Technology*, vol. 153, no. 2, pp. 47-54, 9 March 2006.
11. S. Ohtsuka, T. Teshima, S. Matsumoto and M. Hikita, "Relationship between PD-induced electromagnetic wave measured with UHF method and charge quantity obtained by PD current waveform in model GIS," *2006 IEEE Conference on Electrical Insulation and Dielectric Phenomena*, Kansas City, MO, , pp. 615-618(2006).

An optimized 2D-Robust Adaptive Beamforming algorithm based on Matrix Completion in sparse array

Jiaying Di[1], *Wen* Hu[1], *Mengxia* Li[1] and *Hongtao* Li[2]

[1]Nanjing University of Aeronautics and Astronautics, 210000 Nanjing, Jiangsu, China
[2]Nanjing University of Science and Technology, 210000 Nanjing, Jiangsu, China

Abstract. The sparse arrays can reduce the cost of beamforming, it greatly reduces the number of actual array elements. However, it also brings about the problem of information loss. A 2D-robust adaptive beamforming algorithm in sparse array based on Singular Value Thresholding algorithm is proposed. At first, a signal model of planar array is established based on Matrix Completion, which can be proved to meet Null Space Property. Then the Genetic Algorithm is used to optimize the sparse array, which is determined to reduce the Spectral Norm Error of Matrix Completion and make the array recovered closer to the full array. In the case of sparse array, the missing information is restored by using the theory of Singular Value Thresholding, and then the restored signal is used to design the digital beamformer weights. This algorithm significantly reduces the Spectral Norm Error and forms robust adaptive beam.

1 Introduction

Digital Beam Forming (DBF) technology uses advanced digital signal processing techniques to process signals received by an array antenna. It can greatly improve the anti-jamming capability of radar system, and is one of the key technologies for radar to improve the performance of target detection. In order to implement 2D-robust adaptive beamforming effectively, a large number of antennas and sampling devices are required. There is a problem of low utilization of array elements, which means increased hardware complexity and cost increase in engineering. Since 1960s, sparse arrays have been extensively studied because of their high target resolution and low cost[1]. Sparse arrays can reduce the number of antennas, but the average sidelobe increases with the number of units removed, therefore, the application of it is limited. For low-rank matrix, the Matrix Completion (MC) theory [2],[3],[4],[5] can recover the complete matrix by partial matrix elements, it is a generalization of the theory of compressed sensing on matrix, it is widely used in image processing and pattern recognition. The Singular Value Thresholding (SVT) algorithm is a fast and effective algorithm for matrix completion[6], it brings great convenience to solve the problem above.

Spectral Norm Error(SNE) is an important index to measure the quality of MC. The smaller the SNE, the more accurate the restored matrix. In order to reduce the SNE of MC, we have to find an optimum set of array elements. The Genetic Algorithm(GA) is applied to optimize the set of array elements[7],[8]. It seeks the optimal solution by imitating the mechanism of selection and inheritance of nature. A new 2D robust adaptive beamforming[9] MC model is constructed, the signal matrix is proved to satisfy the Null Space Property(NSP). Thus, an algorithm is proposed to realize adaptive two dimensional robust beamforming[10],[11],[12].

2 Signal model

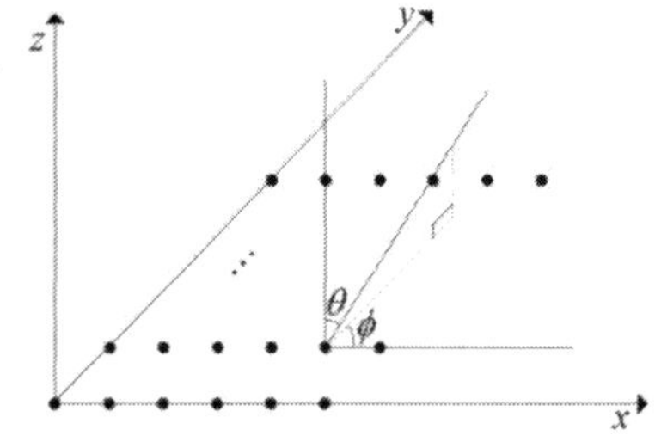

Fig. 1. URA

Now we consider a uniform rectangular array[13] (URA). Fig.1 shows the distribution of the full array elements, where the number of elements in x-direction and y-direction is M_x and M_y. The distance between the neighboring elements in x-direction and y-direction is d_x and d_y. (ϕ,θ) means the pitch and the azimuth angle of the target.

Assume K targets are contained in received signals, , and $s_i(t)$ refers to the waveform of the i th target, (ϕ_i,θ_i) means the elevation and azimuth angle of the i th target and N is the number of snapshot. Suppose the uncorrelated narrow-band signals are received in far field, thus the received signal of M_x elements in x-direction can be written as

$$\mathbf{Y}_x(t) = \mathbf{A}_x\mathbf{S}(t) + \mathbf{N}_x(t) \quad (1)$$

where $S(t) = diag(s_i(t))$ is the diagonal matrix of $K \times K$ dimensions, $\mathbf{N}_x(t)$ is the received noise, $\mathbf{A}_x$ is the steering vector of $K \times K$ dimensions, which can be given by

$$\mathbf{A}_x = [\mathbf{a}_x(\phi_1,\theta_1), \mathbf{a}_x(\phi_2,\theta_2), \cdots, \mathbf{a}_x(\phi_K,\theta_K)] \quad (2)$$

Suppose the wavelength of received signal is λ, then $\mathbf{a}_x(\phi_i,\theta_i)$ is

$$\left[\mathbf{a}_x(\phi_i,\theta_i)\right]_{m_x} = e^{j\frac{2\pi}{\lambda}d_x \sin(\theta_i)\cos(\phi_i)(m_x-1)} \quad (3)$$

where $m_x = 1,2,\cdots,M_x$. Likewise, the received signal of M_y elements in y-direction can be represented as

$$\mathbf{Y}_y(t) = \mathbf{A}_y\mathbf{S}(t) + \mathbf{N}_y(t) \quad (4)$$

where $\mathbf{N}_y(t)$ is the received noise, $\mathbf{A}_y$ is the steering vector of $K \times K$ dimensions, which can be denoted as

$$\mathbf{A}_y = [\mathbf{a}_y(\phi_1,\theta_1), \mathbf{a}_y(\phi_2,\theta_2), \cdots, \mathbf{a}_y(\phi_K,\theta_K)] \quad (5)$$

where $\mathbf{a}_y(\phi_i,\theta_i)$ is defined as

$$\left[\mathbf{a}_y(\phi_i,\theta_i)\right]_{m_y} = e^{j\frac{2\pi}{\lambda}d_y \sin(\theta_i)\cos(\phi_i)(m_y-1)} \quad (6)$$

where $m_y = 1,2,\cdots,M_y$. Therefore, the received signal model of URA can be demonstrated as

$$\mathbf{Y}(t) = \mathbf{A}_x\mathbf{S}(t)\mathbf{A}_y^T + \mathbf{N}(t), t = 1,2,\cdots,N \quad (7)$$

When the noise power is far smaller than the signal powers, we can get the $rank(\mathbf{Y}(t)) \le rank(\mathbf{S}(t)) = \eta$, namely, matrix $\mathbf{Y}(t)$ is low rank[14]. And matrix in (7) meets strong incoherence property, thus it can be recovered with high probability via MC[15].

3 Beaming forming algorithm

3.1 GA sampling

The original matrix cannot be recovered from the sampled matrix by MC when $m < r(n_1 + n_2 - r)$ [16]. Where the n_1, n_2 are the number of rows and columns for the target matrix and r is the rank of it. The smaller the SNE is, the closer the restored matrix is to the full array, and the higher accuracy of the system. The influence of sparse array elements on SNE can not be ignored. GA is used to optimize the sparse array in this paper to reduce the SNE.

$$\mathbf{Y}_S(t) = (d_1, d_2, \cdots, d_m) \quad (8)$$

It is assumed that $\mathbf{Y}_S(t)$ is the sparse array consisting of m array elements $d_1, d_2, \cdots, d_m$. The optimization of spatial sampling array selection based on GA can be expressed as follows:

$$\min SNE = f(d_1, d_2, \cdots, d_m) \quad (9)$$

Take the sparse array $\mathbf{Y}_S(t)$ as the decision variable. A sparse array is chosen as an individual, the fitness function of GA is used to evaluate the merits of each individual. In order to reduce the optimization objective SNE, the fitness function is

$$f(d_1, d_2, \cdots, d_m) = SNE = \|\mathbf{M} - \mathbf{Y}\|_2 / \|\mathbf{M}\|_2 \quad (10)$$

where $\mathbf{Y}_S(t)$ is the full array, $\mathbf{M}$ is the matrix recovered by MC.

The objective function can be written as

$$f(d_1, d_2, \cdots, d_m) = \min\left\{fit\left(d_1, d_2, \cdots, d_m\right)\right\} \quad (11)$$

The optimized spatial sampling array selection algorithm based on GA is as follows:

Step 1 Initial population establishment.
Step 2 The fitness function of individuals in a population is calculated.
Step 3 Determine whether the optimization criteria are met and, if satisfied, turn to step 8, or continue.
Step 4 Select dominant individuals.
Step 5 Generalized cross operation.
Step 6 Generalized mutation operation.
Step 7 Turn to step 2.
Step 8 The end of the output is the best individual.

Full rectangular antenna array can be represented as $\mathbf{Y}_R(t)$, P_Ω is the sparse sampling operator. Thus, a signal model based on MC can be represented as

$$\begin{cases} \min \|\mathbf{Y}_R(t)\|_* \\ s.t.\ P_\Omega(\mathbf{Y}_R(t)) = P_\Omega(\mathbf{Y}_S(t)) \end{cases} \quad (12)$$

Null space of sampling operator is $Null(P_\Omega) = \left\{\mathbf{M} \in \mathbf{R}^{n_1 \times n_2} : P(\mathbf{M}) = 0\right\}$. According to the NSP, the matrix in the null space is hard to recover. As we analyze the array signal, on account of vector $\mathbf{a}_x(\phi_i,\theta_i) \ne 0$, hence, elements in $\mathbf{A}_x$ and $\mathbf{A}_y$ are all non-zero. The elements in $\mathbf{S}(t)$ are non-zero, so the elements in $\mathbf{Y}_R(t) = \mathbf{A}_x\mathbf{S}(t)\mathbf{A}_y^* + \mathbf{N}(t)$ are non-zero. Thus, we can draw the conclusion that $P_\Omega(\mathbf{Y}_R(t)) \ne 0$, which means the matrix meets NSP.

3.2 SVT-EXP-LCMV algorithm

After the establishment of the sparse array, the SVT algorithm is applied to recover the full signal, and the EXP-LCMV algorithm is used to form the robust beam. SVT algorithm combines Bregman iteration with fixed point iteration[17], the algorithm is simple and fast. It is suitable for solving the signal model in this paper.

In SVT algorithm, the signal model of MC is transformed as (11). The corresponding iterative process can be expressed as the following formula, which can be solved by the corresponding iterative algorithm

$$\begin{cases} \mathbf{Y}_R^k(t) = D_\tau\left(\mathbf{Z}^{k-1}\right) \\ \mathbf{Z}^k = \mathbf{Z}^{k-1} + \delta_k P_\Omega\left(\mathbf{Y}_s(t) - \mathbf{Y}_R^k(t)\right) \end{cases} \tag{13}$$

where δ_k is the step size of each iteration. The initial value of the iteration step is

$$\delta_0 = 1.2\frac{m_1 \times m_2}{m} \tag{14}$$

where m_1, m_2 are the number of rows and columns of a matrix, m is the number of data in $\mathbf{Y}_s(t)$, and 1.2 is down sample rate multiple.

Firstly, singular value decomposition of the matrix $\mathbf{Z} \in \mathbf{R}^{m_1 \times m_2}$ is

$$\mathbf{Z} = \mathbf{U}_Z diag(\sigma)\mathbf{V}_Z \tag{15}$$

For each given positive number τ, $D_\tau(\mathbf{Z})$ can be represented as

$$D_\tau(\mathbf{Z}) = \mathbf{U}_Z diag(\{\sigma - \tau\}_+)\mathbf{V}_Z^T \tag{16}$$

After the iterative process, the corresponding left and right singular value vectors $\mathbf{U}_S$ and $\mathbf{V}_S$ of the sparse array. Without the consideration of noise, the sparse array can be written as

$$\mathbf{Y}_R(t) = \mathbf{A}_x\mathbf{S}(t)\mathbf{A}_y^* = \mathbf{U}_S\mathbf{\Sigma}\mathbf{V}_S^* \tag{17}$$

Vectorize(17), it holds that

$$vec\left(\mathbf{Y}_R(t)\right) = vec\left(\mathbf{A}_x\mathbf{S}(t)\mathbf{A}_y^T\right) = vec\left(\mathbf{U}_S\mathbf{\Sigma}\mathbf{V}_S^*\right) \tag{18}$$

Based on Kronecker product property, (17) can be transformed in vector form as

$$\mathbf{y}_R(t) = \mathbf{A}_y \otimes \mathbf{A}_x vec\left(\mathbf{\Sigma}(t)\right) = \bar{\mathbf{V}}_S \otimes \mathbf{U}_S vec\left(\mathbf{\Sigma}\right) \tag{19}$$

Solving the auto-correlation matrix $\mathbf{R}_Y$ of $\mathbf{y}_R(t)$, we get that

$$\mathbf{R}_Y = \mathbf{U}_S\mathbf{\Sigma}_S\mathbf{V}_S^H + \mathbf{U}_n\mathbf{\Sigma}_n\mathbf{V}_n^H \tag{20}$$

where $\mathbf{\Sigma}_S$ is the diagonal matrix of large eigenvalues, $\mathbf{U}_S$ is the interference subspace, $\mathbf{\Sigma}_n$ is the diagonal matrix of small eigenvalues, and $\mathbf{U}_n$ is the noise subspace.

In robust adaptive beamforming, static weight vector of Linear Constrained Minimum Variance(LCMV)[16] is

$$w_{LCMV} = \mathbf{C}\left(\mathbf{C}^H\mathbf{C}\right)^{-1}\mathbf{f} \tag{21}$$

where $\mathbf{C}$ is the constraint matrix of dimension $m_x m_y \times R$, $\mathbf{f}$ is the constraint response vector of dimension $R \times 1$, R is the rank of correlation matrix $\mathbf{R}_Y$. For the purpose of forming a two-dimensional robust adaptive beam, the covariance matrix of input signal of LCMV algorithm is corrected by exponential method to improve the degree of divergence of small eigenvalue in the noise subspace and improve the robustness of beamforming at low snapshot.

We can get $\mathbf{R}_Y^\alpha = \mathbf{U}_S\Lambda_S^\alpha\mathbf{U}_S^H + \mathbf{U}_n\Lambda_n^\alpha\mathbf{U}_n^H$ by correcting $\mathbf{R}_Y$, where α is correction index, and $\alpha \in [0,1]$.

The diagonal matrix of the small eigenvalues of the modified noise subspace is

$$\Lambda_n^\alpha = diag\left\{\sigma_1^\alpha, \sigma_2^\alpha, \cdots, \sigma_n^\alpha\right\} \tag{22}$$

Hence, the small eigenvalues of the noise subspace are updated to $\sigma_i^\alpha, i = 1, 2, \cdots, n$

Since $\alpha \in [0,1]$, the degree of divergence $\sigma_i^\alpha, i = 1, 2, \cdots, n$ is improved.

The small eigenvalue divergence in the noise phonon space decreases with the decrease of the correction exponent α. The weighted vector of the modified EXP-LCMV algorithm is

$$w_{EXP-LCMV} = \mathbf{R}_X^{-\alpha}\mathbf{C}\left(\mathbf{C}^H\mathbf{R}_X^{-\alpha}\mathbf{C}\right)^{-1}\mathbf{f} \tag{23}$$

4 Simulations

The effectiveness of the proposed SVT-EXP-LCMV algorithm is verified by simulations in this section. Let the full array be a 64×64 URA where the total amount of array elements is $M = 4096$ and the array element spacing in x-direction and y-direction are equal. We sample array elements of the URA by GA to build a sparse array. Assume the pitch angle of main gain direction is 45°, and the azimuth angle is 90°. The corresponding pitch angles and azimuth angles of the three incoherent interference signals in the space are $(50°, -120°)$, $(55°, -70°)$, $(45°, 90°)$. Set Interference Noise Ratio (INR) as 20dB, and the number of snapshots in the following experiments remains 50.

4.1 The comparison under different arrays

We select 1200 elements from the URA. In Fig.2-Fig.4, the 2D robust beamforming in different arrays are shown.

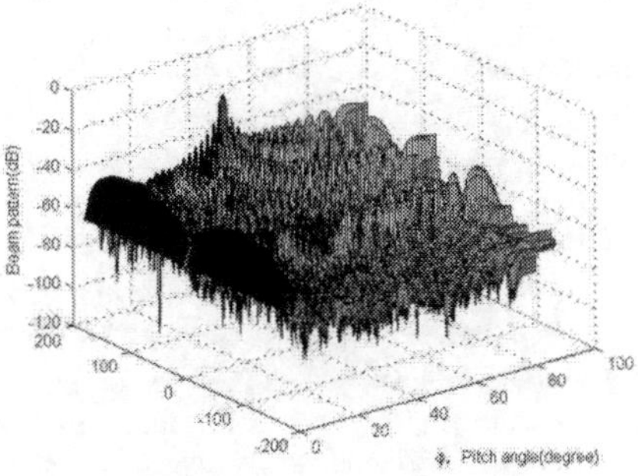

Fig. 2. Beamforming in Full Array

The 2D beamforming in full array is shown in Fig.2. It shows that the main peak value is 0dB and average sidelobes level is -65.49dB . The 2D beamforming in sparse array is shown in Fig.3. It shows that the main peak value is 0dB and average sidelobes level is -32.85dB. In the third experiment, 2D beamforming by SVT-EXP-LCMV algorithm is examined in the same sparse array.

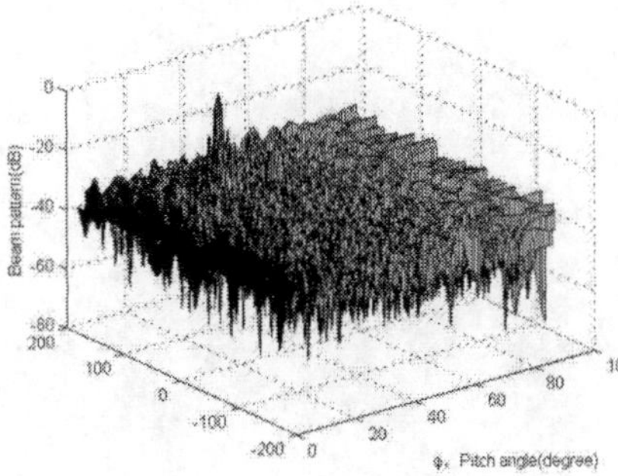

Fig. 3. Beamforming in Sparse Array

From Fig.4, it can be noted that peak value differs by -15.76dB from main peak and main sidelobe, and average sidelobes level is -62.69dB. Compared to the full array, the number of sparse array elements is reduced by 70%. The beam can be effectively formed, and the problem of proportional deterioration of the average sidelobe as the number of units removed is effectively solved at the same time.

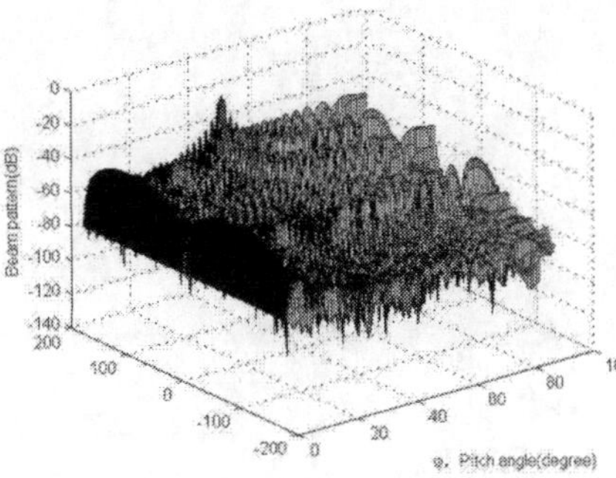

Fig. 4. Beamforming in MC

4.2 The comparison of SNE under different sampling ways

Table 1 is the SNE under the random matrix selection method and the method of element selection based on genetic algorithm.

It can be seen from the table that the number of sparse array elements is the same at the same time, and the SNE of matrix filling under the method of element selection based on Genetic Algorithm is smaller than that of the average matrix selection method. The advantage of the array element selection method based on GA is explained. algorithm. It can be seen from the table that the number of sparse array elements is the same at the same time, and the SNE of matrix filling under the method of element selection based on Genetic Algorithm is smaller than that of the average matrix selection method. The advantage of the array element selection method based on GA is explained.

Table 1. SNE under GA and random sampling

	The number of the sparse elements		
	900	1200	1500
SNE under GA sampling	2.75×10^{-2}	6.68×10^{-3}	4.53×10^{-3}
SNE under random sampling	4.62×10^{-2}	3.55×10^{-2}	6.87×10^{-3}

5 Conclusion

In this paper, an exponential correction beamforming (SVT-EXP-LCMV) algorithm based on singular value threshold (SVT) is proposed, and the genetic algorithm (GA) is used to optimize the sparse array elements. The restoration error of the matrix filling is reduced, and a robust two-dimensional beam forming direction map is generated, which has a good ratio of main lobe to side lobe, small number of array elements and low cost.

References

1. B Li, A Petropulu, *Radar precoding for spectrum sharing between matrix completion based MIMO radars and a MIMO communication SYSTEM*, IEEE Global Conference on Signal and Information, 737-741(2016)
2. Recht B, Re C, *Parallel Stochastic Gradient Algorithms for Large-Scale Matrix Completion*, Mathematical Programming Computation, 201-226(2013)
3. Chen C, He B, Yuan X, *Matrix Completion via an Alternating Direction Method*, IMA Journal of Numerical Analysis, 227-245(2012)
4. Candes E J, Recht B, *Exact Matrix Completion via Convex Optimization*, Foundations of Computational Mathematics, 717(2009)
5. R Tripathi, B Mohan, K Rajawat, *Adaptive Low-Rank Matrix Completion*, IEEE Transactions on Signal Processing, 3603-3616(2017)

6. JF Cai, Cand EJ S, Z Shen, *A Singular Value Thresholding Algorithm for Matrix Completion*, Siam Journal on Optimization, 1956-1982 (2008)
7. Chen Ke Song, Han Chun Lin, He Zi Shu, *An Array Element Spacing Constraint Sparse Array Antenna Synthesis Method*, Journal of Radio Science(2007)
8. Caorsi S., Lommi A., Massa A., Pastorino M., *Peak Sidelobe Level Reduction with a Hybrid Approach Based on GAs and Difference Sets*, IEEE Trans. on antenna and propagation, 1116-1121(2004)
9. Chen Tao, Han Qiang Qiang, Liu Lu Tao, *A Robust Two Dimensional Adaptive Beamforming Algorithm*, Journal of Central South University, 6007-6018(2013)
10. Li Hong Tao, He Ya Peng, Zhu Xiao Hua, Hu Wen, *Robust Adaptive Beamforming Algorithm based on Spectral Analysis*, Journal of Radio Science, 147-151(2012)
11. G.M. Raz, *An Approach to Adaptive Beam-Forming for Wideband Systems Using a Subband Decom Position*, Sensor Array and Multichannel Signal Processing Workshop, 300-305(2000).
12. Frost O L, *An algorithm for linearly Constrained Adaptive Array Processing*, Proceedings of IEEE, X, 926-935(2000)
13. Li Dong Feng, Gong Zhong Lin, *Genetic Algorithm for Pattern Synthesis of Ultra Low Sidelobe Linear Array Antenna*, Journal of Electronics, 123-126(2003)
14. Ding Qian Jun, Wang Yong Liang, *Research on Descending Adaptive Filtering Algorithm*, Radar Science and Technology, 837-840(2005)
15. D Pimentel-Alarcon, G Ongie, L Balzano, *Low algebraic dimension matrix completion*, Allerton Conference on Communication, 790-797(2017)
16. Ma S, Goldfarb, Chen L, *Fixed Point and Bregman Iterative Methods for Matrix Rank Minimization*, Mathematical Programming, 321-353(2011)
17. Li Hong Tao, He Ya Peng, *Adaptive Constrained Minimum Variance Adaptive Directional Pattern Control Based on Eigenvector*, Journal of Nanjing University of Science and Technology, 529-533, (2011)

An APG-MUSIC Algorithm Based-on Optimized Sampling Array

Mengxia Li[1], *Wen* Hu[1], *Jiaying* Di[1] and *Hongtao* Li[2]

[1] Nanjing University of Aeronautics and Astronautics, Nanjing,Jiangsu,China
[2] Nanjing University of Science and Technology, Nanjing,Jiangsu,China

Abstract. This paper proposes a novel two-dimensional direction of arrival (2D-DOA) estimation with optimized sparse sampling array, which is combined with Accelerated Proximal Gradient singular value thresholding(APG) and Multiple Signal Classification(MUSIC). Firstly, a signal model of 2D-DOA estimation in sparse array is established, which is proved to satisfy low rank feature and NULL Space Property(NSP). Then, Genetic algorithm (GA) is applied to a sparse sampling array to optimize the performance of matrix completion(MC). Finally, MUSIC combined with APG is studied to recover received signal matrix and estimate the direction of arrival. The results of computer simulation demonstrate that compared with conventional 2D-DOA algorithms, the proposed algorithm reduces the number of array elements needed dramatically and effectively lowers the average sidelobes level of spatial spectrum.

1 Introduction

Direction of arrival(DOA), an important branch of array signal processing [1,2] has received a lot of research achievements and wide applications in electronic reconnaissance, sensor networks, radar signal processing etc. In recent years, DOA estimation has been facing the urgent demand of high precision, high resolution and high robustness. In response to this trend, two-dimensional(2D) DOA estimation[3,4] has aroused the interest of the majority of scholars and emerged a large number of estimation methods.

Matrix completion(MC)[5–9], a data reconstruction technique, is derived from compressed sensing[10,11] and has obtained potential applications in several fields, i.e., image processing, collaborative filtering, spectrum sensing. For a low rank matrix[12], MC technology can complement missing data in a observation matrix with a high probability by transforming rank minimum problem into nuclear norm optimization problem[13]. In radar applications, the introduction of MC can significantly save processing costs and reduce the number of antennas and sampling devices. In Ref.[14,15], MC is applied to MIMO communication system. In Ref.[16], a method of wideband DOA based on co-prime arrays with sub-Nyquist sampling is proposed. In recent years, some scholars have studied the DOA estimation based on coprime arrays. The method in Ref.[17] improves the degree of freedom of DOA estimation. Then in Ref.[18], the proposed method applies MC in coprime arrays. Nowadays, the theory has also made a great deal of achievements in fast algorithm and accurate reconstruction[19,20] and showed great research value and broad application prospect. However, sampling from time domain is indispensable and 2D DOA for URAs cannot be solved by those above methods.

Based on the above analysis, this paper proposes a 2D-DOA estimation method combining Accelerated Proximal Gradient singular value thresholding with Multiple Signal Classification(APG-MUSIC) algorithm under optimized sampling array by introducing MC into 2D-DOA in planar array. Firstly, a 2D-DOA estimation signal model is built by radiation source data in sparse array, which is proved to meet the properties of low rank and NSP. Then, to improve the performance of received signal matrix reconstruction, we provide a Genetic Algorithm(GA) off-line optimization method which aims at minimizing matrix reconstruction error. Finally, APG and MUSIC are respectively employed to reconstruct received signal matrix and estimate the direction of wave arrived. Compared with conventional 2D-DOA algorithms, the proposed method can make full use of the known array elements and reduce the average sidelobes level.

2 Signal model of 2D-DOA based on MC

2.1 Signal model of 2D-DOA in planar array

A uniform rectangular array(URA) is shown in figure 1. Assume that the planar array is composed of a total number of $M_x \times M_y$ elements, and the number of entries in x-direction and y-direction is M_x and M_y respectively, with the corresponding array spacings d_x and d_y. The elevation and azimuth angles of targets are θ and ϕ respectively.

Suppose that the received signals contain k targets and the number of snapshot is N. $s_i(t)$ $(t = 1, 2, \cdots, N)$

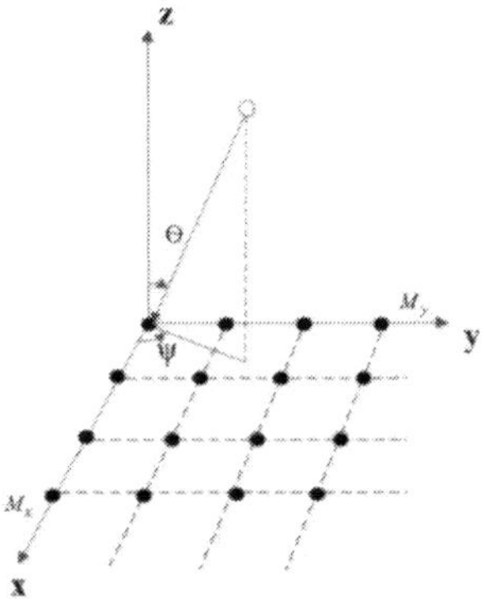

Fig. 1. Structure of uniform rectangular array

represents the waveform of i-th $(i=1,2,\cdots,k)$ target, (θ_i,ϕ_i) refers to the elevation and azimuth angle of i-th target. Considering that the received signals are uncorrelated and narrow-band in far field, thus the received signals of M_x elements in x-direction can be written as follows:

$$\mathbf{X}_x(t)=\mathbf{A}_x\mathbf{S}(t)+\mathbf{N}_x(t) \quad (1)$$

where $\mathbf{S}(t)=diag(s_i(t))$ is a diagonal matrix, $\mathbf{N}_x(t)$ is the noise matrix. $\mathbf{A}_x$ is the steering vector in x-direction, which can be given by

$$\mathbf{A}_x=[\boldsymbol{a}_x(\theta_1,\phi_1),\boldsymbol{a}_x(\theta_2,\phi_2),\ldots,\boldsymbol{a}_x(\theta_k,\phi_k)] \quad (2)$$

Suppose λ is the wavelength of received signal, then $\boldsymbol{a}_x(\theta_i,\phi_i)$ can be represented as a matrix form below:

$$\boldsymbol{a}_x(\theta_i,\phi_i)=\begin{bmatrix}1\\ \exp(j2\pi d_x\sin\theta_i\cos\phi_i/\lambda)\\ \vdots\\ \vdots\\ \exp[j2\pi(M_x-1)d_x\sin\theta_i\cos\phi_i/\lambda]\end{bmatrix} \quad (3)$$

Similar to above analysis, the received signal model of M_y array elements in y-direction can be expressed as:

$$\mathbf{X}_y(t)=\mathbf{A}_y\mathbf{S}(t)+\mathbf{N}_y(t) \quad (4)$$

where $\mathbf{A}_y$ is the steering vector of y-direction which can be represented as

$$\mathbf{A}_y=[\boldsymbol{a}_y(\theta_1,\phi_1),\boldsymbol{a}_y(\theta_2,\phi_2),\cdots,\boldsymbol{a}_y(\theta_k,\phi_k)] \quad (5)$$

where $\boldsymbol{a}_y(\theta_i,\phi_i)$ is defined as the following vector form:

$$\boldsymbol{a}_y(\theta_i,\phi_i)=\begin{bmatrix}1\\ \exp(j2\pi d_y\sin\theta_i\sin\phi_i/\lambda)\\ \vdots\\ \vdots\\ \exp[j2\pi(M_y-1)d_y\sin\theta_i\sin\phi_i/\lambda]\end{bmatrix} \quad (6)$$

Above all, the received signal of this planar array can be written as

$$\mathbf{X}(t)=\mathbf{A}_x\mathbf{S}(t)\mathbf{A}_y^{\ T}+\mathbf{N}(t) \quad t=1,2,\cdots,N \quad (7)$$

In practice, the signal power is usually much higher than the noise power, thus it can be seen that the plane uniform rectangular array has a relatively high received signal-to-noise ratio(SNR) for far-field narrow-band uncorrelated signals. Thus, in the equation (7), $\mathbf{N}(t)$ will have no effect on the rank of $\mathbf{X}(t)$, and $rank(\mathbf{X}(t))\le rank(\mathbf{S}(t))=k$, that is $\mathbf{X}(t)$ is a low-rank matrix.

2.2 NSP property of 2D-DOA estimation model

The received signal in complete array can be recovered from the sparse array signal by MC algorithm. The received signal matrix of full rectangular array is expressed as $\mathbf{X}_R(t)$ and the received signal of sparse array can be $\mathbf{X}_s(t)$. Let a sampling operator be P_Ω, thus 2D-DOA signal model based on MC is

$$\begin{cases}\min \ \|\mathbf{X}_R(t)\|_* \\ s.t. \ \ P_\Omega(\mathbf{X}_R(t))=P_\Omega(\mathbf{X}_s(t))\end{cases} \quad (8)$$

The null space of orthogonal projection is $Null(P_\Omega)=\{\mathbf{M}\in\mathbf{R}^{n_1\times n_2}:P(\mathbf{M})=\mathbf{0}\}$. According to NULL Space Property, it is difficult to recover the matrix which exists in the null space of orthogonal projection by MC. Let us analyse the signal model established in previous section. Since the vector $\boldsymbol{a}_x(\theta_i,\phi_i)\neq\mathbf{0}$, any element in the matrix $\mathbf{A}_x$ consisting of such vectors is non-zero, too. Likewise, any element in the matrix $\mathbf{A}_y$ is also non-zero. And the diagonal elements in the diagonal matrix $\mathbf{S}(t)$ is non-zero. Then we can conclude that elements in $\mathbf{X}_R(t)$ are all non-zero. Consequently, arbitrary orthogonal projection meets $P_\Omega(\mathbf{X}_R(t))\neq\mathbf{0}$, that is, $\mathbf{X}_R(t)$ meets NSP and can be recovered in higher probability by MC.

3 GA-based of optimization of sampling elements

In the signal model based on MC, a small number of elements in planar array need to be selected to construct sparse array. Generally speaking, the array elements can be sampled uniformly at random. In this paper, genetic algorithm(GA) is implemented to optimize the selection of sampling elements, which makes the recovery error of MC smaller. If the number of sampled entries m obeys $m\ge Cr(n_1+n_2-r)$ for some positive constant C, the complete received matrix can be recovered with high probability. Let $m\ge 2r(n_1+n_2-r)$, then we convert the problem to minimizing recovery error, which can be represented as:

$$\min SNE=f(\mathrm{d}_1,\mathrm{d}_2,\cdots,\mathrm{d}_m) \quad (9)$$

where $(\mathrm{d}_1,\mathrm{d}_2,\cdots,\mathrm{d}_m)$ is the position vector set of the array and fit is a fitness function characterizing the reconstruction error of planar array signals, which can be denoted as

$$fit(\mathrm{d}_1,\mathrm{d}_2,\cdots,\mathrm{d}_m)=SNE=\|\mathbf{X}_{full}-\mathbf{X}_{MC}\|_2/\|\mathbf{X}_{full}\|_2 \quad (10)$$

where $\mathbf{X}_{full}$ is the complete received signal matrix and $\mathbf{X}_{MC}$ is the recovered matrix. The flow chart of optimizing the sampling elements by GA is as shown in figure 2.

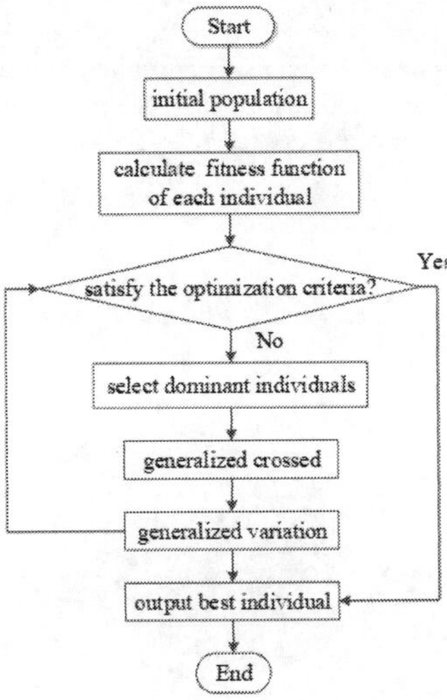

Fig. 2. The flow chart of optimizing the selection by GA

4 APG-MUSIC algorthim

We proposes a APG-MUSIC algorithm to solve the contradiction between the number of array elements and the average sidelobes level. Firstly, the complete received signal matrix $\mathbf{X}_R(t)$ can be achieved by APG algorithm from $\mathbf{X}_s(t)$, then we vectorize the matrix and solve its auto-correlation matrix. Finally, the targeted angles can be estimated from the spatial spectrum by MUSIC.

The established sparse signals model baesd on MC can be transformed into APG problem as follows:

$$\min \quad \mu\|\mathbf{X}_R(t)\|_* + \frac{1}{2}\|P_\Omega(\mathbf{X}_R(t)) - \mathbf{X}_s(t)\|_F^2 \tag{11}$$

The problem mentioned above can be solved by corresponding iterations and each iteration uses formula (12) to update the current matrix:

$$S_\tau(\mathbf{G}) = \mathbf{U}Diag((\sigma - \mu/\tau)_+)\mathbf{V}^T \tag{12}$$

where μ is a positive constant and $\mathbf{U}/\mathbf{V}/\sigma$ are obtained from the singular value decomposition(SVD). Given $\mu > 0$, let $\mathbf{X}^0 = \mathbf{X}^{-1} = \mathbf{X}_s(t)$, $t^0 = t^{-1} = 1$, $k = 0,1,2,\cdots,$ the iteration process from $\mathbf{X}^k$ to $\mathbf{X}^{k+1}$ can be given by the following steps:

1. Set $\mathbf{P}^k = \mathbf{X}^k + \frac{t^{k-1}-1}{t^k}(\mathbf{X}^k - \mathbf{X}^{k-1})$;

2. $\mathbf{G}^k = \mathbf{P}^k - (\tau^k)^{-1}P_\Omega^*(P_\Omega(\mathbf{P}^k) - \mathbf{X}_s(t))$, where $\tau_k > 0$;

3. Obtain $\mathbf{U}$, $\mathbf{V}$, σ from the SVD of $\mathbf{G}^k$ then do contraction operator on it, that is,
$\mathbf{X}^{k+1} = S_{\tau^k}(\mathbf{G}^k) = \mathbf{U}Diag((\sigma - \mu/\tau^k)_+)\mathbf{V}^T$

4. Update $t^{k+1} = \frac{1+\sqrt{1+4(t^k)^2}}{2}$;

5. If the stop condition of iterations is satisfied, then end the loops and the output is turned out to be $\mathbf{X}^{opt}$, that is, $\mathbf{X}^{opt} = \mathbf{X}^k$. Otherwise, go to step 1.

The stop condition can be chosen as shown in equation (13).

$$\frac{\left\|\mathbf{X}^k - \mathbf{X}^{k-1}\right\|_F}{\max\left\{\left\|\mathbf{X}^k\right\|_F, 1\right\}} < \varepsilon \tag{13}$$

where ε is a tolerance for stopping criterion. The contraction operator in Step3 can be denoted as

$$S_a(x) = \begin{cases} x-a & x > a \\ x+a, & x < -a \\ 0, & others \end{cases} \tag{14}$$

where a is a positive constant. From above iterative process, regardless of the noise, the structure of the received signal can be represented as

$$\hat{\mathbf{X}}_R(t) = \mathbf{A}_x\mathbf{S}(t)\mathbf{A}_y^* = \mathbf{U}\mathbf{\Sigma}\mathbf{V}^* \tag{15}$$

Vectorize $\hat{\mathbf{X}}_R(t)$, we can obtain the vector form $\boldsymbol{x}_R(t)$:

$$\boldsymbol{x}_R(t) = \sum_{i=1}^{k} \boldsymbol{a}(\phi_i, \theta_i)\boldsymbol{s}_i(t) + \boldsymbol{n}(t) \qquad t = 1,2,\cdots,N \tag{16}$$

Compute the auto-correlation matrix $\mathbf{R}_x$ of the received signal vector $\mathbf{R}_x$:

$$\mathbf{R}_x = \frac{1}{N}\sum_{n=1}^{N} \boldsymbol{x}(n)\boldsymbol{x}^H(n) \tag{17}$$

where N is the number of snapshots.

Then solve the eigen-decomposition of the matrix $\mathbf{R}_x$, we can get

$$\mathbf{R}_x = \mathbf{U}_S\mathbf{\Sigma}_S\mathbf{U}_S{}^H + \mathbf{U}_N\mathbf{\Sigma}_N\mathbf{U}_N{}^H \tag{18}$$

where $\mathbf{U}_S$ is the signal subspace consisting of eigenvectors of k larger eigenvalues and $\mathbf{U}_N$ is the noise subspace composed of the remaining eigenvectors.

The signal and noise subspace also satisfies:

$$\mathbf{U}_S\mathbf{U}_S^H + \mathbf{U}_N\mathbf{U}_N^H = \mathbf{I} \tag{19}$$

Therefore the 2D spatial spectrum can be obtained to conduct spectral peak search:

$$P_{2D\text{-}MUSIC} = \frac{1}{\boldsymbol{a}^H(\theta,\phi)\mathbf{U}_N\mathbf{U}_N^H\boldsymbol{a}(\theta,\phi)} \tag{20}$$

where $\boldsymbol{a}(\theta,\phi) = \boldsymbol{a}_y(\theta,\phi) \otimes \boldsymbol{a}_x(\theta,\phi)$.

5 Simulations and analysis

In this section, computer simulations are carried out to demonstrate the effectiveness and superiority of the proposed algorithm. Assume that the URA contains 64 × 64=4096 entries totally and elements spanned spacing in x-direction and y-direction equals to half of the wavelength, that is, let $d_x = d_y = \lambda/2$. GA algorithm

 https://doi.org/10.1051/matecconf/201820801004

is employed to optimize the sampling array. We choose m=1800. Consider three incoherent sources in the space. Their elevation angle θ and azimuth angle ϕ are (10°,15°), (30 ,25) and (50°,35°) respectively.

Experiment1 2D-DOA estimation in sparse array is discussed. Set Signal Noise Ratio(SNR) as 10dB and the number of snapshots as 150. The 2D spatial spectrum of randomly sampling array by MUSIC algorithm is shown in figure 3. As shown in the pattern, the average sidelobes level is -2.99dB. It can be concluded that due to the sparsity of the array, the average sidelobes level rises and the spectrum broadens, which will seriously affects the accuracy of 2D-DOA estimation.

Experiment2 2D-DOA estimation based on APG-MUSIC with optimized sampling array is discussed. We utilize optimized sparse array to reconstruct received signal matrix by APG and then implent MUSIC to estimate the spatial spectrum. The result is shown in figure 4. It can be easily seen that the average sidelobes level is -56.31dB. It is easy to find that by comparison with figure 3 the proposed method has better performance in 2D-DOA estimation under the same sampling size. It is mainly due to its ability to reconstruct the received signal matrix with a higher probability, which increases accumulated SNR to some extent. Therefore, the APG-MUSIC 2D-DOA estimation method can effectively suppress sidelobes level.

Experiment3 The relationship between sampling array selection and reconstruction performance is verified. We compare uniformly random sampling with GA optimized selection when the sampling number is 1500, 1800, 2100, 2400 and 2700 respectively. It can be easily seen from figure 5 that GA optimized sampling has better performance in matrix reconstruction and significantly reduces recover error. The experimental results show the importance and research value of sampling array optimization in array signal reconstruction.

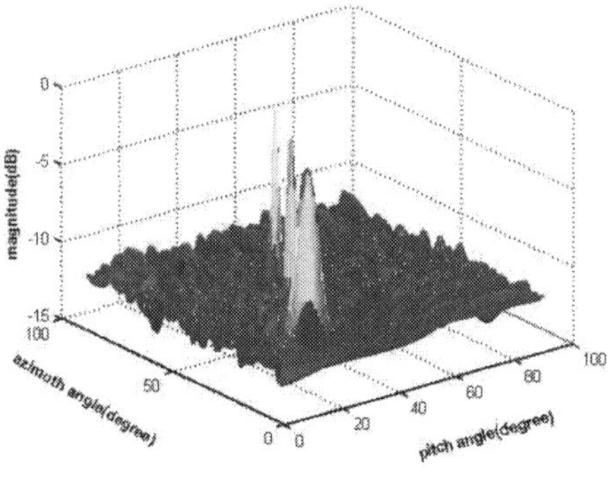

Fig.3. 2D-DOA estimation by MUSIC under randomly sampling array

6 Conclusion

An APG-MUSIC 2D-DOA estimation method based on sparse sampling array optimized is proposed in this paper. Compared with conventional 2D-DOA estimation methods, the proposed method not only achieves low sidelobes level, but also reduces the sampling number dramatically and improves the utilization rate of array, which has high research value and broad application prospect. In my paper, the noise is not taken into consideration. However, how to recover a complete matrix accurately in a noisy environment is an important research direction. Consequently, the related problems of 2D-DOA estimation based on MC under noise need to be further studied.

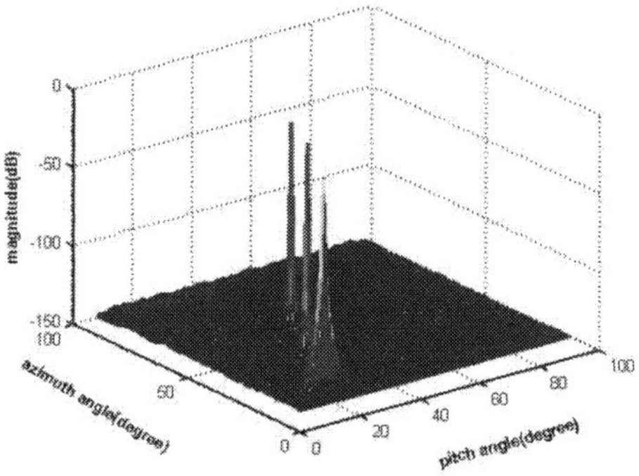

Fig. 4. 2D-DOA estimation by APG-MUSIC under GA optimized selection

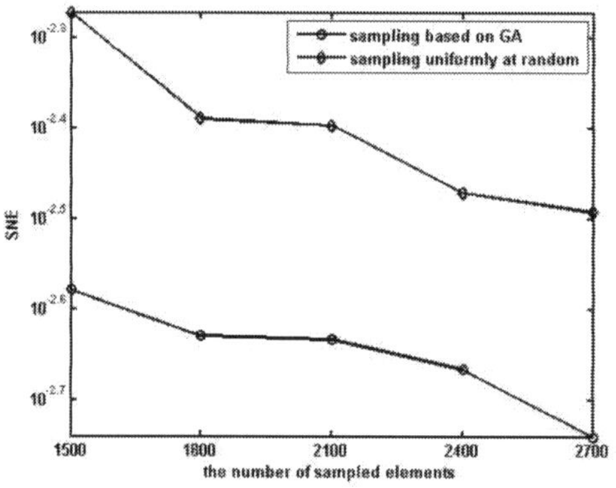

Fig. 5. SNE verus the sampling number in two different methods

References

1. DUDGEON D E. Fundamentals of digital array processing[J]. Proceedings of the IEEE,1977,**65**(6):898-904.
2. Zhang Xiaofei. *Theory and application of array signalprocessing* [M]. Beijing:National Defence Industry Press,2010.
3. Luo Zheng,Zhang Min,Li Pengfei. The new algorithm of two-dimensional DOA based on auto-correlation matrix with high power[J]. Journal of aviation 2012,**33**(4):696-704.

4. Pal P, Vaidyanathan P P.A grid-less approach to underdetermined direction of arrival estimation via low rank matrix denoising[J]. Signal Processing Letters, IEEE,2014,**21**(6): 737-741.
5. Cands E J,Plan Y.Matrix Completion WithNoise[J]. Proceedings of the IEEE,2009.
6. Cands EJ, Eldar YC, Strohmer T, Voroninski V. Phase retrieval via matrix completion. SIAM J Imaging Sci 2011;**6**(1):199225.
7. Cands EJ, Emmanuel S, Recht B. Exact matrix completion via convex optimization. Commun ACM 2012;**55**(6):1119.
8. Chen C, He B, Yuan X. Matrix completion via an alternating direction method.IMA J Numer Anal 2012;**32**(1):22745.
9. Toh Kim-Chuan, Yun Sangwoon. An accelerated proximal gradient algorithm for nuclear norm regularized linear least squares problems. Pacific J Optim 2010;**6**(3):61540.
10. Donoho DL.Compressed sensing. IEEE Trans Inform Theory 2006;**52**(4):1289306.
11. Duarte MF, Baraniuk RG. Spectral compressive sensing. Appl Comput Harmon Anal 2013;**35**(1):11129.
12. Yuan X, Yang J.Sparse and low-rank matrix decomposition via alternating direction methods[J]. preprint,2009,12.
13. Recht B, Xu W, Hassibi B. Necessary and sufficient conditions for success of the nuclear norm heuristic for rank minimization.Eprint Arxiv 2008;**16**(5):306570.
14. Li B, Petropulu A. Spectrum sharing between matrix completion based MIMO radars and a MIMO communication system[C]. *IEEE International Conference on Acoustics, Speech and Signal Processing. IEEE*,2015:2444-2448.
15. Zhang N, Xie C L, Tang J, et al. Three-dimensional extensions of matrix completion based MIMO radar[C]//*Iet International Radar Conference*. 2015:4. -4.
16. Wanghan L V, Wang H, Liu F, et al. Wideband DOA Estimation Based on Co-Prime Arrays with Sub-Nyquist Sampling[J]. Ieice Transactions on Fundamentals of Electronics Communications and Computer Sciences,2016, E99.A(9):1717-1720.
17. Chen Chuan, Guo Yong, Jia Yong. A DOA estimation method based on dual frequency coprime array[J]. Computer Application Research, 2017,**34**(05):1459-1462.
18. Wu Chenxi, Zhang Min, Wang Ke. A underdetermined DOA estimation method of coprime array based on matrix completion[J]. Engineering Science and Technology,2017,**49** (05): 156-163.
19. Hu Y, Zhang D, Ye J, et al. Fast and accurate matrix completion via truncated nuclear norm regularization. *IEEE Trans Pattern Anal Mach Intell* 2013;**35**(9):211730.
20. Recht B. A simpler approach to matrix completion. J Mach Learn Res 2011;**12**(4):3413-30.

Ultrasonic Image Enhancement to Internal Defect Detection during Material Inspection

Thouraya Merazi-Meksen, *Malika* Boudraa and *Bachir* Boudraa

USTHB, Electronics & Computer Science Dpt, Algiers, Algeria

Abstract. Ultrasonic waves are efficiently exploited to detect internal defects such as cracks and occlusions in non-destructive testing (NDT) of materials. In some automated procedures, emitter and receiver probes are displaced step by step and signals are displayed to form images that make diagnosis easier to interpret. However, too much images are unnecessarily stored when they correspond to safe zones (without presented defects). In this paper, we propose a method that enhances ultrasonic images quality in order to improve the discrimination between images containing or not a shape regarding a detected defect. The method is based on the adapted histogram equalization technique followed by morphological operations that erase all useless shapes on processed images.

1 Introduction

Non Destructive Testing of Materials consists in inspecting the integrity of a structure without causing damages in order to preserve its future use. Ultrasonic waves are exploited for this purpose because of their safety and their low cost [1]. The principle of ultrasonic inspection is based on analyzing produced waves when an incident one interacts with a discontinuity (defect) presented in the structure under control. Principally, three types of waves are generated: reflected one, refracted one and diffracted one [2]. Diffracted waves present the advantage to propagate all around the defective zone. This makes their detection less dependent of the receiving probe position. However this type of waves is generated only by acute tips. That is why crack defect detection benefits by this wave exploitation in the Time of Flight Diffraction named technique (TOFD) [3].

Nowadays, the graphic performances allow quick image formation and visualization. So ultrasonic signals are digitalized and stored to form ultrasonic images easier to interpret and shapes regarding internal detected defects may be analyzed. Locating, dimensioning as well as characterizing are thus deduced by exploiting existing image processing techniques [4,5,6]. This facilitates the decision making and allows more robustness and reliability in the defect recognition.

In the case of TOFD image, as diffracted wave propagates in all directions around the discontinuities, its energy is significantly dispersed. Thus, resulting observed signal amplitude is relatively low involving low contrast in the formed images. The forms regarding presented defects may be difficult to observe. In order to increases the capacities of the detection [7, 8, 9].

Even though the mass storage is becoming increasingly more important, we think that classifying formed images in "defective zone " and "non defective zone" as their formation is suitable and will avoid the storage of a huge of useless images. This will allow saving memory and time. Pattern recognition techniques also give a substantial help in this domain when they are applied to defect shape recognition [10,11]. For this purpose, the first processing step consists in discriminating the different shapes presented in the analyzed image and eliminating background noise.

This paper presents a proposed method to extract principal feature presented in a TOFD image by histogram analysis and mathematical morphology application. In a first step, adaptive histogram equalizing technique is exploited in order to enhance the contrast and to recognize gray levels regarding significant presented shapes. Thus, the image is region segmented consequently and erosion and dilatation morphological operators can be successively applied to eliminate background irregularities while maintaining the resulting shapes at their effective sizes.

Next section of this paper describes briefly the TOFD inspection technique. In section 3, the technique to enhance image contrast by exploiting adapted histogram and morphological operators is developed. The results obtained on a TOFD image presenting a crack defect are shown in section 4. Section 5 summarizes the conclusions.

2 TOFD image formation

TOFD images are synthesized during ultrasonic inspection techniques by exploiting diffracted waves. Each row of these images is constituted by the samples of a received signal during the scanning of the structure under control [3]. Two probes are used: one emitting ultrasonic pulses, the other receiving diffracted waves that are generated at tips of presented discontinuities. Those two probes are maintained at equal distances and are displaced step by step, according to a straight line. At each position, the received signal is digitized and his samples will constitute the pixels of one row in the formed image. The amplitudes are displayed in grey levels, black for negative values, grey near the zero and white for maximal ones. Thus, images obtained scanning safe zones (no defect) will be uniformly grey while a white shape will appear in the image in a defective zone. Indeed, in safe zones, signals are constituted only by noise, while signal amplitude will increase in case of a presented discontinuity that has refracted the emitted ultrasonic wave. Figure 1 shows an example of a TOFD image. As both surface and the back-wall of the controlled structure produce at each probe position reflected waves (Fig. 1 (a), on each row of the image it will appear at the same columns white and black pixels regarding maximum and minimum amplitudes, forming thus the corresponding lines in the image (Fig. 1. (b), The pulse regarding a detected defect (between surface and back-wall echoes) produces a shape that can give

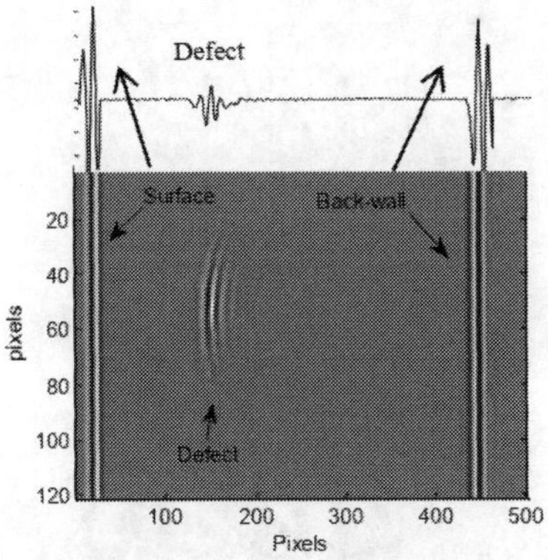

Fig 1. At each position, the samples of the received A-scan signal are displayed as pixels of a row in the formed image.The level of the grey depends of the sample amplitude. Vertical white and black lines result from surface and back-wall echoes.

The defect location is proportionally deduced by the position of its corresponding shape between surface and back-wall echoes.

3 Image enhancement

Because diffracted waves propagate all around the defective area, their energy is dispersed also all around. That implies low amplitudes in representative signals, so low contrast in the formed images. The defect detection and characterization accuracies depend largely of an enhancement pre-processing phase. In this section, the histogram equalization based technique to image enhancement is described as well as its improved version.

3.1 Histogram equalization

A histogram is a graph that shows frequency of occurring of pixels intensities in the whole image [8, 9]. When a region is constituted only by noise, each pixel value has the same probability to occur. If the gray-level histogram corresponds to an image I(x,y), composed of light objects in a dark background, object and background pixels will have grey levels grouped into two dominant modes.

Extracting objects from background returns to select a threshold *th* that separates these modes. Then any point *(i,j)* for which $I(i,j) > th$ is called an object point, otherwise, the point is called a background point.

Histogram equalization allows to enhance image quality by removing superimposed dark or light fog. Equalizing the histogram leads to making it as flat as possible and apportioning initial image grey levels all over the grey level band *L*. (L=256 in most of cases). Resulting equalized histogram *p* is calculated as follows:

$$P_n = \frac{\text{number of pixels with intensity } n}{\text{total number of pixels}} \quad (1)$$

$n = 1, 2,3,\dots.L\text{-}1$

And the resulting image

$$g(i,j) = (L-1)\sum_{n=0}^{I(i,j)} p_n \quad (2)$$

The histogram equalized image *g* will be defined by

$$g(i,j) = (L-1)\sum_{n=0}^{I(i,j)} p_n \quad (3)$$

where floor() rounds down to the nearest integer. This is equivalent to transforming the pixel intensities, *k*, of *f* by the function

$$T(k) = floor((L-1)\sum_{n=0}^{k} p_n \quad (4)$$

The motivation for this transformation comes from thinking of the intensities of f and g as continuous random variables *X*, *Y* on [0, L − 1] with *Y* defined by:

$$Y = T(X) = (L-1)\int_0^x p_x(x)dx \quad (5)$$

where p_x is the probability density function. *T* the cumulative distributive function of *X* multiplied by *(L−1)*.

Figure 2 illustrates an example of envelopes of an histogram, before and after equalizing (respectively in (a) and (b)).

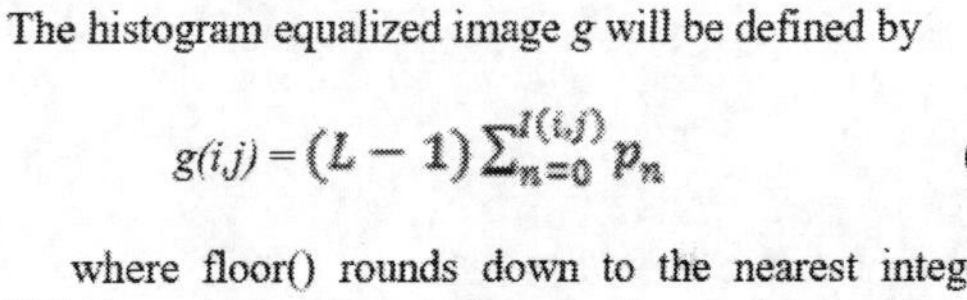

Fig 2. Example of histogram before equalization (a), and after equalization (b).

While the histogram operates on the entire image, adaptive histogram operates on small regions. Thus, each region's threshold is determined depending on local image characteristics. This allows to increase contrast and facilitates detection and recognition of presented shapes. The neighboring regions are then combined using interpolation in order to eliminate induced boundaries. The contrast, especially in homogeneous areas, can be limited in order to avoid amplifying the noise which might be present in the image.

3.2 Morphological operations

Erosion is a basic morphological operation that is exploited to eliminate noise in binary images [12]. Its principle consists in scanning the image by using a structuring element with a specific form that depends on the shape to select or to suppress. The structuring element is constituted by a group of pixels and has its origin at the center. It is shifted over the image and at each pixel of the image its elements are compared with the set of the underlying pixels. If the two sets of elements match the condition defined by the set operator (*e.g.* if the set of pixels in the structuring element is a subset of the underlying image pixels), the pixel underneath the origin of the structuring element is set to a pre-defined value (0 or 1 for binary images). This operation, while eliminating artifacts, reduces processed shapes. That is why it is often followed by the dilatation operation that uses the same structuring element to enlarge the boundaries.

4 Application and results

In this work a TOFD image has been formed by displacing the tandem of emitting and receiving probes on a structure where a crack is simulated by a drilled hole as illustrated in figure 3. Indeed, the circular sides reflects incident waves all around the axis just like diffracted ones. This is why this type of waves are generaly simulated by drilled holes.

The probes are moved step by step, 0.5mm 120 times producing 120 ultrasonic signals that have been used to synthetize the image constituted thus by 120 rows.

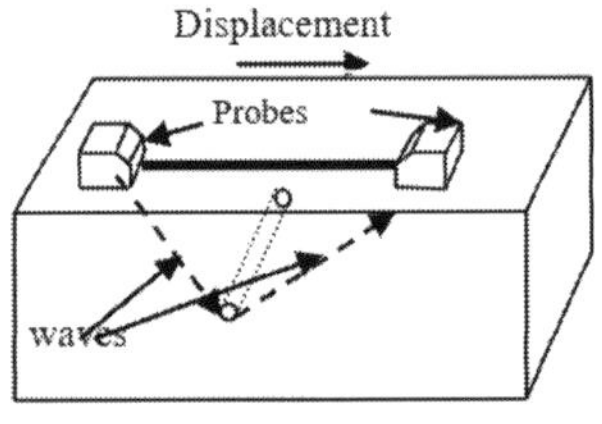

Fig 3. A TOFD inspection description. The receiving and emitting probes are moved step by step according to a staright line as shown by the arrow.

Figure 4(a) shows the formed image where surface echoes have been ignored. The arc of barabola corresponds to reflected waves by the drilled hole.

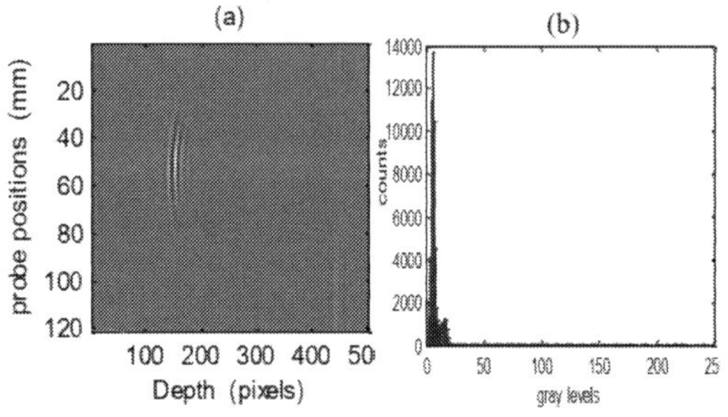

Figure 4. (a): Tofd image, formed by simulating a crack defect by a drilled hole. (b): Histogram of the image represented in (a).

The histogram of this image has been equalized (Fig.5 (a)) and the resulting image is shown in figure 5(b). In addition to defect shape (set of parabolas), the drilled hole as well as back wall echoes are visible, confirming the efficiency in image enhancing by equalizing histogram.

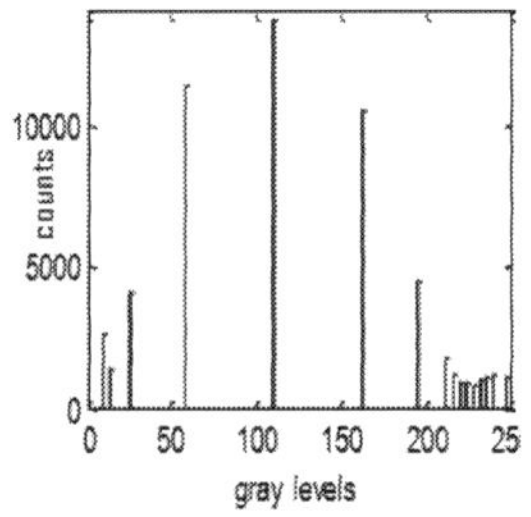

Fig 5 (a). Equalized histogram previously processed.

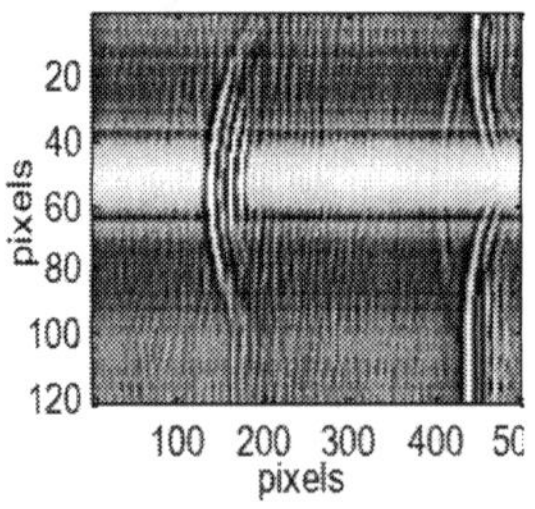

Fig 5 (b). Resulting image.

Considering the resulting histogram, it appears that significant gray levels are peaks between 50 and 200. So, the binarization can be performed by cancelling insignificant pixels (gray levels greater than 200 and less than 50). Figure 6(a) shows obtained binary image where the pertinent features are discriminated: namely the drilled hole, the reflected waves that simulate the diffracted ones and the backwall echoes (at the right of the image).

Background noise has been removed by exploiting erosion technique as explained in previous section. The structuring element has been chosen as a disk of 2 pixels radius because the most important shape (that is regarding reflected waves) is curvature. As all shapes are thus eroded, a dilatation phase has to be performed in order to recover effective sizes of the features. Figure 6(b) illustrates resulting processed image.

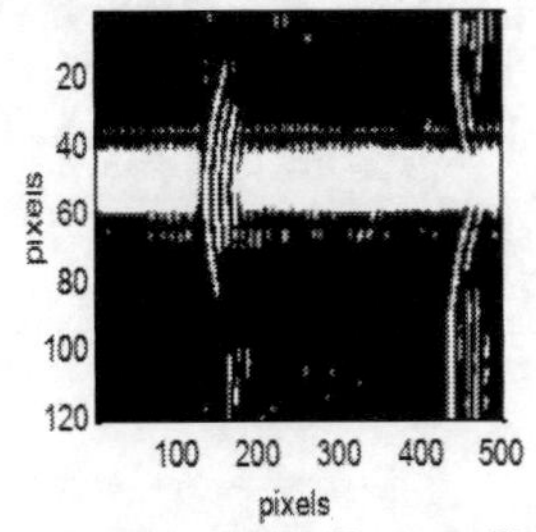

Fig 6 (a). Binarized image obtained by cancelling all pixels of gray levels that are different from those selected by the peaks of the equalized histogram.

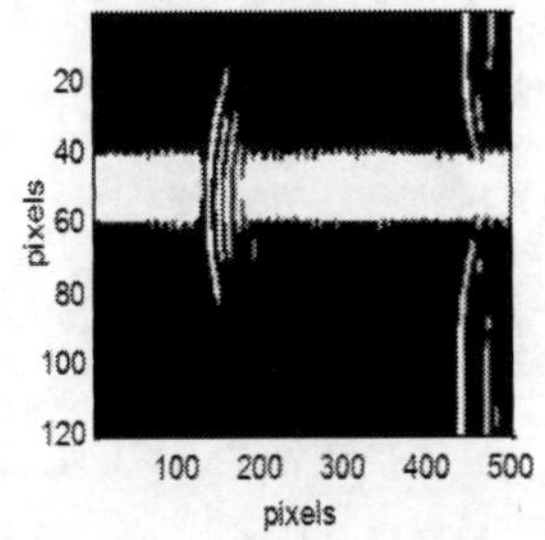

Fig 6 (b). Resulting image after the erosion/dilatation phases.

At this stage, it becomes easier to exploit pattern recognition techniques to parabolas detection in order to automate diffracted waves detection. Indeed, parameterized forms may be detected by least mean square algorithm, as well as RANSAC filter [6]. This phase will totally automate this type of control.

5 Conclusion

During automatic procedures of non destructive testing of materials, a large amount of images are synthetised. Their interpretation is widely dependant of their quality, especially when diffracted waves are exploited. Indeed, these waves are of low amplitude because of their energy dispersion all around a presented discontinuity. However, making the detection independant of the probes positions, diffracted waves are suitable for crack detection. In this paper, it has been shown how the exploitation of histogram equalization and morphological operations enhances contrasts and discriminates efficiently presented shapes in the image. By improving the quality of produced TOFD images in an automated material inspection facilitates control diagnosis.

References

1. R.A. Blake, & J.P. Allebach, Digital ultrasonic image construction using electronic ordered dither techniques, J Nondestruct Eval. 1981 pp 2- 75.
2. J. Krautkramer and G.Simpson. Ultrasonic Testing of Materials. Springer Science and Business Media. 2013.
3. M.Silk, The potential of scattered or diffracted ultrasound in the determination of crack depth", Research techniques in NDT, vol 3, June 1975.
4. X. Gao, Y. Shen, & Luo, L. Noise Reduction Method for Ultrasonic TOFD Image Based on Image Registration ,J Nondestruct Eval 325-330. 2013 doi:10.1007/s10921-013-0185-9.
5. R. Gonzalez, C.and. R.E. Woods Digital image Processing ", Third Edition, 2008.
6. Hornberg A.Handbook of Machine vision. Wiley VCH (2006)
7. P. J. Janani Premaladha and K. S. Ravichandran Image enhancement techniques: A study Indian Journal of Science and Technology, vol. 8 issue 22. doi: 10.17485/ijst/2015/v8i22/79318
8. R.Maini andd A. Himanshu A Comprehensive Review of Image Enhancement Techniques. journal of computing, volume 2, issue 3, march 2010.
9. P. Janani, J. Premaladha and K. S. Ravichandran Image enhancement techniques: A study, Indian Journal of Science and Technology, vol. 8, issue 22, doi: 10.17485/ijst/2015/v8i22/79318
10. P.A Petcher. and S. Dixon Parabolas Detection using Mached Filtering for Ultrasound B-scans. Ultrasonics.vol. 52 pp 138-144. (2012).
11. S. Singh, & M. Kaur, Machine vision system for automated visual inspection of tile's surface quality. *IOSR J Eng*, vol. 2issue 3, pp 429-432, 2012.
12. J. Serra P. Soie Morphological mathematics and its applications to image processing. Springer Science and Business Media. 2012.

Optimization of Probe Train Size for Available Bandwidth Estimation in High-speed Networks

Veronika Kirova[1], *Eduard* Siemens[1], *Dmitry* Kachan[1], *Oksana* Vasylenko[2] and *Kirill* Karpov[1]

[1]Anhalt University of Applied Sciences, Department of Electrical, Mechanical and Industrial Engineering, Bernburgerstr. 55, Köthen, Germany
[2]O. S. Popov Odessa National Academy of Telecommunications, Department of Higher Mathematics, Kovalska Str. 1, Odessa, Ukraine

Abstract. Available bandwidth parameter is a crucial characteristic in terms of networking and data transmission. The beforehand knowledge of its value and use of this parameter in various traffic engineering algorithms and QoS calculations is a key for high-efficient multigigabit data transport in nowadays networks. The challenge in available bandwidth estimations is not only in its accuracy and processing speed but also in the reduction of the amount of probe traffic injected into the network by keeping an adequate level of estimation accuracy. In this paper we extend existing active probing measurement algorithms for end-to-end available bandwidth estimation along with methods to reduce estimation times and amount of injected traffic while keeping measurement accuracy constant and even reducing the uncertainty of estimations. The main goal of this research was to detect a sufficient ratio of MTU, packet train size with the link capacity and available bandwidth (AvB) in up to 10 Gbps networks. In order to explore measurement accuracy under different conditions, a new tool for the AvB estimation named Kite2 has been developed and is presented in the paper. Comparative performance of AvB estimations using Kite2, Kite and Yaz is presented. Finally we calculate with statistical means dependency between the estimation error probability, measurement probing overhead and the measurement time.

1 Introduction

Accuracy of available bandwidth (AvB) estimation heavily relies on the method of packet probing and the mathematical assumptions concluded from the delay distribution of the probe packets [1].

Available bandwidth of the end-to-end path defines the instantly unused capacity of the path in a contrary to the overall capacity, which is defined by the physical parameters of the link, so AvB depends on the instant traffic load and has almost unpredictable character.

Since the traffic load in the link has a burst character, the task of defining spare capacity in a certain time is challenging, because a process of measurement is affected by continuous intervention by a diversity of cross traffic, context switching of transmission/reception processes, small packet buffers, interrupt coalescence mechanisms, etc. [2].

The knowledge of AvB is especially important for contemporary network protocols and applications, as it enables them to adapt the transmission data rate tightly to the abrupt changes in the communication channel, which are typical, for example, in SDN networks. Also fast and highly accurate AvB estimations lay the basis for efficient operations of SDNs with regard to re-routing and switching of high-priority paths through high-speed networks. Furthermore, non-intrusive AvB measurements are used for an embedding of machine learning and artificial intelligence algorithms into applications for data exchange. Hereby the precise estimation of instantly available bandwidth serves as main input parameter for machine learning algorithms of transport protocols.

It is worth noting that active measurement methodologies rely on the emission of probe traffic in the network and afterwards processing of the impact of intermediate network devices on the injected traffic: loss rate, delays, jitters, packet reordering, etc. [3]. Consequently, as active measurement tools raise traffic load in the network, one of its significant flaws is related to the intrusion caused by the measurement traffic which can result in a QoS deterioration, and thus provide biased available bandwidth measuring results. Also, the massive network occupation by active probes intermediate devices can falsely identify as DoS attacks or flooding. It can lead to the probe traffic rejection and cause the measurement process failure. The level of disturbance by active probes is then one of the key features which must be taken into account by the active probing tool design.

One of the approaches to reduce the bias caused by the active probing is based on the minimization of the samples number in its sending train sizes. On the other hand, if an active measurement tool operates with a small probing amount, it provides shorter execution times, which are significant for the highly dynamic traffic of the most networks. However, it also increases the risk to underestimate the available resources.

In this paper we evaluate the accuracy of active measurement algorithms of AvB estimations and optimize the packet sizes in probe trains by finding a trade-off between the amount data injected by active probing, estimation times and suitable accuracy of AvB estimation. In order to investigate measurement performance under the different probing intensity, a new tool called **Kite2** is proposed here. It extends the capabilities of **Kite** which was used for AvB estimations before [4]. With the presented improvements, Kite2 achieves faster convergence of AvB measurements and a significantly lower uncertainty of the estimations than Kite. The algorithms used in Kite2 along with its performance results are also described in this paper. Finally the limitations of measurements with Kite2, Kite and the older tool **Yaz** [5] are shown in section 4.2.

2 Background and related work

Several software tools for the available bandwidth estimation were developed in recent years which are using different methods and models of estimation network metrics, such as *pathload* [6] and *pathChirp* [7] which are based on evaluation of one-way-delay properties changes while sending periodic streams of packets; algorithm IGI [8], which detects proportional correlation of initial packet gaps changes and cross traffic in the path, thus detect a tight link; already mentioned Yaz which makes calculations based on difference of interpacket intervals on receiver and sender side, *Abing* [9] which evaluates packet pair dispersion; *ASSOLO* [10] which is based on enhanced algorithm of *pathChirp* and increase the accuracy of measurement by increasing of the probing stream density; *ttcp* [11] which measures achievable TCP throughput; *pathrate* [12] which is based on packet pairs dispersion method; *Netperf* that uses large TCP transfer [9]; and others which were described in papers like [13].

A summary of existing tools and comparison of their performance is given in a few papers like [14] and [15]. It is important to distinguish between different network metrics which are measured by these tools: link capacity, available bandwidth or achievable TCP throughput as each of these parameters define different characteristics.

Despite the fact of such amount of related investigations on AvB estimations, authors of this paper found only a few research related to applicability of these tools in the high-speed networking of 10 Gbps. A proposed methodology can be found in [4] and a pilot investigation in real networks in [16].

3 Testbed topology

The results of the paper have been collected from experiments performed on a testbed described in *Figure 1*.

The laboratory environment of 10Gbps network consists of equipment:

- 2 instances of Netropy 10G WAN Emulators, which emulate WAN links using variable inferences like packet loss ratio up to 100%, delays of up to 100s, diversified distributions of packet corruption and duplication laws;

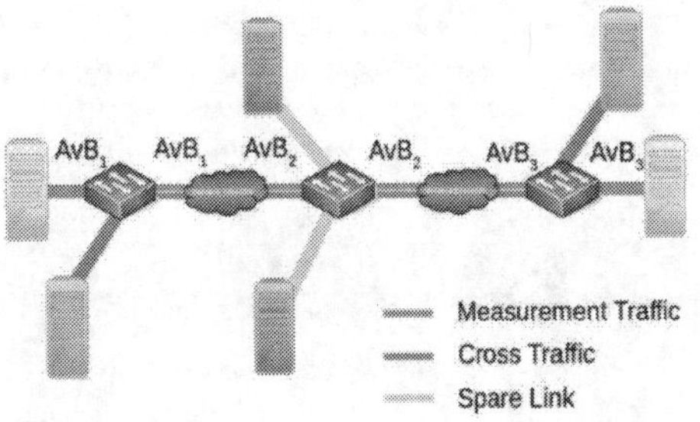

Fig 1. Setup of a measurement testbed.

- 2 instances of end systems one on which is sender with 12 x Intel(R) Xeon(R) CPUX5690 3.47 GHz processors, 32 GB RAM and equipped with 10G Chelsio NIC (driver cxgb4-2.0.0-ko), another one is receiver with 6 x Intel(R) Xeon(R) CPU X5690 3.47GHz processors, 42GB of RAM and 10G Chelsio NIC. Both machines run Linux distribution – Ubuntu 16.04 system with the 4.4.0-97-low-latency kernel.
- 10GE network switches Summit x650 from Extreme Networks were used as network devices.

In order to emulate fluidity of AvB under real world, one of the network emulators cyclically has been changed its capacity with such steps: 2 Gbps, 7 Gbps, 10 Gbps during the whole measurement.

Also, measurements have been performed on servers with different NICs - Intel (ixgbe-4.2.1-k) with an eye to confirm results tendency.

The RTT on the link between sender and receiver without emulation is equal to 250 µs. Additional emulated delay was set to a fixed value of 20 ms.

4 Measurement tool description

In terms of this research a new tool named Kite2 has been developed. It borrows its name from Kite application [4] estimation methodology of which had formed the basis of the developed tool.

Approach of Kite2 is based on the Probe Rate Model (PRM) of the active probing measurement and basically uses analysis of inter-packet interval deviation on the receiver side (IpIr) in order to detect whether the sending rate of probe packets meets the available bandwidth limits of the link between the sender and the receiver.

4.1. Kite2 workflow

The test application was developed with performance and simplicity in mind. It is written in C++ programming language using HighPerTimer library for precise time measurements [17], has a client-server architecture and supports 3 probing modes:

- Sequential (train mode);
- Batch (packet pairs);
- Packet pairs inverted.

These modes are described below. The measurement process is comprises of short send sessions (probing), according to specified probing mode:

- sequential, when the probe samples are sent with a constant interval, and the receiving side measures interval between each successive packets, called "train mode" [18] *(Figure 2)*;

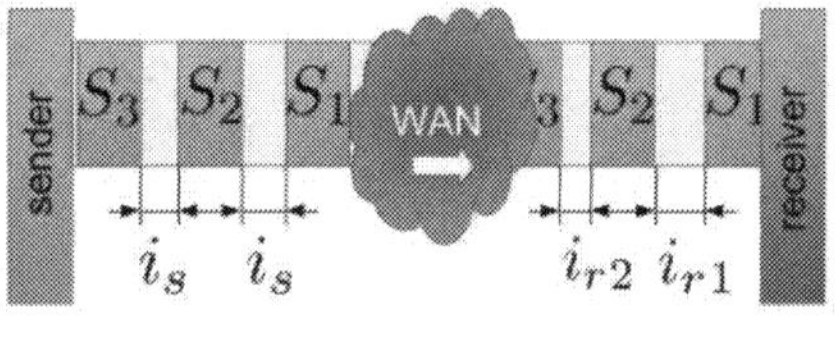

Fig 2. Train probing mode measurement.

- batch, when two consecutive samples are sent with the shortest possible interval and after that pair of packet, a pause of a specific inter-packet interval is inserted. It results in coupled packets. Receiver measures then intervals between pairs, which is defined as inter-send packet interval on the sending side, e.g. between 2nd and 3rd, 4th and 5th packets. This operational mode is called "packet pairs" (*Figure 3*).

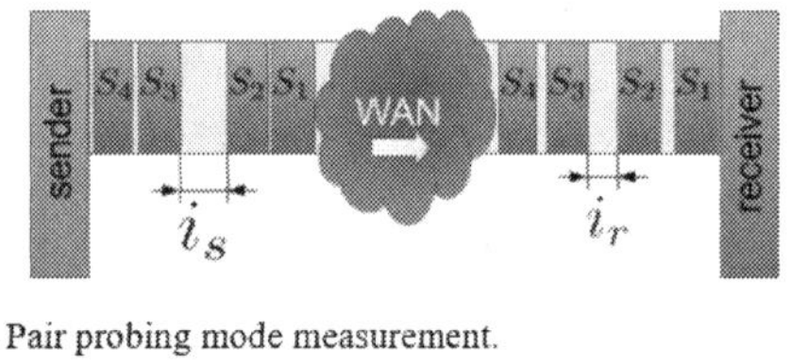

Fig 3. Pair probing mode measurement.

- "Packet pairs inverted" mode means the same operation as "packet pairs" one, but the receiver instead measures interval within a pair, which corresponds to the shortest possible send interval, e.g. between 1 and 2, 3 and 4 packets. This mode was developed in a contradiction to the previous ones in order to analyze the efficiency of well-known modes and a new one in the high-speed network environment.

In order to be able to investigate contemporary high capacity links, it is necessary to have a reliable time source and precise time measuring functions to keep a constant interval between probing samples while sending ones and to obtain accurate results during their reception. Linux API does not provide the sufficient accuracy [19], suitable for 1 Gbps and above networks, from this point HighPerTimer library provides up to 20 nanoseconds accuracy for time acquisition and waiting operations, thus allowing to investigate 10 Gbps and 40 Gbps networks, provided by FILA at Anhalt University of Applied Sciences.

Kite2 has an option to specify sending rate or approximate utilization of the bandwidth of a link, as the main goal of Kite2 is to implement an effective approach of bandwidth estimation with the smallest possible impact on the channel. The idea behind this estimation, as described above, is based on the observation of inter-packet interval deviations between sending and receiving sides.

However, despite the method simplicity there are pitfalls introduced by different features of network devices, such as undetermined buffering methods of intermediate network devices or interrupt coalescence on a host running Kite2 which is an objective of the current research.

Under the cover, client and server exchange UDP messages. The first one sends probing samples, structure of which is shown in *Figure 4*, with the defined interval.

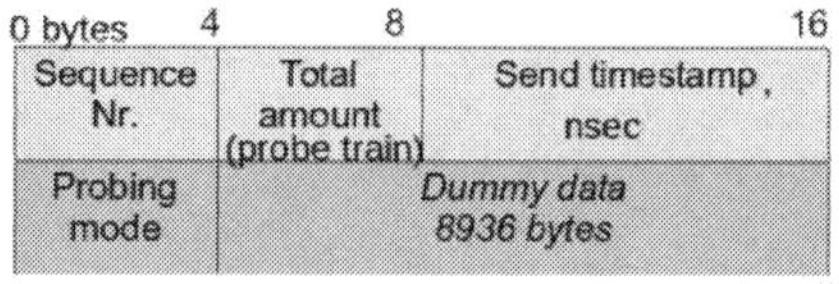

Fig 4. Kite2 sample structure (UDP payload).

The interval can be specified by a user as the transmission data rate option. Server waits for the first sample and then begins a probing session estimation, initiated by this first sample. During the single probe, server gathers samples calculating required intervals obtained from timestamps. On the sending side the sending time stamp is set by the Kite2 in the application level while on the receiving side the time stamp is set by the kernel on reception of the data packet. The sender keeps most recent samples according to measurement type: adjacent messages for "train" or "packet pairs inverted" modes, odd for "packet pairs". Two packets are enough for current probing modes.

When all issued probes are received, the server calculates an average inter-packet interval and reception data rate based on this average. In case of packet losses receiver waits for 3 sec and prepares a report message to provide the results to the client application. This result calculations are done after the probes session for reduction of calculation load in order to obtain accurate timestamps.

At the server side processes the result of probes calculating a running-average Iplr for probes, forms the report with the estimated AvB and send it to the sender. For each reported AvB value client recalculates the data rate for transmission and the number of samples to be sent:

$$R_{s_i} = 1.1 \cdot R_{r_{i-1}} \quad (1)$$

where R_s – send data rate, R_r – receive data rate, i – number of samples in train;

and the amount of samples:

$$N = \frac{c \cdot R_r}{8L} \quad (2)$$

where N – amount of samples for the probe, c is the maximal portion of occupied bandwidth (as a ratio), L – size of the packet.

The client sends probes repeatedly increasing send data rate on each iteration until there is no increment of receiving data rate in response at least 3 times to ensure it is a current boundary of a link and not a random measurement error. The last value of receiving data rate is assumed the available bandwidth.

4.2 AvB tools performance comparison

Performance and accuracy of estimations have been approved tests under two general conditions:

1. On a testbed of an emulated 10 Gbps link with the variable network parameters (capacity, delay and loss);
2. In an environment between two dedicated servers with the physical link capacity of 1Gbps and background network load of TCP traffic approx. 150 Mbps (measured by *iperf3* tool).

The application tests in emulated 10G Network have been performed with the duration of the 80s. During one measurement cycle of the 20s, 4 variations of capacity has been configured in order to emulate an instability of AvB in a real environment and two comparative tools are used:

- **Yaz** or calibrated *pathload* tool based on the *Pathload* algorithm of self-load periodic streams but uses only one probe packets train opposed to fleets of trains in *Pathload* evaluates inter-packet interval (and corresponding to it constant bit rate - CBR) for each iteration of measurement as an increased value of the previous inter-packet interval on the sender side by its difference with the inter-packet interval on the receiver side. Yaz terminates the process of measurements as soon as difference of inter-packet intervals gains predefined thresholds. The previous study of this tool shows high performance in networks up to 1 Gps. Thus, Yaz is used as a reference tool in this research.

- **Kite** uses the same technique but with enhanced features like automatic adjustment of a number of samples which allows using the tool in high-speed networks, so they so called Step Decreasing Factor – rule of increasing the probe send rate. The adjustment of the number of samples in a train reduces significantly the data overhead for the bandwidth estimation while the SDF offers an enhances algorithm of increase of probe data rate which leads to significantly faster convergence of estimated data rates in comparison to Yaz.

Emulated delay while test case was continually enlarged to the threshold of 0.5 ms. Results of experiments are shown in *Figure 5*.

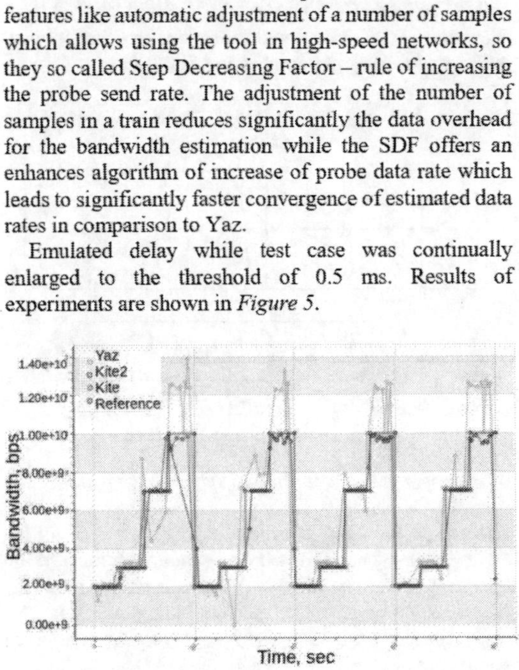

Fig 5. Accuracy of the AvB tools measurements in network with 0.5 ms of delay.

The light-blue graph corresponds to the reference bandwidth which is equal to the capacity in an idle link, the dark blue presents results of measurement by Kite2 tool, and results of experiments done by other two tool - Yaz and Kite are shown as red and yellow lines accordingly. Dots on each graph corresponds to the estimated values by each tool per one measurement iteration (so-called 'instant bandwidth value').

On the idle link when the capacity is limited only by emulators, all three tools show low overhead for the estimation of AvB up to 3 Gbps. With the increase of capacity overestimations of Yaz grow constantly, as well as the number of estimation outliers [20].

The most precise estimation for the range of 7-10 Gbps shows Kite2 tool, despite the fact that it is tended to overestimate the AvB at the high speeds. It might be related to the amount of occupied capacity by this tool while estimation. For the 10G link, it consumes approx. 1,6 Gbps for calculation each finite AvB value in a contradiction to 300 Mbps consumed by Kite tool.

Estimation error during the experiments is shown in *Figure 6*. Bars of the graphs corresponds to a minimal and maximal estimated error value, green and red areas show lower and upper quartiles of the mean. As it is represented by the boxplots for each examined tool, the lowest range of error values corresponds to Kite and Kite2 measurement with a tendency of underestimation up to 100 Mbps for the 10G network. Peak outliers of Yaz measurements in *Figure 6* are detected in the wide range of the boxplot, which spreads 400 Mbps of overestimation.

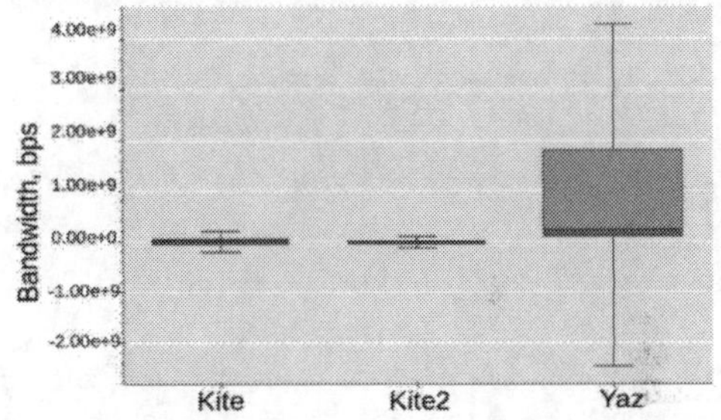

Fig 6. Estimation error of the AvB tools measurements in network with 0.5 ms delay.

In order to track the changes of measurement accuracy in the deterioration of conditions by the network, 1% constant Packet Losses has been configured on the link. Results of experiments are shown in *Figure 7*.

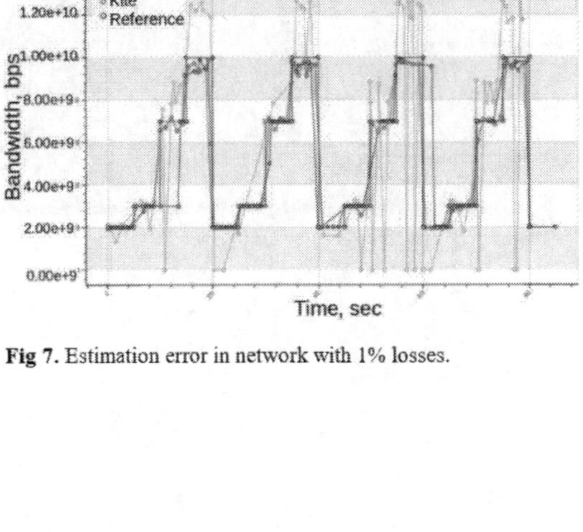

Fig 7. Estimation error in network with 1% losses.

Kite and Kite2 present the error scatter shown in *Figure 8*.

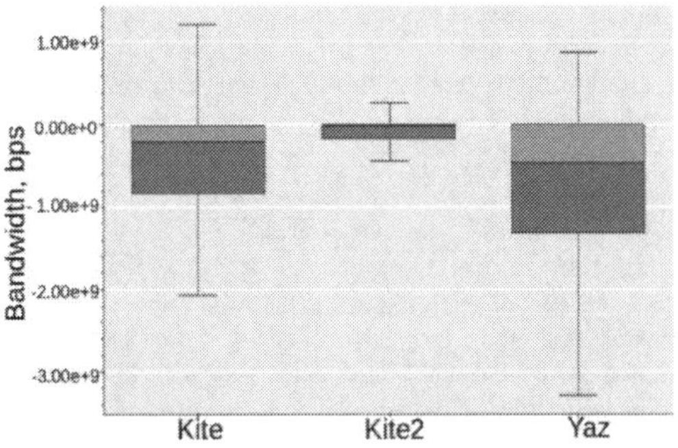

Fig 8. Estimation error of the AvB tools in a network with 1% of losses.

A tendency of AvB underestimation observed in all investigated tools and varies from 100Mbps for Kite2 to 1.3Gbps for Yaz.

Results of a measurement in environment are shown in *Figure 9*. Test have been performed on machines which run Web and synchronization operations and accumulate a valid network load by clients' requests and data transmission. A physical link at each instance has a capacity of 1 Gbps, RTT value corresponds to 10 ms.

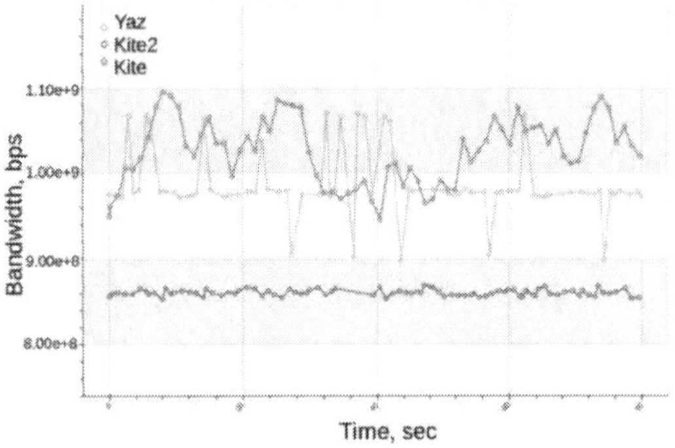

Fig 9. Accuracy of the AvB measurements in a real network.

Deviation of measurement results is shown in *Figure 10* and represents stability of processed values by Kite2 tended to underestimate 150 Mbps of AvB at the first glance.

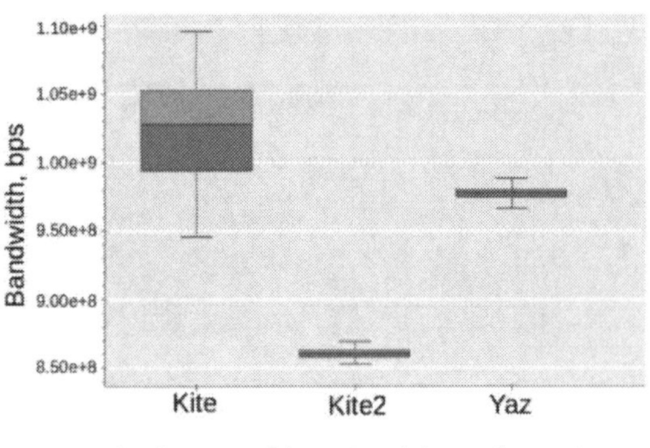

Fig 10. Estimation error of the AvB tools in a real network.

In order to determine an actual AvB *iprerf3* tool which is based on link congestion has been used. Its results during of 1 min operation with a send data rate speed of 1 Gbps showed that AvB value is within 740 Mbps to 875 Mbps on the investigated link with the background bursty web traffic.

Results of the tests show competitiveness of Kite2 measurement tool and its potential to investigate high-speed networks.

5 Experimental research

This section describes the correlation between the measurement accuracy and the probing packet size. The threshold of a network load for a defined measurement accuracy has been chosen experimentally.

5.1 Test scenarios

Measurements have been performed on the emulated network testbed with spare link and constant AvB equal to its capacity of 10 Gbps. Duration of each test corresponded to 360 s and had been repeated 10 times.

In addition, each test was executed with an eye for two Maximal Transmission Unit (MTU) size values: 1500 bytes (the most common value in Internet) and 9000 bytes (common in 10 Gbps and above networks) which corresponds to 1472 bytes of data in UDP packet payload and 8972 bytes accordingly, further referred to as payload. The steps of ratio variation for the network occupation by probing samples are presented in a *Table 1*.

Table 1. Number of estimated results (N) and bandwidth usage ratio (p).

p	0.015	0.025	0.035	0.05	0.065
N	773	438	250	159	150

p	0.095	0.15	0.35
N	120	122	89

An observed result has been taken with the payload size of 8972, while tests with the lower value have been rejected due to the high error rate (more that 60% for the 1472 bytes). Therefore, the research was focused on higher MTU size, besides, for bigger payload the measurement performance increased and processing time was reduced to 0.75 s in average but causes more significant network load, though.

5.2 Statistical analysis of the test results

With the help of statistical software for analyzing of empirical data *SPSS* [21] we define the correlation values for observations: measurement errors for available bandwidth estimation (*X*), the ratio of bandwidth utilization by measurement probes (*Y*), the amount of estimation results (*Z*).

Correlation coefficients corresponds to such values:

- **Pearson's**

$$r_{xy} = \frac{(\overline{xy} - \bar{x} \cdot \bar{y})}{S_x \cdot S_y} \approx -0.396, \quad (3)$$

where:
- n is the sample size;
- x_i, y_i are the single samples indexed with i;
- $\bar{x}, \bar{y}, \overline{xy}$ are the sample mean;
- S_x, S_y are the standard deviations.

- **Spearman's**

$$r_s = 1 - \frac{6\sum_{i=1}^{n} d_i^2 + T_X + T_Y}{n \cdot (n^2 - 1)} \approx -0.671 \quad (4)$$

where:
- $d_i = rg(X_i) - rg(Y_i)$ is the difference between the two ranks of each observation;
- n is the number of observations;
- T_X, T_Y – corrections of related ranks.

Because the value of relation between *X* and *Y* is evaluated on the data set, thus its statistical significance should be tested, i.e. estimation of possibility to expand the obtained results over population universe.

Let's calculate the *t*-statistics for the correlation coefficients:

- Pearson's

$$t(X, Y) = r\frac{\sqrt{n-2}}{\sqrt{1-r^2}} \approx -19.76 \quad (5)$$

- Spearman's

$$t(X, Y) = r\frac{\sqrt{n-2}}{\sqrt{1-r^2}} \approx -41.46 \quad (6)$$

and compare both with a critical value t_{crit}:

$$t_{\alpha,l} = \begin{cases} 2.578 & for\ \alpha \le 0.01 \\ 1.961 & for\ \alpha \le 0.05 \end{cases} \quad (7)$$

Since the absolute value of *t*-statistics are greater than critical ones $|t| > t_{0.01,l} = 2.578 > t_{0.05,l} = 1.961$ for $\alpha \le 0.05$ and $\alpha \le 0.01$ we can assume that the correlation coefficient is significant for $\alpha = 0.01$ level.

Hence, the relationship *exists* between measurement error of available bandwidth estimations (*X*) and the bandwidth occupation by probing train (*Y*) for significance value $\alpha \le 0.01$. Pearson's correlation coefficient is equal *-0.396* which means the average inverse relation. I. e. while increasing the value of *X* feature, the value of *Y* one is decreasing. The corresponding results is possible to expand over all the data of the same type, because the value of correlation coefficient is significant for level $\alpha \le 0.01$. Spearman's coefficient is equal to *-0.671* which confirms the average reverse relation between the mentioned feature sets. The test for significance of Spearman's coefficient also confirms that the results can be expanded over the general population for level $\alpha \le 0.01$ ($|41.24| > 2.578_{0.01}$).

Correlation coefficients for *X* and *Z* according to *Equations (3, 4)*:

Pearson's $r_{xz} \approx 0.598$ and Spearman's $r_s \approx 0.669$.

To check statistical significance of correlation value between *X* and *Z* we provided calculation of *t*-statistics:

Pearson's $t(X, Y) \approx 34.18$ and Spearman's $t(X, Y) \approx 41.24$

and compared the results with a critical value t_{crit} according to *Equitation (7)*.

Hence, the relationship *exists* between measurement errors of available bandwidth estimation (*X*) and the number of estimated results (*Z*) for significance value $\alpha \le 0.01$. Pearson's correlation coefficient is equal *0.598* which means the average direct relation. I. e. while increasing the value of *X* feature, the value of *Z* one increasing as well. The corresponding results is possible to expand over all the data of the same type (the value of correlation coefficient is significant for level $\alpha \le 0.01$. Spearman's coefficient is equal to *0.669* which confirms the average direct relation between the mentioned feature sets. The test for significance of Spearman's coefficient also confirms that the results can be expanded over the general population within $\alpha \le 0.01$ ($|41.24| > 2.578_{0.01}$).

6 Conclusion and future work

The research presented here shows that there is a direct correlation between the probing size in available bandwidth estimation and measurement errors, as described in Section 5.2. It has been shown that the methods, implemented in Kite2 enables us to reduce the amount of samples by 20%, while the bandwidth estimation deviation was reduced to 100 Mbps for 10 Gbps links. Tests with payload sizes of 1472 and 8972 bytes have shown that MTU size essentially impact the estimation time. While estimation with smaller packets takes up to 6.5s measurements with jumbo frames need only 0.75s for stable results. Moreover, estimation with normal IP packet sizes is limited by 3.4 Gbps while with jumbo frames up 9.98 Gbps where estimated with sufficient accuracy. Statistical investigations on tests show that 5% probing load ratio gives up to 94% measurement accuracy in 10 Gbps networks with 20 ms of delay, compared to the 97% of accuracy when probing samples occupy quarter of the path's AvB. That will be taken into account in further experiments.

Another significant finding was, that the network delay does not affect the accuracy of estimations and processing time. As next research tasks we consider adjustment of probing size in high-speed networks. While in emulated networks the estimation accuracy was very accurate in real-world networks achieved results have shown much higher uncertainty. So algorithm adoptions for real-world networks have to be applied. Also some more sophisticated test scenarios with presence of heavy cross traffic must be elaborated and executed. Also, an algorithm for the sending rate adaptation to the changing network conditions could lead for further improvements.

Acknowledgement

This work has been funded by Volkswagen Foundation for trilateral partnership between scholars and scientists from Ukraine, Russia and Germany within the project CloudBDT: Algorithms and Methods for Big Data Transport in Cloud Environments.

References

1. R. Prasad, C. Dovrolis, M. Murray and K. Claffy, "Bandwidth estimation: metrics, measurement techniques, and tools," IEEE Network, vol. 17, no. 6, 2003, pp. 27-35.
2. K. M. Salehin and R. Rojas-Cessa, "Method for measuring the packet processing time of Internet workstations with the detection of interrupt coalescence", 2016 IEEE 17th International Conference on High Performance Switching and Routing (HPSR), Yokohama, 2016, pp. 142-147.
3. N. Hu and P. Steenkiste, "Evaluation and characterization of 33 Areas in Communications, vol. 21, no. 6, 2003, pp. 879-894.
4. D. Kachan, E. Siemens and V. Shuvalov, "Available bandwidth measurement for 10 Gbps networks," 2015 International Siberian Conference on Control and Communications (SIBCON), Omsk, 2015, pp. 1-10.
5. J. Sommers, P. Barford, W. Willinger "A Proposed Framework for Calibration of Available Bandwidth Estimation Tools," Proc. of the 11th IEEE Symposium on Computers and Communications, 2006, pp. 709-718.
6. M. Jain and C. Dovrolis, "Pathload: a measurement tool for end-to-end available bandwidth," in Proceedings of the 3rd Passive and Active Measurements Workshop, 2002.
7. V. Ribeiro, R. Riedi, R. Baraniuk, J. Navratil and L. Cottrell, "pathChirp: efficient available bandwidth estimation for network paths," in Proceedings of the 4th Passive and Active Measurements Workshop, 2003.
8. Ningning Hu and P. Steenkiste, "Evaluation and characterization of available bandwidth probing techniques," in IEEE Journal on Selected Areas in Communications, vol. 21, no. 6, pp. 879-894, Aug. 2003.
9. L. Angrisani, S. DAntonio, M. Vadursi and G. Ventre, "Performance comparison of different techniques for available bandwidth measurement in packet switched networks," in VECIMS
10. 2003 - International Symposium on Virtual Environments, Human-Computer Interfaces, and Measurement Systems, 2003.
11. E. Goldoni, G. Rossi and A. Torelli, "Assolo, a New Method for Available Bandwidth Estimation," 2009 Fourth International Conference on Internet Monitoring and Protection, Venice/Mestre, 2009, pp. 130-136.
12. E. T. Saulnier and B. Bortscheller, "Data transfer bottlenecks over SPARC-based computer networks," Proceedings of 20th Conference on Local Computer Networks, Minneapolis, MN, 1995, pp. 289-295.
13. C. Dovrolis, P. Ramanathan and D. Moore, "Packet-dispersion techniques and a capacity-estimation methodology," in IEEE/ACM Transactions on Networking, vol. 12, no. 6, pp. 963-977, Dec. 2004.
14. Y Labit, Powezarski, N. Larrieu, "Evaluation of Active Measurement Tools for Bandwidth Estimation in Real Environment", Workshop on End-to-End Monitoring Techniques and Services, IEEE 2005
15. A. A. Ali, F. Michaut and F. Lepage, "End-to-end available bandwidth measurement tools : a comparative evaluation of performances," Proc. 4th International Workshop on Internet Performance,IPS-MoMe 2006.
16. C.M.Leong, T.S. Wei, A.T. Samsudin "Performance evaluation of optimized available bandwidth estimation tools in FTTH networks" Communications (APCC), 2012 18th Asia-Pacific Conference.
17. I. Fedotova, E. Siemens, and H. Hu, "A high-precision time handling library," J. of Comm. and Computer, vol. 10, pp. 1076-1086, 2013.
18. C. Dovrolis, P. Ramanathan, and D. Moore, "What do Packet Dispersion Techniques Measure?," Proc. IEEE INFOCOM, Apr. 2001, pp. 905–14.
19. D. Kachan, E. Siemens, H. Hu. Tools for the high-accuracy time measurement in computer systems, (Russian) // 6th Industrial Scientific Conference «Information Society Technologies». Moscow, 2012
20. Kirova V., Korniichuk M. "Performance Comparison of PRM-based Available Bandwidth Estimation Tools in 10G Emulated Networks", 72d scientific and technical conference of professors faculty, academics, postgraduate students and students, 2017.
21. "IBM SPSS Statistics v23.0 documentation." [Online] Available: https://www.ibm.com/support/knowledgecenter/en/SSLVMB_23.0.0/spss/product_landing.html. [Accessed 04-Jan-2018].

An Evolving Hypernetwork Model to Quantify Progress Potential of Emerging Research Topic

Jia Liu, *Kewei* Yang, *Jianguo* Xu, *Yingying* Gao, and *Qingqing* Yang

Department of Systems Engineering, National University of Defense TechnologyChangsha, Hunan, China

Abstract. There is considerable and growing interest in the emergence of research topics. However, current methods to detect the emergence are still problematic mainly due to information loss and aging effect. In this study, we show three intrinsic mechanisms including preferential attachment, exponentially growth and heterogeneous fitness values that decay with time. Depending on the input assumptions, all topics tend to follow a universal temporal pattern according to our model which results in strongly sufficiency to quantify progress potential.

1 Introduction

There is considerable and growing interest in the emergence of research topics, both from the policy-making and academic perspective. Numerous studies and projects have proposed definitions and models to detect, track, and predict emerging topics. For instance, the Comprehensive Strategy on Science, Technology and Innovation of Japan in 2017 (STI 2017) and the Annual Report on the Top 10 Emerging Technologies of the World Economic Forum [1]. Evaluating innovation potential is essential, not only for promoting efficiency of scientific research but also for detecting emerging research topics with strategic significance [2]. The intrinsic improvability of a new research area allows decision makers to track its evolutionary trajectory, or progress potential. It mainly includes two reasons. For one, the studies always attempt to deal with a practical or scientific problem. For another, the current study is affirmably carried out on the basis of existing one, and trying to improve the science and technology level through cooperation.

We contrast this problem of quantifying the innovation potential of emerging research topic based on its intrinsic improvability, which generally focuses on exogenous, market-driven factors or experts factors, for which bibliometrics and Delphi have played a key role. Many researchers utilize various types of relations among publications to establish network based models of bibliometrics, such as bibliographic coupling, co-citation analysis and others [3]. Different citation relations can all be used to aggregate publications. In the case of text-based approaches [4], some studies have established clusters based on the co-occurrence of terms [5], using a large-scale database [6]. However, the effectiveness of any approach for identification is extremely difficult to verify. And such approaches tend to focus on measuring the attributes of novelty and fast growth, and to ignore the consideration of other potentially important attributes, such as impact [7]. After the small-world model proposed by Watts and Strogatz, Barabási and Albert (BA) proposed the scale-free network [8]. The academic community has set off an upsurge of research on complex networks. Despite its different applications, it is still hard to depict some real-life systems.

All in all, traditional homogeneous complex networks have following limitations. There is no consensus on the concept of emergence especially the impact of co-operation. What's more, bibliometrics indicators such as the number of papers or citations are poor prediction of topic's future [9] in Fig 1. The number of scientific literature for each research point relies on the item's age [10]. As a result, the older it is, the more likely it is to be favored, which violates to the intuition that people prefer the emerging things [11]. In the experiment, we regarded Medical Subject Headings (MeSH) term as research topic. PubMed databases of the NCBI（National Center for Biotechnology Information）provide freely source in a well formatted structure. (which is publicly available at http://www.ncbi.nlm.nih.gov/pubmed/advanced) [12]. Moreover, it is also short of predictability: a subset of research points that collect nearly 500 papers in 1966 own widely different evolutionary potential (Fig. 1, inset). To avoid information loss and aging effect, it naturally leads to the emergence of hypernetworks. The concept of hypernetworks is a natural multidimensional generalization of networks and represents n-dimensional relations. The evolving hypernetworks in the existing literatures are almost all uniform, except Guo and Zhu [13] considering the characteristics of non-uniformity and weight of hyperedges. In the non-uniform model, at each time step, both the size of new nodes and the randomly selected existing nodes in one hyperedge are random variables.

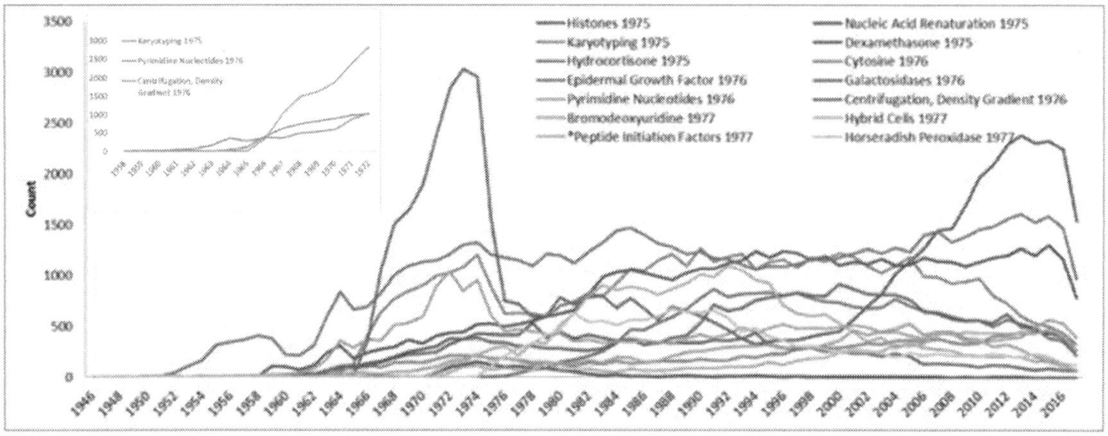

Fig. 1. The life curves of 14 topics selected from 1975 to 1977 in 'CELL'.

2 The evolving hypernetwork model

2.1 Fundamental mechanisms

We started by hackle intrinsic mechanisms to drive new research topic evolving. Intuitively, the life curves of different MeSH terms vary widely in Fig. 1. Some topics are highly valued when they are first proposed. While, some of them have been dormant for a long time before they catch the attention of people, that is, the curve rises out rapidly after a period of time. Some have been developing steadily, that is, the curve flattens out. In addition, in terms of total quantity, some of the hot topics exist in dozens or even hundreds of times of literature, compared to the general topic.

Preferential attachment captures the fat-tailed nature of the publication distribution which several hotspots are more likely to be used and finally result in 'the rich get richer' 14. Despite its rationality in comparing the empirical data and theoretical models, the first-mover advantage of preferential attachment results from strong time bias in the system.

Secondly, unlike the traditional uniform network, the total amount of publications each year is not the same, but increases exponentially year by year15.

$$N(t) \propto \exp(\beta t) \quad (1)$$

where $N(t)$ is the amount of documents at the time t, β is the speed of the literature growth. Then the scholars can noticed that the proportion of the literature must be reduced rapidly. Furthermore, the evolving network in the existing publications is non-uniform. That is, the number of research points of each paper is not a constant which is determined by a given distribution function like Poisson function. There are approximately 20 MeSH terms in each publications of 'CELL', new nodes entering into the network or previously existing nodes selected may not be the same in Fig 2.

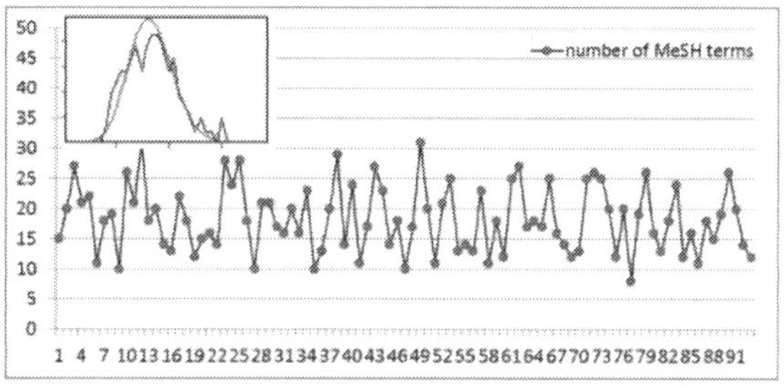

Fig. 2. The number of MeSH terms of 100 papers.

Old research points will be replaced by new ones eventually. Therefore, in the long run, the new research subject will dies slowly and the total amount of relevant literature will also tend to be stabilized, such that the trend is similar to the shape of the normal distribution. But it is difficult to determine when this distribution begins, due to the difference in the innovative characteristic of the new research point, i. e. Is it a standard normal distribution, left or right?

In the beginning, the research pioneers creatively propose a new research idea and relative researchers with the early derivation of theories. Once forming the theory and receiving more attention of experts in the field, which usually takes a long time, much more researchers, on the basis, start a large number of extension and application researches, until the hotspot is replaced by another. For instance, there is a new MeSH term in 2015 called latent autoimmune diabetes of adults in Fig. 3(New appearance in 2015 in left subfigure and emerging in 2016 on the right).

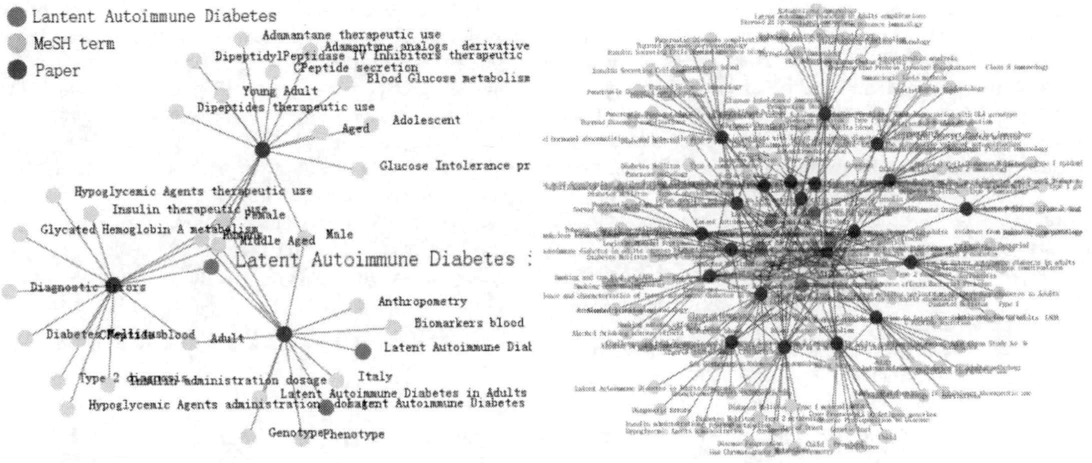

Fig. 3. A sample hypernetwork of latent autoimmune diabetes of adults.

2.2 Theoretical analysis

Now we proceed to an evolving hypernetwork model based on the above-reported empirical observations. Initially, we assume the hyperdegree h_i^{t0} initially equals to 1. The hyperdegree is the number of hyperedges that connect to the node, and the number of related papers in the real world. At time t, $\exp(\beta t)$ new papers are introduced. The number of each paper k is determined by probability $p(k)$. The time evolution of expected number of h_i^t satisfies

$$\frac{dh_i}{dN} = (p_k k)\frac{h_i R(\Delta t_i)}{\sum_{j=1}^{N} h_j R(\Delta t_j)} \quad (2)$$

where $R(\Delta t_j)$ is the relevance of time which is similar to normal distribution.

$$R(\Delta t_i) = \frac{1}{\sqrt{2\pi}\sigma_i}\exp\left(-\frac{(\Delta t_i^{\omega} - u_i)^2}{2\sigma_i^2}\right) \quad (3)$$

where u indicates the time for a paper to reach its citation peak, σ is longevity, capturing the decay rate, ω is the degree of skewness.

The equation represents the evolution of each node's hyperdegree depends not only on itself but also on the current hyperdegree and relevance of all other nodes. Two simplifications could be introduced to solve this partial differential equation. At first, the number of MeSH terms each new paper contains is subordinated to an extremely narrow Poisson distribution

$$p(k) = \frac{\lambda^k}{k!}e^{-\lambda}, (k = 0, 1, \cdots) \quad (4)$$

where its expectation and variance are λ. According to the Poisson theorem, $\lim_{t\to\infty} p(k)k \to \lambda$. Secondly, there are millions of nodes with non-negligible infects of infinite series $\sum_{j=1}^{N} h_j R(\Delta t_j)$. Combining with (1) leading to $\Delta t_i = t - t_i = \ln(N/i)$, we get $dN = \beta N d\Delta t_i$ .While $R(\Delta t_j)$ decays faster than $\frac{1}{\Delta t_j}$ and $\lim_{\Delta t_i \to \infty} R(\Delta t_j) \to 0$, the finally growth of the infinite series converges to a certain value $\Omega^* = \frac{1}{N}\sum_{j=1}^{N} h_j R(\Delta t_j)$. This allows us to approximate (3) with

$$\frac{dh_i^t}{d\Delta t_i} = \beta\lambda \frac{h_i R(\Delta t_i)}{\Omega^*} \quad (5)$$

With the initial condition $h_i^0 = 1$, assuming $\alpha_i = \frac{\lambda \beta h_i}{\Omega^*}$, thus

$$h_i^t = \exp\left(\alpha_i \Phi\left(\frac{\Delta t_i^{\omega} - u_i}{\sigma_i}\right)\right) \quad (6)$$

When $\omega > 0$ and $\Delta t_i \to \infty$, $\Phi\left(\frac{\Delta t_i^{\omega} - u_i}{\sigma_i}\right) \to 1$. Therefore the ultimate number related MeSH term i equals to a remarkably value $h_i^{\infty} = \exp(\alpha_i)$. Else if $\omega < 0$ and $\Delta t_i \to \infty$, then, $\Delta t_i^{\omega} \to 0$, $h_i^{\infty} = \exp\left(\alpha_i \Phi\left(-\frac{u_i}{\sigma_i}\right)\right)$, indicating that the ultimate impact of a research topic is independent with ω.

3 Experiment

3.1 Results

For the purpose of this study, we selects the scientific data of medical field from 1946 to 2017. PubMed databases in the National Center for Biotechnology Information (NCBI) provide free publications data.

Moreover, PubMed comprises more than 28 million records for biomedical literature from MEDLINE, life science journals, and online books. Without loss of generality, we randomly choose fifty papers per year as long as 72 years utility data, granted between 1946 and 2017, downloaded from the NCBI website. 15 articles and their MeSH terms are randomly selected each year, such as ("Dinoflagellida" [MeSH Terms]). It finally provided us (on April 1, 2018) with a retrieval of 75 records.

To illustrate the universality of the model, we choose four data with completely different characteristics in Fig. 4. First, as illustrated, the red line, corresponding to the key word 'Pyrimidine Nucleotides', has grown rapidly in the early period and has also declined significantly in the later period. The blue line, corresponding to the key word 'Hexosyltransferases', has been steadily rising, only to decline sharply in the end. The entire life curve of the purple line is relatively stable, which corresponds to 'Leukemia Virus, Murine', initially maintaining a small increase in the early period, and then gradually decaying. The green line, corresponding to the key word 'Epidermal Growth Factor', is not the same as the above three. There was a long period of incubation in the early period, followed by a rapid increase and then a slow decline. They can basically summarize the developing characteristics of all various data. Simultaneously, the large amount of these four data further enhances their persuasiveness and representation. The Fig.4 Inset shows the cumulative amount of the four life curves.

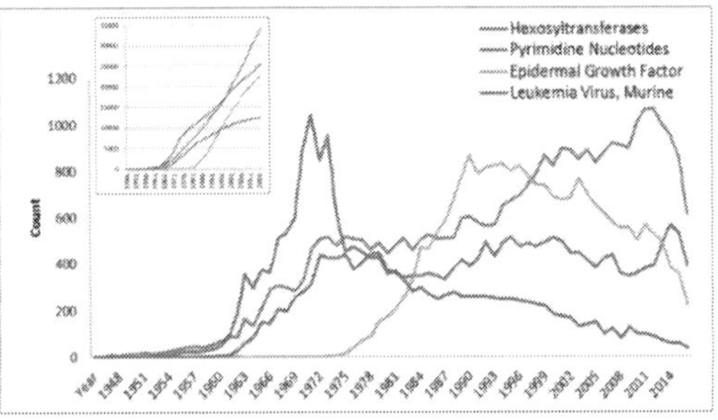

Fig. 4. Four life curves with typical characteristics.

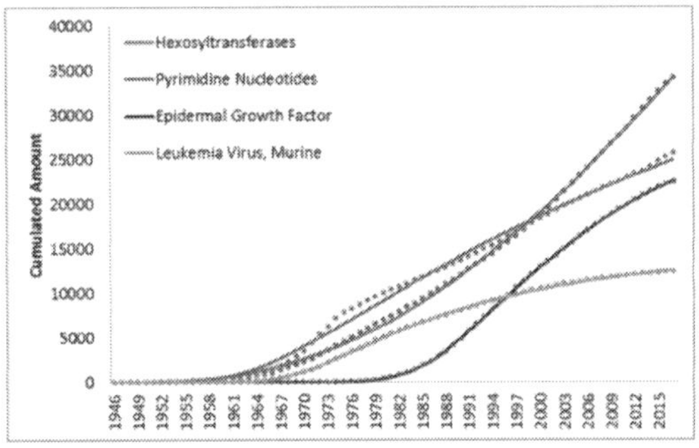

Fig. 5. The fitting results of model Eq. 5.

We have obtained a model suitable for this paper. The fitting results of this model we got are shown in Fig. 5. The solid line corresponds to our model fitted with the data. The dashed line denotes the life curve of the four real data, respectively. Obviously, the degree of fit between the solid and dashed lines is high, especially the blue line (corresponding to 'Epidermal Growth Factor') and the green line (corresponding to 'Leukemia Virus, Murine'), which are almost perfectly fitted. The second is the red line(corresponding to 'Hexosyltransferases').

The α and δ value of 'Hexosyltransferases' are highest for its continuously innovative ability, It has been rising from 1950 and taking a long time. 'Epidermal Growth Factor' has the smallest ω value, and its influence is mainly concentrated in the early stage since its appearance.

Table 1. The results of parameter estimates

	CASE1	CASE2	CASE3	CASE4	AVG
α	27.96	12.9	10.71	9.77	15.335
ω	-0.22	-0.76	-1.75	-0.51	-0.81
u	0.11	0.14	0.002	0.3	0.138
σ	0.41	0.13	0.001	0.1	0.16025

CASE1: Hexosyltransferases; CASE2: Pyrimidine Nucleotides; CASE3:Epidermal Growth Factor; CASE4: Leukemia Virus Murine

3.2 Comparison and analysis

The observed accuracy prompts us to compare with other models. We therefore identified several models that the others have been used in the past to fit histories the Model (Wang, 2013), the Logistic model and Gompertz model as shown in Table 2.

Table 2. Three models for comparison.

Model	Formula	Instruction
Model(Wang) 13 (Science, 2013)	$h_i^t = m(e^{\lambda_i \Phi(\frac{\ln t - \mu_i}{\sigma_i})} - 1)$	m measures the average references each new paper contains. λ_i is the relative fitness, μ indicates the time for a paper to reach its citation peak, σ is longevity.
Logistic Model	$h_i^t = \frac{a}{1+be^{-cx}}$	a , b is two constants. c is longevity. x corresponds to immediacy of paper.
Gompertz Model	$h_i^t = ae^{-e^{(p+qt)}}$	a is a constant. p sets the displacement in h_i^t . q characterizes the growth rate of citations.

Correlation coefficient is the state or relation of being correlated. Specifically, a relation existing between phenomena or things or between mathematical or statistical variables which tend to vary, be associated, or occur together in a way not expected on the basis of chance alone. In Fig. 6, the correlation is between the model fitting and the actual trends of data. That reflects the approximate degree of the model and the actual situation. This shows that the stronger the correlation is, indicating that the model is more accurate. From the Fig.6, we can clearly see that the correlation of our model, Model(Wang) and Gompertz Model is relatively close, but our model is still better than the other two models. And, the effect of Logistic Model is not satisfactory.

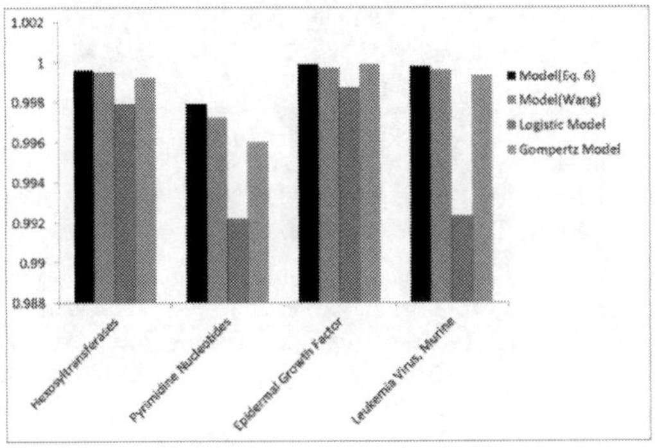

Fig. 6. Correlation coefficient among four cases.

The root-mean-square deviation (RMSD) or root-mean-square error (RMSE) is a frequently used measure of the differences between values (sample and population values) predicted by a model or an estimator and the values actually observed. The RMSD represents the sample standard deviation of the differences between predicted values and observed values. So the smaller the RMSD, the better. As illustrated from Fig.7, the RMSD among our model on the four samples (H,P,E&L) are obviously smaller than those of the other three models (Model(Wang), Logistic Mode and Gompertz Model). That means, our model clearly outperforms the other three. At the same time, we can also see that the second-optimal model is Wang's model. The effects of the remaining two models are slightly worse.

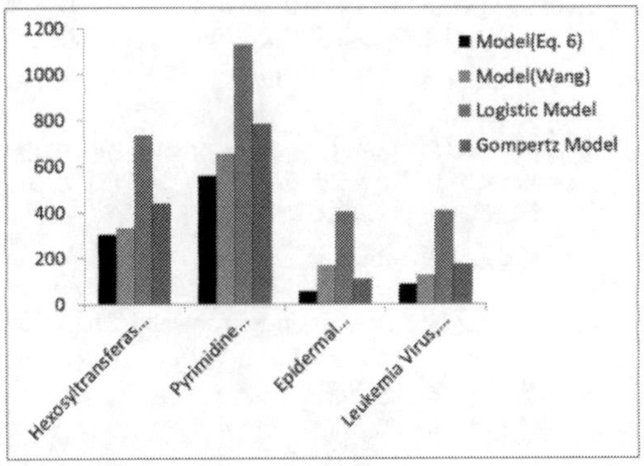

Fig. 7. RMSD among four cases.

4 Discussion

Under the premise of the same number of parameters, the fitting effect of our model is generally superior to others. It benefits from the understanding of the evolutionary mechanism of research topics and the restoration of information as possible. The innovation α, duration σ, and interval of the topic u will affect the ultimate development of itself. Therefore, the best choice for new researchers is the research point with both the larger α and σ, which means large research scale and continuous innovation.

And it is worth emphasizing that although ω does not affect the total number of final documents, it still is of great significance. ω can be a measure of the time efficiency of this topic being recognized. The bigger ω is, the shorter time it takes. When ω is more than 1, it can be thought that a number of researchers followed suit as soon as proposed. The smaller ω is, the more gradually it increases at first. When σ=0.2 and μ=0, the effect of ω is shown in the Fig. 8. Based on comprehensively analyzing of the literature amount of Moore's law, the influence of time affect is neither exponential decline nor left normal distribution.

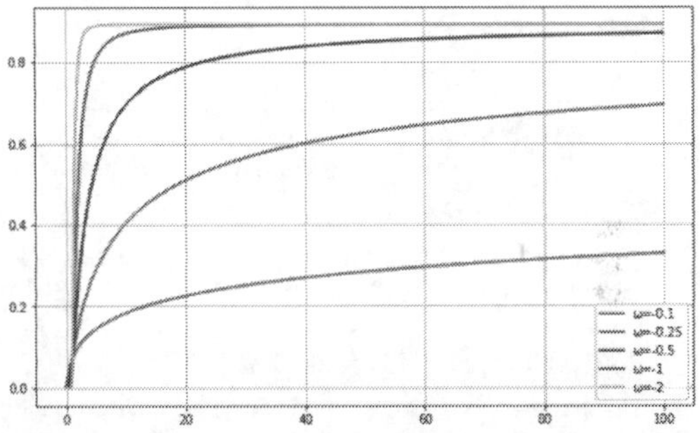

Fig. 8. The impact analysis of the parameter ω.

Citation-based measurement, which used to gauge impact, including analysis of impact factors to citation of short-term papers, always lacks predictability. In this paper, we propose a citation model with long-term predictability. Meanwhile due to little is known about the mechanism of time evolution of individual papers in the past research. Here, we derive a mechanistic model for the dynamics of citations in individual papers, allowing us to collapse the citation history of papers from different journals and disciplines into a single curve, indicating that all papers tend to follow the same universal time model.

The observed patterns help us to discover the basic mechanisms of management science impact. We can use this universal model to quantify progress potential of emerging research topic to predict the emergence of emerging technologies within a certain discipline. In combination with expert experience, we will select emerging technologies that will have the most development impact in the future. Through the government's vigorous policy support and social resources support, it will lead to the emergence of emerging technologies in advance. Therefore, it is doubtless to have tremendous boost to the development of the discipline, and even the entire history of science and technology.

The model proposed in this paper can fit the data of different characteristics well. In particular, it solves the problem of 'Sleeping Beauties' mentioned by Wang, whose reason is that the Model (Wang) limit the time to the early stages of publication with the Log function. The fundamental reason why logistic regression fails to fit well is that it's characteristics of 'fast growth and fast disappearing'. Although our model is highly applicable, it is not completely universal, such as neural network, which experienced several degrees of sinking and developing. At the same time, we also observed that the early prediction is not good enough when the study was not followed for a long time. Therefore, it's better to analyze the data which has been accumulated for a period of time. This paper involves only the validation and analysis of the model, and uses only one type of data. Next step, the model will continue to be expanded in terms of data types. At the same time, we will focus on the predictability of the model to explore the future research.

References

1. M. Halaweh, "Emerging technology: What is it," Journal of technology management & innovation, vol.8, nov. 2013, pp. 108-115, doi:10.4067/S0718-27242013000400010.
2. D. Rotolo, D. Hicks, and B.R. Martin, "What is an emerging technology?" Research Policy, Dec.2015, vol. 44, p.1827-1843, doi:10.1016/j.respol.2015.06.006.
3. H. Noh, Y.K. Song, and S. Lee, "Identifying emerging core technologies for the future: Case study of patents published by leading telecommunication organizations ," Telecommunications Policy, vol. 40, Oct. 2016, pp. 956-970, doi:10.1016/j.telpol.2016.04.003.
4. H. Small, H. Tseng, and M. Patek, "Discovering discoveries: Identifying biomedical discoveries using citation contexts", Journal of Informetrics,vol. 11, Feb. 2017, p. 46-62, doi:10.1016/j.joi.2016.11.001.
5. J. Joung, and K. Kim, "Monitoring emerging technologies for technology planning using technical keyword based analysis from patent data," Technological Forecasting & Social Change, Jan. 2017, vol. 114, pp. 281-292, doi:10.1016/j.techfore.2016.08.020.
6. Zhang Y, Zhang G, Chen H, et al. Topic analysis and forecasting for science, technology and innovation: Methodology with a case study focusing on big data research[J]. Technological Forecasting & Social Change, April 2016, vol. 105, pp.179-19, doi:10.1016/j.techfore.2016.01.015 1.
7. S. Zhang, and F. Han, " Identifying emerging topics in a technological domain," Journal of Intelligent & Fuzzy Systems, vol. 31, Sep. 2016, pp.2147-2157, doi: 10.3233/JIFS-169054.
8. M. Medo, G. Cimini, and S. Gualdi, "Temporal effects in the growth of networks," Physical review letters, vol. 107, Dec 2011, pp. 956-970, doi: 10.1103/PhysRevLett.107.238701.
9. I. Park, K. Lee, and B. Yoon, "Exploring Promising Research Frontiers Based on Knowledge Maps in the Solar Cell Technology Field" Sustainability, vol. 7, Oct. 2015, pp. 13660-13689, doi:10.3390/su71013660.
10. D. Rotolo, I. Rafols, M.M. Hopkins, and L. Leydesdorff, "Strategic intelligence on emerging technologies: Scientometric overlay mapping," Journal of the Association for Information Science & Technology, vol. 68, Dec. 2015, pp. 214-233, doi:/10.1002/asi.23631.
11. J.L. Guo, X.Y. Zhu, Q. Suo,and J. Forrest , "Non-uniform evolving hypergraphs and weighted evolving hypergraphs," Scientific reports, Nov. 2016, doi:10.1038/srep36648.
12. Leydesdorff L, Comins J A, Sorensen A A, et al. "Cited references and Medical Subject Headings (MeSH) as two different knowledge representations: clustering and mappings at the paper level", Scientometrics, vol.109, March 2016, pp. 2077-2091,doi: 10.1007/s11192-016-2119-7.
13. D. Wang, C. Song, and A.L. Barabási, "Quantifying long-term scientific impact," Science, vol. 342, Oct. 2013, pp.127-132, doi: 10.1126/science.1237825.
14. G. González-Alcaide, P. Llorente, and J.M. Ramos, "Bibliometric indicators to identify emerging research fields: publications on mass gatherings," Scientometrics, Nov. 2016,vol. 109, pp. 1283–1298, doi: 10.1007/s11192-016-2083-2.
15. Q. Wang, "A bibliometric model for identifying emerging research topics," Journal of the Association for Information Science and Technology, vol. 69, Feb. 2018, pp. 290-304, doi:10.1002/asi.23930.

Unimodular waveform design for MIMO radar transmit beamforming

Hongtao Li[1, 2], *Cheng* Chen[1] and *Xiaohua* Zhu[1]

[1]Nanjing University of Science and Technology, Nanjing, China
[2]CETC Key laboratory of aerospace information applications, Shijiazhuang, China

Abstract. In this paper, we consider unimodular waveform optimization for MIMO radar transmit beamforming with prior information on the directions of target and interferences. A phase-only conjugate gradient method is presented to optimize the unimodular waveform by maximizing the ratio of power at the target location to the power at the interference locations. Simulations show that the transmit beampattern obtained from the optimized waveform achieves a coherent gain of the target location while minimizing the radiation powers around the interference locations.

1 Introduction

Compared to the traditional phased-array radar, the multiple-input multiple-output (MIMO) radar allows each transmit array element to transmit arbitrary waveform [1-5]. By exploiting waveform diversity, we can design the transmit beampattern for colocated MIMO radar flexibly. Traditionally, the waveforms transmitted from each antenna are designed to maximize the radiation power of the target location or to match a given transmit beampattern [6-9]. However, these methods fail to meet the demand for designing transmit waveforms with a beampttern which places nulls at the interference locations. Recently, a phase-only variable metric method (POVMM) is proposed to design waveforms for MIMO radar transmit beamforming with nulls in the directions of the interferences [3]. However, optimizing transmit waveforms, which match the beampattern with a significant gain at the target location and nulls at the interference locations, still remains an unsolved problem. In this paper, a phase-only conjugate gradient method is proposed to optimize unimodular waveforms to obtain such a beampattern by maximizing the ratio of power at the target location to the power at the locations of interferences.

2 Signal model

Consider a MIMO radar system with a uniform linear array (ULA) of M transmit antennas. Let the waveform sequence with sample length L transmitted from the ith transmit antenna be $s_i \in \mathbb{C}^{L\times 1}$, where $i \in \{1,2,\ldots,M\}$. Then the covariance matrix of the transmit signal can be given by $\boldsymbol{R} = \boldsymbol{S}\boldsymbol{S}^{\mathrm{H}}/L$, where $\boldsymbol{S} = [\boldsymbol{s}_1, \boldsymbol{s}_2, \ldots, \boldsymbol{s}_M]^{\mathrm{T}}$ denotes the transmit waveform matrix. Suppose the inter element spacing between two adjacent antennas is half of the carrier wavelength, and the transmit steering vector is given by:

$$\boldsymbol{a}(\theta) = \frac{1}{\sqrt{M}}[1, e^{-j\pi\sin\theta}, \ldots, e^{-j\pi(M-1)\sin\theta}]^{\mathrm{T}} \tag{1}$$

The radiation power at location θ can be written as:

$$\begin{aligned} P(\theta) &= \boldsymbol{a}^{\mathrm{T}}(\theta)\boldsymbol{R}\boldsymbol{a}^{*}(\theta) \\ &= \boldsymbol{a}^{\mathrm{T}}(\theta)\boldsymbol{S}\boldsymbol{S}^{\mathrm{H}}\boldsymbol{a}^{*}(\theta)/L \end{aligned} \tag{2}$$

Let s be a $LM\times 1$ vector obtained from stacking the columns of transmit waveform matrix $\boldsymbol{S}$. Then, Eq. (2) can be rewritten as

$$\begin{aligned} P(\theta) &= \boldsymbol{s}^{\mathrm{H}}(\boldsymbol{I}_L \otimes \boldsymbol{a}^{*}(\theta))(\boldsymbol{I}_L \otimes \boldsymbol{a}^{\mathrm{T}}(\theta))\boldsymbol{s}/L \\ &= \boldsymbol{s}^{\mathrm{H}}(\boldsymbol{I}_L \otimes \boldsymbol{a}^{*}(\theta)\boldsymbol{a}^{\mathrm{T}}(\theta))\boldsymbol{s}/L \\ &= \boldsymbol{s}^{\mathrm{H}}\boldsymbol{A}(\theta)\boldsymbol{s}/L \end{aligned} \tag{3}$$

where

$$\boldsymbol{A}(\theta) = \boldsymbol{I}_L \otimes \boldsymbol{a}^{*}(\theta)\boldsymbol{a}^{\mathrm{T}}(\theta) \tag{4}$$

Where $\boldsymbol{I}_L$ is a $L\times L$ identity matrix and $\otimes$ denotes the Kronecker product operator.

In the scenario of interest, a target is located at the angle θ_t and K interferences are located at the angles of $\{\theta_1, \theta_2, \ldots, \theta_K\}$. To design a beampattern which concentrates the transmit power at the target location and places nulls at the interference locations, we formulate the constrained optimization problem as follows:

$$J = \max_{s} \frac{s^{\mathrm{H}} A(\theta_t) s}{\sum_{k=1}^{K} s^{\mathrm{H}} A(\theta_k) s}$$
$$s.t.\ |s(i)| = 1,\ i = 1,2,\ldots,LM \tag{5}$$

where the constant modulus constraint is considered due to the limited dynamic range of the radar amplifiers [10].

3 Optimal unimodular waveform design

Since s is an unimodular sequence, it can be written as:

$$s = [e^{j\varphi_1}, e^{j\varphi_2}, \ldots, e^{j\varphi_{LM}}]^{\mathrm{T}} \tag{6}$$

Therefore, the waveform optimization problem is equivalent to the phase-only optimization problem and a conjugate gradient method can be used to obtain the optimal waveform.

To compute the gradient of the phase of the waveform, we consider extracting the first-order term of Taylor series expansion of J [3, 11]. Let $\Delta = \mathrm{Diag}(\delta_1, \delta_2, \ldots, \delta_{LM})$ be a diagonal matrix where $\delta_i (i = 1,2,\ldots,LM)$ is the small perturbation of the phase code φ_i. After the unimodular sequence is perturbed as $s \rightarrow e^{j\Delta} s$, the Taylor series expansion of $J(e^{jx\Delta} s)$ is expressed as follows:

$$J(e^{jx\Delta} s) = J(s) + x\mathrm{d}J(s)(\Delta) + \cdots \tag{7}$$

where x is a scalar and

$$\mathrm{d}J(s)(\Delta) = \left.\frac{\mathrm{d}}{\mathrm{d}x}\right|_{x=0} J(e^{jx\Delta} s) = \sum_{1 \le i \le LM} \frac{\partial J}{\partial \varphi_i} \delta_i \tag{8}$$

Define

$$N(e^{jx\Delta} s) = s^{\mathrm{H}} e^{-jx\Delta} A(\theta_t) e^{jx\Delta} s \tag{9}$$

and

$$D(e^{jx\Delta} s) = s^{\mathrm{H}} e^{-jx\Delta} \sum_{k=1}^{K} A(\theta_k) e^{jx\Delta} s \tag{10}$$

The Taylor series expansion of $N(e^{jx\Delta} s)$ and $1/D(e^{jx\Delta} s)$ are respectively given by:

$$N(e^{jx\Delta} s) = s^{\mathrm{H}} (A(\theta_t) - j[\Delta, A(\theta_t)] + \cdots) s \tag{11}$$

$$1/D(e^{jx\Delta} s) = 1/(s^{\mathrm{H}} \sum_{k=1}^{K} A(\theta_k) s) + s^{\mathrm{H}} (j[\Delta, \sum_{k=1}^{K} A(\theta_k)] + \frac{1}{2}[\Delta, [\Delta, \sum_{k=1}^{K} A(\theta_k)]]) s / (s^{\mathrm{H}} \sum_{k=1}^{K} A(\theta_k) s)^2 + \cdots \tag{12}$$

where the Lie bracket in Eqs. (11) and (12) is calculated as $[A, B] = AB - BA$. Then the first-order term $\mathrm{d}J(s)(\Delta)$ can be computed from the product of the Taylor series expansion of $N(e^{jx\Delta} s)$ and $1/D(e^{jx\Delta} s)$, as follows:

$$\mathrm{d}J(s)(\Delta) = \frac{-j}{s^{\mathrm{H}} \sum_{k=1}^{K} A(\theta_k) s} \mathrm{Tr}\{(\Delta[A(\theta_t) - J(s) \sum_{k=1}^{K} A(\theta_k), ss^{\mathrm{H}}]\} \tag{13}$$

where $\mathrm{Tr}(\cdot)$ denotes the trace of a matrix.

By comparing Eqs. (8) and (13), the phase-only gradient of J is given by the follow expression:

$$\nabla J = \frac{1}{s^{\mathrm{H}} \sum_{k=1}^{K} A(\theta_k) s} \mathrm{Im}(\mathrm{diag}[A(\theta_t) - J(s) \sum_{k=1}^{K} A(\theta_k), ss^{\mathrm{H}}]) \tag{14}$$

where $\mathrm{Im}(\cdot)$ and $\mathrm{diag}(\cdot)$ denote imaginary part and diagonal part of a matrix respectively.

As the phase-only gradient ∇J is given by Eq. (14), a conjugate gradient method [12] can be used to obtain the optimal waveform. Let φ be a $LM \times 1$ vector of the phase code $\varphi_i (i = 1,2,\ldots,LM)$, and the basic steps of the algorithm is summarized as follow:

a) For $i = 0$, select an initial phase code vector φ_0 for transmit waveform, compute $g_0 = h_0 = \nabla J(\varphi_0)$.
b) Let $i = i+1$, for all $x_i \ge 0$, find t_i such that $J(\varphi_{i-1} + x_i h_{i-1}) \ge J(\varphi_{i-1})$.
c) Let $\varphi_i = \varphi_{i-1} + x_i h_{i-1}$.
d) Compute $g_i = \nabla J(\varphi_i)$, $h_i = g_i + a_{i-1} h_{i-1}$, $a_{i-1} = \frac{(g_i - g_{i-1})^{\mathrm{T}} g_i}{\|g_{i-1}\|_2^2}$.
e) If $|J(\varphi_i) - J(\varphi_{i-1})| \le \varepsilon$, where $\varepsilon \ge 0$ is a parameter to control the convergence, let φ_i be the optimal phase code vector; otherwise, go to step b).

4 Simulation

In this section, the performance of transmit beampattern obtained from the proposed method is compared to the

https://doi.org/10.1051/matecconf/201820802003

method in [3]. We consider two cases which have different interference distributions. In case 1, the interference locations are discretely distributed at $\{-40°, -5°, 45°, 50°\}$; in case 2, the interferences are continuously distributed between $-25°$ and $-10°$. Suppose the target location of interest is $20°$ and the transmit array consists of 16 and 32 antennas.

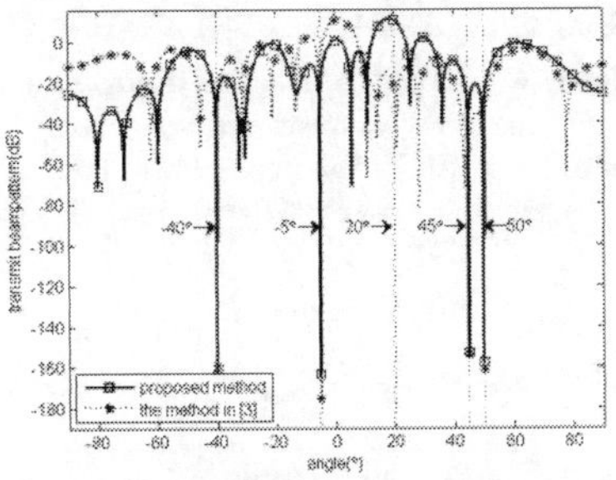

a Transmit beampatterns with 16 transmit array elements

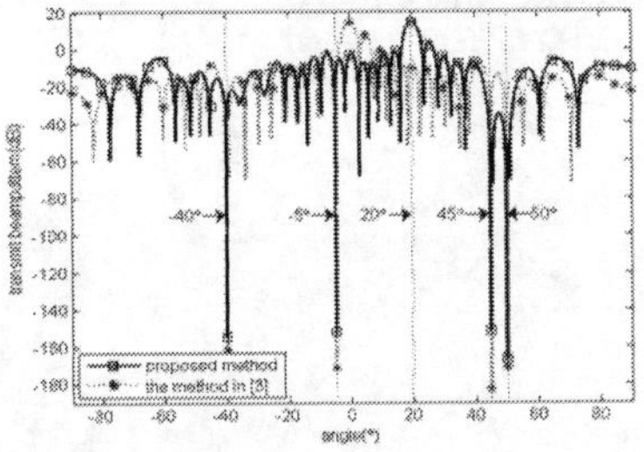

b Transmit beampatterns with 32 transmit array elements

Fig 1. Comparisons of transmit beampattern

Fig. 1 and Fig. 2 depict the transmit beampatterns for the interferences of discrete distribution and continuous distribution respectively. Among the four plots, the proposed method can achieve a coherent gain at the target location, while the method in [3] forms a gain at arbitrary locations. In Fig. 1a and 1b, the null depths of at least 150 dB are obtained from both the methods with 16 and 32 transmit array elements. It seems that the number of the array elements has little effect on null depths obtained from the proposed method and the method in [3] shows better performance in interference suppression when the number of the array element is 32. In Fig. 2, the method in [3] shows almost 8 dB deeper null depth than the proposed method both in the case of 16 and 32 transmit array elements, without concentrating the power in the target direction. We can see that a transmit beampattern with 32 transmit array elements shows better null performance than the one with 16 elements. This is due to the fact that more degrees of freedom provided by 32 transmit array elements are used to reject interferences.

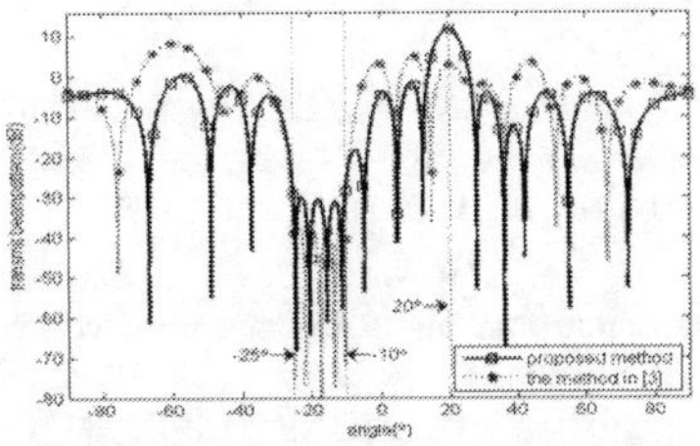

a Transmit beampatterns with 16 transmit array elements

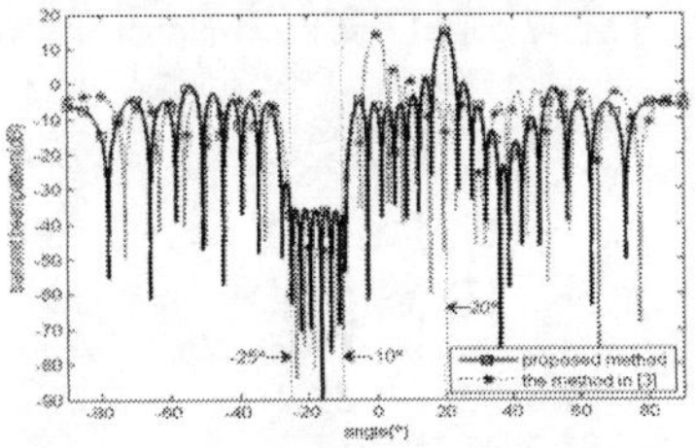

b Transmit beampatterns with 32 transmit array elements

Fig 2. Comparisons of transmit beampattern

5 Conclusion

In this paper, unimodular waveforms have been optimized for MIMO radar transmit beamforming by maximizing the ratio of power at the target location to the power at the interference locations. Numerical results show that the beampatterns can achieve a coherent gain in the target location while forming nulls in the interference locations.

Conflict of interests

The authors declare that there is no conflict of interests regarding the publication of this paper.

Acknowledgment

This work was supported by the National Natural Science Foundation of China (Grant no. 61401204), Postdoctoral Science Foundation of Jiangsu Province (Grant no. 1501104C), Technology Research and Development Program of Jiangsu Province (Grant no. BY2015004-03), China Postdoctoral Science Foundation (Grant no. 155676), and the Fundamental Research Funds for the Central Universities (Grant no. 30916011319).

References

1. Jian, L., Stoica, P. ‘MIMO radar with colocated antennas’, *IEEE Signal Porcess. Mag.*, (2007), **24**, (5), pp. 106-114, doi: 10.1109/MSP.2007.904812
2. Roberts W., He H., Stoica, P., and Jian, L.: ‘Probing waveform synthesis and receiver filter design’, *IEEE Signal Porcess. Mag.*, (2010), **27**, (4), pp. 99-112, doi: 10.1109/TSP.2010.936724

3. Lilin, G., Hai, D., and Himed, B.: 'Waveform optimization for transmit beamforming with MIMO radar antenna arrays', *IEEE Trans. Antennas and Propagation,* (2015), **63**, (2), pp. 543-552, doi: 10.1109/TAP.2014.2382637
4. Zhang, X., He, Z., Rayman-Bacchus, L., and Yan, J.: 'MIMO radar transmit beampattern matching design', *IEEE Trans. Signal Porcess.,* (2015), **63**, (8), pp. 2049-2056, doi: 10.1109/TSP.2015.2398841
5. Jardak, S., Ahmed, S., and Alouini, M., S.: 'Generation of correlated finite alphabet waveforms using gaussian random variables', *IEEE Trans. Signal Porcess.,* (2014), **62**, (17), pp. 4587-4596, doi: 10.1109/TSP.2014.2339800
6. Ahmed, S., Thompson, J. S., Petillot, Y., R., and Mulgrew, B.: 'Finite alphabet constant-envelope waveform design for MIMO radar', *IEEE Trans. Signal Porcess.,* (2011), **59**, (11), pp. 5326-5337, doi: 10.1109/TSP.2011.2163067
7. Stoica, P., Jian, L., and Yao, X.: 'On probing signal design for MIMO radar', *IEEE Trans. Signal Porcess.,* (2007), **55**, (8), pp. 4151-4161, doi: 10.1109/TSP.2007.894398
8. Stoica, P., Jian, L., and Xumin, Z.: 'Waveform synthesis for diversity-based transmit beampattern design', *IEEE Trans. Signal Porcess.,* (2008), **56**, (6), pp. 2593-2598, doi: 10.1109/TSP.2007.916139
9. Xu, H., Blum, R., S., Wang, J. and Yuan J.: 'Colocated MIMO radar waveform design for transmit beampattern formation', *IEEE Trans. Aerosp. Electron. Syst.,* (2015), **51**, (2), pp. 1558-1568, doi: 10.1109/TAES.2014.140249
10. Guolong, C., Hongbin, L., and Rangaswamy, M.: 'MIMO radar waveform design with constant modulus and similarity constraints', *IEEE Trans. Signal Porcess.,* (2014), **62**, (2), pp. 343-353, doi: 10.1109/TSP.2013.2288086
11. Zhang, J.D., Zhu, X.H., and Wang, H.Q.: 'Adaptive radar phase-coded waveform design', *Electron. Lett.*, (2009), **45**, (20), pp. 1052-1053, doi: 10.1049/el.2009.1099
12. Conjugate gradient method [online]. Available: http://en.wikipedia.org/wiki/ conjugate_gradient_method

Monitoring System of Shaft Based on ZigBee Wireless Sensor Network

Man-hong FAN, Sheng-qian MA

Key Laboratory of Atomic and Molecular Physics & Functional Materials of Gansu Province, College of Physics and Electronic Engineering, Northwest Normal University. Anning East Road 967, Lanzhou, China

Abstract. Traditional shaft temperature and humidity detection system is wired mode, the transmission distance is limited. In order to overcome this shortcoming, this paper designed a detection system, which is based on ZigBee wireless sensor network and digital temperature-humidity sensor. In the system, the data's acquisition and transmission of master-slave nodes is realized by the star network, and the data is displayed on the host computer through serial port. The system can realize multi-point detection and real-time monitoring.

1 Introduction

The function of the monitoring system mainly includes collection, master, alarming, information transmitting and host computer displayed. Sensor is the perception of the monitoring system, which can transform the information into electrical signals, and then the system can collect, transmit and analyze these signals, which can indirectly reflect the state of the underground environment; The main system not only control to extract the information of sensor and transmit the data to the channel of communication, but also need to determine whether the collected information is normal. If the collected information out of range, the master will carry out the alarm. The part of information transmitting is responsible for sending the collected data to the host computer smoothly and displayed on it. Make sure, transmit information of data is effectively and timely. On the PC will display the data received in real time. In order to monitor the state of the underground environment all the time. The host computer by judging the machine over the information of data, displays that the state of underground environment is normal or abnormal, the alarm will rang when the data exceed the normal range.

The application of wireless environmental monitoring system is very extensive. Dohare(2015) uses the wireless Communication and Environment Monitoring in Underground Coal Mines. According to MASHALKAR and KAZI (2015), learning more about Smart Dust. Jha (2014) puts the networked Sensors on the architecture. Egan(2005) is used the ZigBee to automation and industrial controlling. Lee and SuA(2007) studies the wireless in Bluetooth. Alliance(2009) introduces the ZigBee standard.

2 Hardware design of the system

This system mainly involves the following modules: the master control of STM32, temperature and humidity sensor, LCD screen, JTAG interface, Clock and reset, wireless communication of ZigBee and power supply. Figure 1 is diagram of the system.

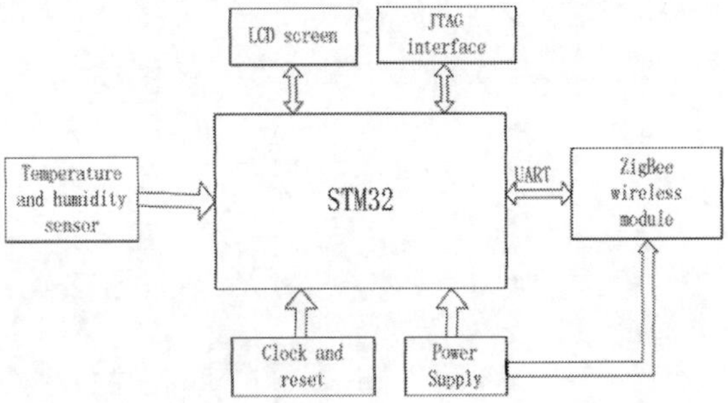

Fig 1. diagram of the system

2.1 The processor

STM32 is a microprocessor chip which is based on ARM Cortex-M3 kernel. The chip has the advantages of strong compatibility, high integration, strong real-time, which is easy to hardware development. Figure 2 is the STM32 Minimum system schematic. Table 1 is STM32 standard performance.

Before the STM32F105 and STM32F107 interconnected series of micro controllers, Italy semiconductor has introduced the STM32 basic series, the enhanced series, the USB basic series, the complementary series; the new series products follow the 72MHz processing frequency of the enhanced series. The memory includes 64KB to 256KB flash memory and 20KB to 64KB embedded SRAM. The new series uses three packages of LQFP64, LQFP100 and

LFBGA100. Different packages keep the consistency of pin arrangement. In combination with the design concept of STM32 platform, the developer can reoptimize function, memory, performance and pin number by selecting the product to meet the personalized application requirements with minimal hardware change.

The STM32L series added low power consumption and low power sleep two low power modes. By using the ultra-low power voltage regulator and oscillator, the micro controller can greatly reduce the power consumption at low frequency. The regulator can satisfy the current requirement without depending on the supply voltage. STM32L also provides dynamic voltage lifting, which is a successful application of energy saving technology for many years, which can further reduce the internal working voltage of the chip in the middle and low frequency operation. Under normal operation mode, the minimum current consumption of flash memory is 230 A/MHz, and the power consumption / performance ratio of STM32L is the lowest 185 A/DMIPS.

Table 1. STM32 standard performance

Number	Function	Parameter
1	Clock frequency	72MHz
2	Interrupt mode	interrupt nesting
3	Voltage range	2.0~3.6V
4	Lowest power consumption in standby mode	2uA

2.2 ZigBee wireless sensor network

CC2530 is a real system on chip (SoC) solution for 2.4-GHz IEEE 802.15.4, ZigBee and RF4CE applications. It can build strong network nodes with very low total material cost.

CC2530 combines the excellent performance of the leading RF transceiver, industry standard enhanced 8051 CPU, programmable flash memory in the system, 8-KB RAM and many other powerful functions. CC2530 has four different flash versions: CC2530F32/64/128/256,

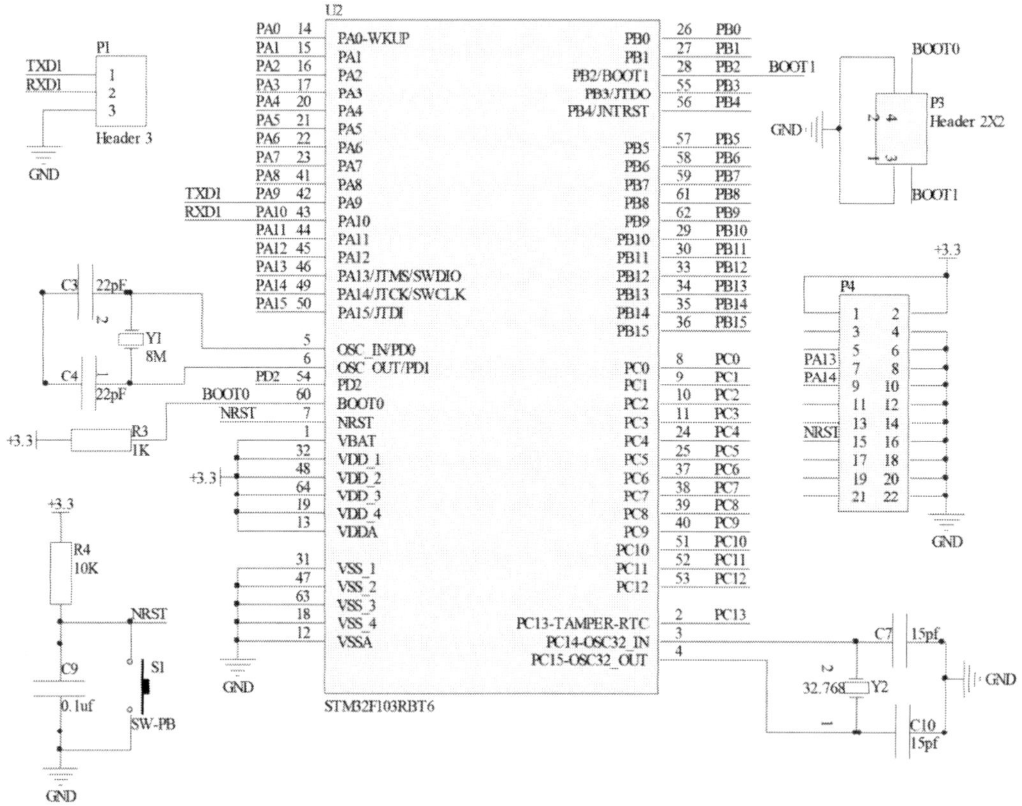

Fig 2. STM32 Minimum system schematic

64/128/256KB respectively. CC2530 has different modes of operation, making it especially suitable for ultra-low power consumption systems. Short switching time between operation modes further ensures low energy consumption.

CC2530 has a IEEE 802.15.4 compatible wireless transceiver. The RF kernel controls the analog wireless module. In addition, it provides an interface between the MCU and the wireless device, which makes it possible to send commands, read States, automatically operate and determine the order of wireless device events. The wireless device also includes a packet filter and address recognition module.

The communication of data between the nodes of network in the sensor must be carried out through the wireless communication. This design is based on CC2530 [7-8]. Although low-cost, it can build a powerful wireless sensor network node. CC2530 can be switched to a different mode in a short period of time, particularly suitable for low power consumption requirements of the application. CC2530 includes the abundant resources of hardware in the internal, the main function module is: the enhanced 8051 core, the memory, the RF transceiver, the A/D, the timer, USART, I/O, DMA controller, AES security coprocessor. The connection pin of CC2530 and STM32 is shown in Figure 3.

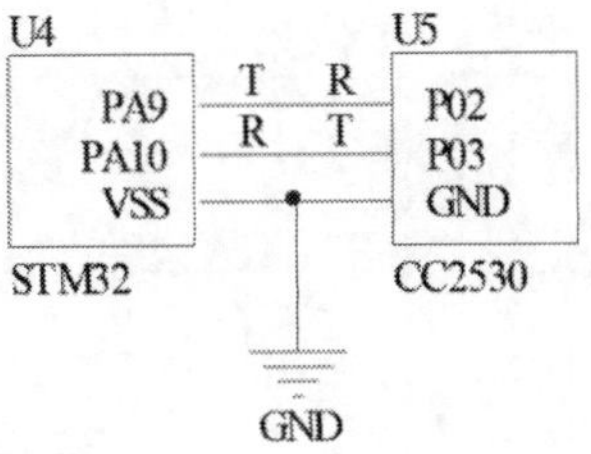

Fig 3. Connection of CC2530 and STM32

2.3 Temperature and humidity sensor

DHT22 is a type of high precision digital temperature and humidity sensor. The sensor includes capacitive humidity sensitive element and a high precision temperature measuring element. Each sensor has completed the temperature calibration before delivery, and will be stored the calibration coefficient in the 8 bits microcontroller chip. Each sensor uses internal form calling these calibration coefficients internal to process collected signal. Figure 4 is the diagram of pins' connection of DHT22.

In Han and Li(2009), point out the design of Temperature And Humidity Detection System for Intelligent Community. Wei(2012) studies the wireless network sensing system bases on ZigBee.

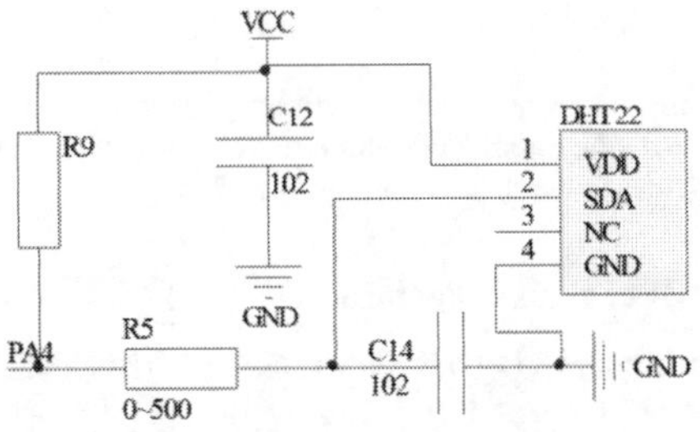

Fig 4. DHT22 connection

3 The software design of the system

3.1 Main program design

The software design for STM32 and sensor acquisition uses the MDK5.10. Keil MDK integrates industry-leading technology, including the integrated developing environment of Keil Vision5 with RealView compiler, which supports ARM7, ARM9 and the latest Cortex-M3 core processor. The software design includes data's acquisition of the route and the sensor of terminal node, transmission of serial and transmission of ZigBee, and the coordinator node by the RS485 that will transfer the data to the host computer. The flow chart of main program is shown in Figure 5.

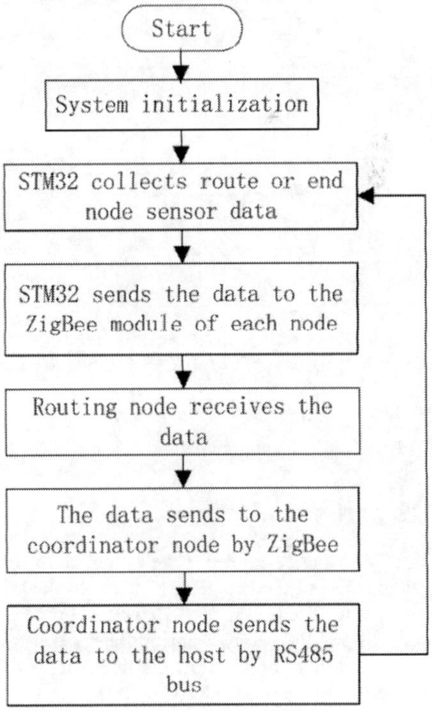

Fig 5. Main program flow chart

STM32 controls the sensor's 5 parameters cycle and time-share collection. If the range is out of the alarm, the result will be displayed by the LCD, and the data that has read is transmitted to the ZigBee module through the serial port, after the data of ZigBee network transmission come to the node of coordinator, and it will be transferred to the host computer through RS485.

3.2 DHT22 data collection

When in the idle state, the bus is high level. STM32 pull bus level to low, when the signal starting, then it will release bus. Time delay about 20～40 μ s, loop detecting DHT22 signal will begin. This signal is about 80 μ s low level response. When received the signal, STM32 will collect the 40 bit data. Finally, LCD and serial port is opened. Figure 6 is the DHT22 Software process.

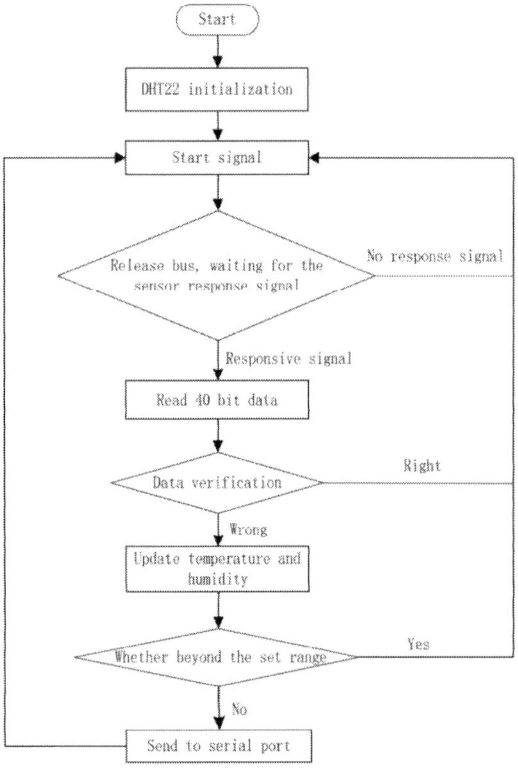

Fig 6. DHT22 Software process

4 Experimental test

The system can be applied to the 17 floors' experimental building. The host computer is in the 17th floor. In order to facilitate testing, the part of power use the mobile power of 5V, the specific situation may need to use the battery of voltage or 9~12 DC power of supply regulator, or directly use the computer USB power to supply. The data collection location is on the first floor and the 9th floor. Each floor was collected 10 different points of the temperature and humidity. Table 2 shows the information of data to the temperature and humidity in the first floor, in Table 3 for the data of information to the temperature and humidity in the 9th floor.

Table 2. Data information in the first floor

Test number	**1**	**2**	**3**	**4**	**5**
Temperature (℃)	17.5	18.9	17.1	20.4	21.2
Humidity (%RH)	32.9	35.7	36.1	35.9	35.4
Test number	**6**	**7**	**8**	**9**	**10**
Temperature (℃)	19.5	18.9	21.4	22.0	21.7
Humidity (%RH)	34.2	36.1	35.8	37.1	34.2

Table 3. Data information in the 9th floor

Test number	**1**	**2**	**3**	**4**	**5**
Temperature (℃)	27.9	29.3	29.5	28.6	27.5
Humidity (%RH)	39.1	38.7	39.4	38.9	40.0
Test number	**6**	**7**	**8**	**9**	**10**
Temperature (℃)	28.9	28.4	28.9	29.1	28.0
Humidity (%RH)	39.9	38.1	39.8	37.9	37.6

5 Conclusion

In this paper, the detected system of the CC2530 wireless temperature and humidity is based on ZigBee technology, and the 20 sensor nodes of the temperature and humidity are studied. In the experimental building, the detection of multipoint data wireless transmission are realized. The results meet the expected requirements. The system has a simple structure, low power, high effective detection, and good ability of anti-interference, which has the higher practical and economic benefit.

Acknowledgment

In this paper, the research was sponsored by the Natural Science Foundation of China (61162017), Department of Education-fund projects in Gansu(1101-03).

References

1. Alliance Z.B. IEEE 802.15. 4, ZigBee standard. http://www. zigbee. org. (2009)
2. Dohare Y.S., Maity T., Das P.S., et al. Wireless Communication and Environment Monitoring in Underground Coal Mines–Review. IETE Technical Review.1-11. (2015)
3. Egan D. The Emergence of ZigBee in building automation and industrial controls. Computing and Control Engineering.16(2), 14-19. (2005)
4. Jha L.N. System Architecture Directions for Networked Sensors. International Journal of Research. 1(11), 450-465. (2014)
5. Lee J.S., Su Y.W.and Shen C.C. A comparative study of wireless protocols: Bluetooth, UWB, ZigBee, and Wi-Fi. Industrial Electronics Society, 2007. IECON 2007. 33rd Annual Conference of the IEEE. Taipei, Taiwan, Nov.05-08, IEEE. 46-51. (2007)
6. MASHALKAR Y.S. and KAZI A.S.M. Wirless Sensor Networks (MOTES): Smart Dust. IJCER, 3(6), 308-312. (2015)
7. Han Chenghao, Li Baifeng, et al. Design of Temperature And Humidity Detection Systemfor Intelligent Community. Journal of Changchun University of Science and Technology(Natural Science Edition). 32(3), 494-497. (2009)
8. Wei Feifei. Research of The Wireless Network Sensing System Bases on ZigBee. Dalian Maritime University. (2012)

An Efficient Hardware-Based Fault-Tolerant Method for SMS4

Hanguang Luo, *Guangjun* Wen and *Jian* Su

University of Electronic Science and Technology of China, 611731 West Hi-Tech Zone, China

Abstract. The SMS4 cryptosystem has been used in the Wireless LAN Authentication and Privacy Infrastructure (WAPI) standard for providing data confidentiality in China. So far, reliability has not been considered a primary objective in original version. However, a single fault in the encryption/decryption process can completely change the result of the cryptosystem no matter the natural or malicious injected faults. In this paper, we proposed low-cost structure-independent fault detection scheme for SMS4 cryptosystem which is capable of performing online error detection and can detect a single bit fault or odd multiple bit faults in coverage of 100 percent. Finally, the proposed techniques have been validated on Virtex-7 families FPGA platform to analyze its power consumption, overhead and time delay. It only needs 85 occupied Slices and 8.72mW to run a fault-tolerant scheme of SMS4 cryptosystem with 0.735ns of detection delay. Our new scheme increases in minimum redundancy to enhance cryptosystem's reliability and achieve a better performance compared with the previous scheme.

1 Introduction

With the rapid development of network and digitization, the information protection has nowadays became an absolutely requirement. Encryption technology is not only be used for computer network, but also used on some special digital equipment and chip cryptographic services to protect information. Due to the limited resources and highly-constrained of the equipment or embedded chip, time constraint, electricity constraint etc., lightweight encryption algorithm is widely used in numerous data privacy places. SMS4 is one of the lightweight block cipher using a Unbalance Feistel Network(UFN) structure, designing for the resource-constrained platforms, which is mandatory in wireless networks by China in 2006[1] .

As the hardware implementations of SMS4 algorithm, stresses, temperature changes and electromagnetic fields affect electronic devices causing it abnormal operation. Furthermore, as a result of rapid diffusivity of its security function, even a bit fault can cause complete decryption failure. In addition, it had not been known that different error output could be used to carry out an attack on the cryptographic algorithm until 1996 by Boneh, DeMillo, and Lipton [2] from Bellcore. Soon afterwards, Differential Fault Analysis(DFA) was extended to DES-like secret key cryptosystems by Biham and Shamir[3].Since SMS4 was published, several authors mounted DFA attacks on it[4,5,6]. The main idea of these attacks is induced several byte fault into the register of the round function, then use DFA attack to retrieve secret key. It is reported in [6] that a successful attack on SMS4 requires only a single random byte faulty responses.

Therefore, an effective error detection scheme is required to improve SMS4 algorithm of fault tolerance. There have been some traditional methods applied to the fault detection of other cipher, e.g. AES. A scheme is put forward in [7], which used reverse functions to recover the original input for the encryption/decryption process; [8,9] can applied to the internal function, round function, or entire encryption process levels, based on the hardware-redundant-based concurrent error detection schemes. Another category of fault detection methods is based on parity codes [10,11,12]. This kind of methods utilized the parity check codes to each byte of the state matrix. It can also be applied to each transformation, each round, or the encryption process levels. To the best of our knowledge, [13] is the only one article which worked to investigate fault-tolerant methods for SMS4 and used reverse functions in the process levels.

The organization of this paper is as follows: In Section 2, we review the basics of SMS4. In Section 3, we introduce the traditional simple method of faults detection and implemented it to the SMS4. Then we present the proposed less redundancy faults detection scheme. Section 4 presents the experimental results and hardware implementation. Finally, conclusions are made in Section 5.

2 Description of SMS4 algorithm

SMS4 is a 128-bit block cipher, which supports 128-bit block size and 128-bit key lengths. It iterates a simple round function 32 times. The 128-bits plaintext is divided into 4 words each has 32-bit length, $P=(P_0,P_1,P_2,P_3)\in(\mathbb{Z}_2^{32})^4$. The ciphertext and round subkeys are also divided into 4 words denoted as $C=(C_0,C_1,C_2,C_3)\in(¡\,_2^{32})^4$ and $RK_i\in¡\,_2^{32}\,(i=0,1,2,...,31)$, respectively. Note that the first round is referred to Round 0.

The one round structure of SMS4 is depicted in Figure.1 and the encryption procedure of SMS4 is as follows:

1) For i =0 to 31, the words are updated as $X_{i+4}=X_i\oplus F(X_{i+1}\oplus X_{i+2}\oplus X_{i+3},\ RK_i)$.

2) The 128-bit ciphertext with output of the last round is generated by applying the switch transformation R denoted as

$$\begin{aligned}(Y_0,Y_1,Y_2,Y_3)&=R(X_{32},X_{33},X_{34},X_{35})\\&=(X_{35},X_{34},X_{33},X_{32})\end{aligned} \quad (1)$$

The Addroundkey is a process that let three input X_{i+1}, X_{i+2}, X_{i+3} do XOR operation, then put the 32-bit output XOR with the round key RK_i. The non-linear transformation S is constituted of four equal S-box in parallel to the 32-bit input. The transformation S is defined as follows.

$$S(B)=(Sbox(a_0),Sbox(a_1),Sbox(a_2),Sbox(a_3)) \quad (2)$$

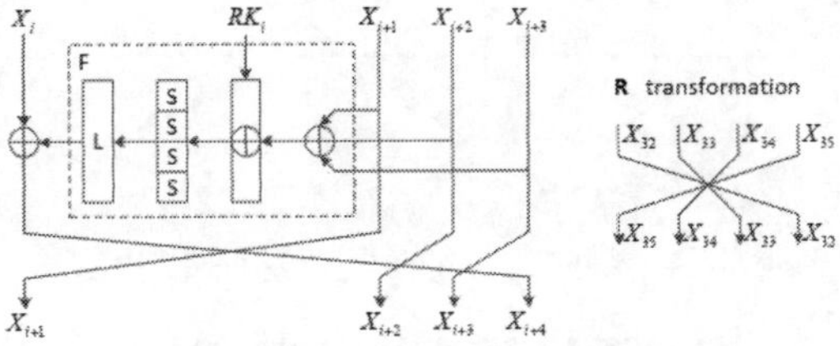

Figure 1. The i -th round of SMS4

Where $a_i\in¡\,_2^8, i\in 0,1,2,3$ denotes the four 8-bit input of S-box which achieved by the output of the Addroundkey process and divided into four equal parts. $B\in¡\,_2^{32}$ is the output of the transformation S.

The diffusion transformation L is a simple linear function which follows the transformation S. Its input is the output of the transformation S and defined as follows.

$$\begin{aligned}L(B)=B\oplus(B<<<2)\oplus(B<<<10)\\\oplus(B<<<18)\oplus(B<<<24)\end{aligned} \quad (3)$$

The Key schedule of the SMS4 is similar to the encryption function and it only has 128-bit key version. The main idea of the Key schedule is to generate 32-bit round key by a 128-bit master key in 32 rounds. As the key schedule algorithm is not involved in our research we omit the concrete process. And interested readers may refer to [14] for the details.

3 Error detection methods

3.1. Error model

The pathway of discussed error model in this paper can be divided into two categories: deliberately introduced error in a malicious attempt and internal error. The first one is usually inserted by an adversary for using Fault Analysis (FA) or Differential Fault Analysis (DFA) in order to retrieve secret key of the algorithm. And the faults can be inserted into the register unit in each round. The latter one occurs during the process of encryption/decryption. This class of fault happens in a natural consequence of the module's normal operation, duration of use or environment, which takes into account a global, physical and electrical failure approach. Both of them can be temporary or permanent faults.

Although SMS4 algorithm allows the 32 iteration round unfold into line structure, the hardware overhead is not suitable for most of the embedded application. Due to the different applications, the demand for encryption/decryption speed, and power consumption, area resources, the structure of SMS4 algorithm is also different. Without loss of generality, we take the 32 iteration round structure into account. We also assume that an error can both be inducted into the register and the state between transformations in each round.

Since most of embedded systems, data are based on bytes of storage and processing, we also assume that the error happened only in a single bit or odd multiplies of bits, whether it is a malicious attempt insert error or internal occur error. Errors, however, the number of bytes and location are random.

3.2 Fault detection techniques

1) Redundancy-Based Techniques (RBT). Traditional fault-tolerant methods have been widely investigated for the implementation of cipher. One of the most simple methods is Redundancy-based Techniques (RBT) [15]. The main idea of the method is as follows. Consider an arbitrary function $y=f(x)$. If $f(*)$ exists an invertible function $z=f^{-1}(y)$ in the domain. We can safely come to the conclusion that must equal to z. i.e. $x=z=f^{-1}(y)$. We can also apply this method to SMS4 fault detection. It not only can be applied at the algorithm level but also at the round function level. The detail can be seen in the Figure.2. at (a) and (b), respectively.

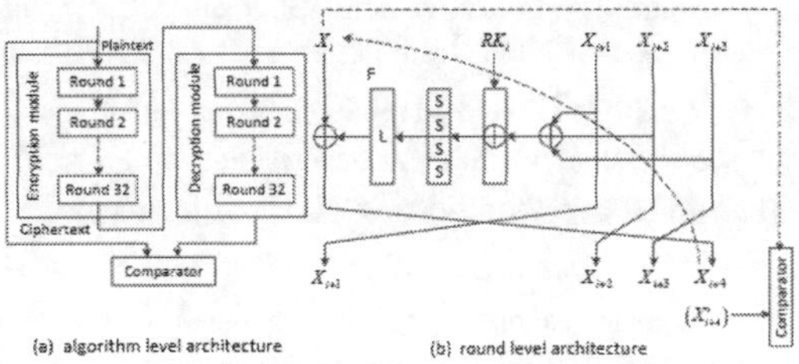

Figure 2. Method of Redundancy-based Techniques

As shown in Figure 2. (a), when RBT is employed at the algorithm level, the ciphertext is decrypted to the plaintext then compared with the original plaintext. If faults are produced in the process of encryption, the result will not equal. When it comes to the round level, due to the structure of SMS4 round function, the encryption and decryption are almost the same. The four input data $X_i, X_{i+1}, X_{i+2}, X_{i+3} \in \mathbb{F}_2^{32}$ which divided from the plaintext, is operated with round key RK_i result in $X_{i+1}, X_{i+2}, X_{i+3}, X_{i+4} \in \mathbb{F}_2^{32}$ (see Figure 2.(b) with the encryption). The imaginary line describes the change data of decryption process in a round, i.e. the round key is also be RK_i, the input of decryption is $X_{i+4}, X_{i+1}, X_{i+2}, X_{i+3} \in \mathbb{F}_2^{32}$ and the output is $X_{i+1}, X_{i+2}, X_{i+3}, X'_{i+4}, \in \mathbb{F}_2^{32}$. If no error occurs in the encryption process the decryption output data X'_{i+4} must equal to X_i, or it is incorrect.

Although the principle of RBT method is quite simple and easy to be implemented, the time expense on the fault detective and the power consumption is almost equivalent to the encryption process. This can't be tolerated with some of the resources limited embedded device, e.g. RFID. At the same time we cannot guarantee, as a part of error detection process, the decryption is exactly correct. If some faults happen to the detection process, the whole error detection progress fails. Therefore to find a suitable new redundancy scheme is necessary.

2) Error Detecting Codes (EDC). Bertoni et al. proposed in [16] in AES algorithm used a parity bit to resist fault analysis. They use a single parity bit to represent the 8 bit word in the middle of the status or register unit.

$$P_{i,j} = \bigoplus_{k=0}^{7} S_{i,j}^{(k)} \tag{4}$$

The same method can also be used in SMS4 algorithm. We check on the single parity bit in order to know whether there is a bit or odd bit fault in each register unit. Meanwhile, it can be extended to arbitrarily bit situation to check more bits in each time. i.e.

$$P = \bigoplus_{i=0}^{n} W^i \tag{5}$$

For the convenience of description and easy to understand the fault detection scheme, the round function is divided into four parts: 1) "AddRK" short for Addroundkey: $(X_{i+1} \oplus X_{i+2} \oplus X_{i+3} \oplus RK_i)$; 2) "Sbox" short for S-box transformation; 3) " Ltranf " short for L transformation; 4) " OUT-X_{i+4}" represented the output X_{i+4} : $X_{i+4} = X_i \oplus L(B) \in \mathbb{F}_2^{32}$. And we define the corresponding parity bit denoted with symbol "I ∗". e.g. The corresponding parity bit of X_i denoted as $\mathrm{I}\ X_i = \bigoplus_{i=0}^{3} p^i \in \mathbb{F}_2$, where p^i is the parity bit to the corresponding register unit, i.e. $p^0 = \bigoplus_{k=0}^{7} x_i^k, \ldots, p^3 = \bigoplus_{k=24}^{31} x_i^k$.

To our scheme, the detection can be applied at round function level or at the operation level. Here we only show the round function level scheme, see Figure 3. The operation level scheme is similar and omitted. The S-boxes of SMS4 algorithm are used in a look-up table (LUT) method.

Initialization, each storage unit is upgraded from the original 8 bit to 9 bit. The actual data is kept in the first 8 bit, and the parity bit is deposited in the last bit. To the look-up table of the S-box, the same as the input data, each corresponding parity bit is calculated and reserved behind the actual data in advance. The round key is also be calculated by the secret key through the key schedule beforehand. Then calculated the corresponding parity bit and reserved it.

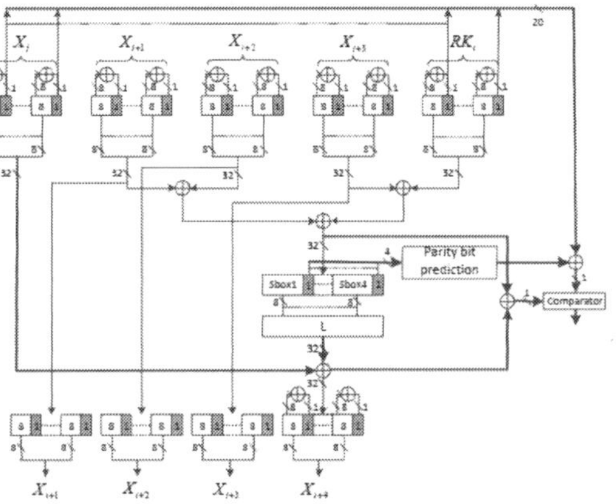

Figure 3. Round function level of parity bit prediction

To the Addroundkey part, as shown in figure 3, we can obtained that

$$\mathrm{I}\ (AddRK) = (\mathrm{I}\ X_{i+1} \oplus \mathrm{I}\ X_{i+2} \oplus \mathrm{I}\ X_{i+3} \oplus \mathrm{I}\ RK_i) \tag{6}$$

The S-box is the only nonlinear device in the SMS4 algorithm. When comes to the S-box transformation, the output and the corresponding parity bit can both be obtained from the look-up table.

L transformation is a linear device denoted in (3). It is obtained by the S-box output with cyclic shift several times, then XOR each other. As the left rotation operations do not affect the parity bit of its outputs, meanwhile the XOR operation times is odd. The parity bit of L transformation is equal to S-box's. So the parity bit prediction of the L transformation is as follow:

$$\begin{aligned}
\mathrm{I}\ (\text{Ltranf}) &= \mathrm{I}\ L(B) \\
&= \mathrm{I} \begin{Bmatrix} B \oplus (B <<< 2) \oplus (B <<< 10) \oplus \\ (B <<< 18) \oplus (B <<< 24) \end{Bmatrix} \\
&= \mathrm{I}\ B \oplus \mathrm{I}\ (B <<< 2) \oplus \mathrm{I}\ (B <<< 10) \\
&\oplus \mathrm{I}\ (B <<< 18) \oplus \mathrm{I}\ (B <<< 24) \\
&= \mathrm{I}\ (Sbox) \in \mathbb{F}_2
\end{aligned} \tag{7}$$

Finally, the X_{i+4} is achieved by X_i XOR with L transformation output. The parity bit is

$$\mathrm{I}\left(OUT-X_{i+4}\right)=\mathrm{I}\left(X_i\oplus L(B)\right)=\mathrm{I}\ X_i\oplus\mathrm{I}\ L(B) \quad (8)$$

From what has been discussed above, if no faults happen in the encryption process, all parts of the parity bits will equal to all the parts of the actual data XOR. i.e.

$$\begin{aligned}
&\bigoplus_{k=0}^{31}\left[\left(X_{i+1}\oplus X_{i+2}\oplus X_{i+3}\oplus RK_i\right)\oplus X_{i+4}\right]^k\\
&=\bigoplus_{k=0}^{31}\left[\left(X_{i+1}\oplus X_{i+2}\oplus X_{i+3}\oplus RK_i\right)\oplus X_i\oplus L(B)\right]^k\\
&=\mathrm{I}\left(X_i\oplus X_{i+1}\oplus X_{i+2}\oplus X_{i+3}\oplus RK_i\right)\oplus\mathrm{I}\ Sbox\\
&=\mathrm{I}\ X_i\oplus\mathrm{I}\ X_{i+1}\oplus\mathrm{I}\ X_{i+2}\oplus\mathrm{I}\ X_{i+3}\oplus\mathrm{I}\ RK_i\oplus\mathrm{I}\ Sbox\in ?_2
\end{aligned} \quad (9)$$

where $\mathrm{I}\ X_i, \mathrm{I}\ X_{i+1}, \mathrm{I}\ X_{i+2}, \mathrm{I}\ X_{i+3}, \mathrm{I}\ RK_i, \mathrm{I}\ Sbox$ are the parity bit of the input date and Keys saved before in the last bit of storage unit.

Finally, before the output X_{i+4} is stored, the data of X_{i+4} should be calculated to the corresponding parity bit then preserved. The fault detection scheme of operation level version is to detect of the above four parts, respectively. From what has been discussed above, both the malicious faults and internal faults can be detected.

4 Simulation results

4.1. Fault detection rate

In order to test the faults detection rate of the EDC scheme, we implemented the detection scheme in C++ and tested the fault detective scheme from two types:

i) Randomly selected 10000 plaintexts data blocks and injected a single bit error (into round key or round plaintexts) into random position and random rounds. However, the secret key of key schedule was remained the same. Though our error detection scheme, all the faults were detected.

ii) Randomly selected 10000 plaintexts data blocks with the same secret key. An odd bit error (into round key or round plaintexts) was injected into random position and random rounds. Here too, all the faults were detected.

Through the experiment above, the proposed scheme can accurately detected the single and multiple odd bit error in SMS4. In order to determine the power consumption and overhead, further hardware experiment should be taken.

4.2. FPGA implementations and comparison

In this section, we used FPGA to verify practicability of the proposed schemes. The SMS4 algorithm and fault detective scheme are both designed in 32 iteration round structure. The S-boxes is designed in LUT-based method, parity bits of the S-boxes were calculated and preserved beforehand. The round keys and corresponding parity bits were also calculated and preserved beforehand. This scheme not only saved the time spent by extension of round keys but also reduced the overhead. We have used Verilog HDL as the design entry for ISE version 14.1 and the synthesis is performed using Xilinx Synthesis Tool on Virtex-7 families (XC7V585T).

As shown in Table 1, the used numbers of Slice LUTs in this paper were 169 compared with the 118 to the original scheme. The Slice Registers were 168 compared with the 129 to the original one. Although the total percentage of increased overhead is approximately 51.79%, the cut down of round key extension must have saved a large part of the overhead compared with the one which has the extension of key schedule. And then we compared with fault detection scheme of [13], the total occupied logic of resources were less than one third of [13].

Table 1. Overhead of different scheme of the SMS4 algorithm on Xilinx Virtex-7 FPGAs

Structure	overhead				Increase Overhead (%)
	Slice LUTs	*Slice Registers*	*IOBs*	*occupied Slices*	
Original	118	129	259	56	-
This paper	169	168	261	85	51.79%
RBT[13]	-	-	-	270	382.14%

Table 2. Power consumption of different scheme of SMS4 algorithm on Xilinx Virtex-7 FPGAs

Structure	Power(mw)				Increase (%)	Delay (ns)
	Logic	*Signals*	*IOs*	*Total*		
Original	1.15	3.07	1.74	5.96	-	-
This paper	2.86	4.12	1.74	8.72	46.30%	0.735

With the increase of overhead, the corresponding power consumption was also increased. We compared the original scheme (don't have the extension of key schedule) with our scheme of SMS4 cryptosystem's power consumption in Table 2. Total power consumption of our scheme was 8.72mW, increased 2.76mW compared with the original scheme. Although [13] didn't show the power consumption of the fault detection scheme, the cost of the power consumption of our scheme must be lower than the cost of [13], since the method of it only did the inverse process of encryption and the S-boxes process is the most power consuming part. From what has been discussed above, considering the contribution of faults detection, the little bit of increasing redundancy is completely acceptable. At the same time, the delay time of 0.735ns which caused by the additional circuit is also acceptable.

5 Conclusion

In this paper we did some research on the SMS4 cryptosystem of fault detection. We showed the traditional scheme of error detective method on SMS4 and presented a new fault detection scheme based on parity code. The simulations showed that our new fault detection structure reached the error detected coverage of 100 percent.

In addition, the proposed fault detection scheme has been implemented on the recent Xilinx Virtex-7 families FPGAs. The overhead, delay and power consumption have all achieved very good performance compared with the original and prevenient scheme. The reason why new scheme of fault detecting structure used so less resource is that we only utilized XOR operation to checkout fault. And the time delay is also in an absolutely acceptable range since XOR operation is easy and fast to be implemented in hardware. At the same time, the reduced of the round key extension part also greatly reduces the cost. Therefore, we can conclude that from our proposed scheme it can well done on the fault detection and defensed Fault Attack on SMS4 cryptosystem.

Acknowledgment

This work was supported in part by the National Natural Science Foundation of China under project contracts No.61601093, No.61791082, No. 61701116 and No.61371047, in part by Sichuan Provincial Science and Technology Planning Program of China under project contracts No.2016GZ0061, 2017GZ0336 and No.2018HH0034, in part by Guangdong Provincial Science and Technology Planning Program of China under project contracts No.2015B090909004 andNo.2016A010101036, in part by the fundamental research funds for the Central Universities under project contract No.ZYGX2016Z011, and in part by Science and Technology on Electronic Information Control Laboratory.

References

1. Office of State Commercial Cipher Administration. "Block Cipher for WLAN Products-SMS4", http://www.oscca.gov.cn/UpFile/200621016423197990.pdf.
2. Dan Boneh, Richard A. DeMillo, Richard J. Lipton, "On the importance of checking cryptographic protocols for faults", in: EUROCRYPT'97, in: LNCS, vol. 1233, Springer 1997, pp. 37–51.
3. Eli Biham, Adi Shamir, "Differential fault analysis of secret key cryptosystems", in: CRYPTO 1997, in: LNCS, vol. 1294, Springer, 1997, pp. 513–525.
4. Wei Li, Dawu Gu, "An improved method of differential fault analysis on the SMS4 cryptosystem", in: ISDPE 2007, IEEE Computer Society, 2007, pp. 175–180.
5. Wei Li, Dawu Gu, "Differential fault analysis on the SMS4 cipher by inducing faults to the key schedule", Chinese Journal on Communications 29 (10) , 2008, pp.135–142.
6. Ruilin Li, Bing Sun, Chao Li, Jianxiong You, "Differential fault analysis on SMS4 using a single fault, Information Processing Letters", vol.111, 2011, pp.156–163.
7. R. Karri, W. Kaijie, P. Mishra, and K. Yongkook, "FaultBased Side-Channel Cryptanalysis Tolerant Rijndael Symmetric Block Cipher Architecture," in Proc. DFT'01, 2001, pp. 418–426.
8. R. Karri and X. Guo "Invariance-based concurrent error detection for advanced encryption standard," in Proc. DAC, 2012, pp. 573–578.
9. M. Mozaffari-Kermani and A. Reyhani-Masoleh, "Concurrent Structure-Independent Fault Detection Schemes for the Advanced Encryption Standard," IEEE Trans. Computers, vol. 59, no. 5, May 2010, pp. 608–622.
10. J. Mathew, et al., "On the design of different concurrent EDC schemes for s-box and gf(p)," in Proc. ISQED'10, 2010, pp. 211-218.
11. Ting An, Lirida Alves de Barros Naviner and Philippe Matherat, "Evaluation of Fault-tolerant Composite Field AES S-Boxes under Multiple Transient Faults," in Proc. NEWCAS'13, 2013, pp 1–4.
12. A. Barenghi, L. Breveglieri, I. Koren, and D. Naccache e, "Fault injection attacks on cryptographic devices: Theory, practice, and countermeasures," Proc. IEEE, vol. 100, no. 11, Nov. 2012,pp. 3056–3076.
13. Xin Xiaoxia, Wang Yi, Li Renfa, "FPGA-based implementation for fault detection of SMS4", Journal of Computer Applications, vol. 35, no. 2, pp. 420-3, 10 Feb. 2015.
14. Specification of SMS4, Block cipher for WLAN products–SMS4, http://www.oscca.gov.cn/UpFile/200621016423197990.pdf
15. Abdel Alim Kamal, Amr M. Youssef, "An FPGA Implementation of AES with Fault Analysis Countermeasures Analysis Countermeasures", in: ICM 2009, Vol.978, May 2009, pp.217-220.
16. G. Bertoni, L. Breveglieri,I. Koren, P. Maistri, and V. Piuri, "Error Analysis and Detection Procedures for a Hardware Implementation of the Advanced Encryption Standard", IEEE Transactions on Computers, Vol. 52, April 2003, pp. 492-505.

Impact of Xen and Virtual Box Virtualization Environments on Timing Precision under Stressful Conditions

Mykola Korniichuk , *Kirill* Karpov · *Irina* Fedotova, *Veronika* Kirova, *Nikolay* Mareev, *Dmytro* Syzov and *Eduard* Siemens

Anhalt University of Applied Sciences - Faculty of Electrical, Mechanical and Industrial Engineering, Bernburger Str. 57, 06366, Köthen, Germany

Abstract. Sharing physical resources among virtual instances introduces time overhead in comparison with direct access to hardware. Such lag is not significant for most of the everyday tasks however it can influence much more in dealing with time-critical applications and especially in case of reliable network services. The purpose of this paper is to compare time overhead on timing operations such as time acquisition and sleep introduced by different virtualization environments: Xen, VMWare ESXi, QEMU. The current research focus to establish possibility of using such platforms in real-time applications with high resource utilization. In terms of present work, there are several load types to be used to simulate real conditions. Measurement performance includes different methods of time measurements and waiting operations. Considering results, the certain recommendations about timing mechanisms using different virtualization environment have been offered.

1 Introduction

Nowadays operating systems are often virtualized due to increasing the hardware utilization and decreasing maintenance costs. However, a lot of different challenges occur due to virtualization. Hypervisor has direct access to hardware and is responsible for sharing all the resources among virtual machines, referred as a guest operating system. This additional software layer, which is influences on the performance and can break time-related functionality.

The precision of time-related operations such as time acquisition and waiting operations are crucial for any networking application. The time overhead on such operations can cause the data rate losses in high performance network applications, such as RMDT transport protocol, where on each sent packet there are 187 time acquisition operations and one sleep procedure.

The goal of this paper is to investigate how precise timing operations behave in different virtual environment under different loads. These results allow to perform an analysis of timing behavior under different conditions and to estimate reliability, provided by TSC and improved sleep methods [4,5]. Such analysis may be helpful for deploying time-related software in a virtual environment.

Measurements are performed on the same hardware setup in so called Enterprise scenario. In this scenario, virtualization is used as a mechanism for fast migration and support of system reliability under conditions of high-performance computation needs. In the experiments presented below, only one guest operating system is being virtualized, and all the measurements performed within the virtualized operating system.

The remainder of this paper is structured as follows: Section 2 reviews related work. The results of experiments with time acquisition and sleep function measurements is presented in Section 4 and 5 respectively. Conclusions and ideas for further work are offered in Section 6.

2 Related work

This work is an extension of our previous paper [1], where QEMU and VMware ESXi hypervisors have been studied. In this paper the same analysis of Xen and Virtual Box virtualization environments will be provided.

Similar analyses for the case of, so called Hosting scenario have been conducted in [2], where impact of different number of virtual machines running in parallel on timing operations inside one of them have been estimated.

S. G. Langer and T. French [3] have conducted research of influence of virtualization on I/O operations. It has been performed on Red Hat, Xen, ESXi, kvm_redhat64 as hypervisors, VMWare player and VirtualBox as hosted-based VMMs.

A detailed performance analysis of Xen Linux was made by P. Barham [4]. The authors made performance measurements in the hosting scenario on the Xen hypervisor-based VMM with different loads to simulate real life servers and compared it with different virtualization software.

The current paper is the extension of paper [1] with two new virtual environments: Citrix XenServer Host 7.0.0-125380c and Virtual Box 4.3.36. All hardware and software equipment, load types and methodologies are the same. With the current paper we intend to finalize the survey of the most valuable virtual environments from the perspective of time-related functionality.

3 Experimental results for time measurements

Time measurements consist of timer data collected with the RDTSC machine instruction, further referred as TSC and timer data collected with clock_gettime() further referred as system call. Experiments is composed of 3 configuration scenarios: Host OS (operating system is installed directly on hardware), Xen hypervisor and Virtual Box hypervisor.

3.1 Idle tested

Figure 1 shows the collected data in an idle state, which further helps to better estimate the results under heavy loads. Bold lines represent the medians, which for Host OS and Xen virtualized OS with TSC Timers are almost the same. However, the results from the Virtual Box significantly differ. As expected, host OS has the lowest time overhead (figure 1 (a)) about 17.2 nsec with TSC timer and 27 nsec with a system call. Xen virtualized OS (figure 1 (b)) shows almost the same median value as for Host OS in case of TSC timer, 18.6 nsec, but has greater overhead for a system call, 119 nsec. Virtual Box (figure 1 (c)) has a greater median overhead in comparison to Xen virtualized OS and host OS – 2.12 μsec for TSC timer and almost the same 2.18 μsec for the system call. Additionally, Virtual Box has the largest spread of values, up to 1 msec for the max value, and 1 nsec for the minimum value. On other platforms (Host OS, Xen, ESXi) minimal difference between time cost values from TSC timer is always 1.4 nsec, while here we have 0.29 nsec.

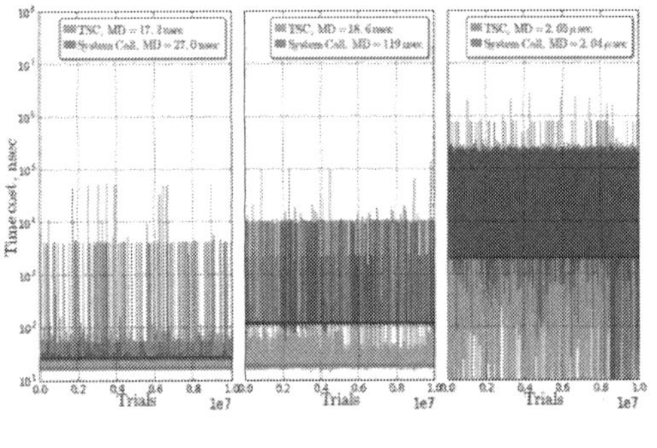

Fig. 1. Measurements of timer cost for the Idle testbed on: (a) Host OS (b) Xen (c) VirtualBox

In Figure 2 the representation of Complementary Cumulative Distribution Function (CCDF) with dashed lines as a means, are showed. Considering the given results, it can be observed that the overhead values which are less than of median values occurs with probability 99%. In idle mode, means and medians have almost the same values. CCDF for Virtual Box has significantly greater tails than for other results, which tells about greater results range in comparison to Host OS and Xen virtualized OS. Considering QEMU CCDF and VMware ESXi, tails are longer on all others test machines, except Virtual Box. Virtual Box tails end on a value up to 10 times greater than all others. These observation allow to assume that Virtual Box uses different way to obtain TSC values.

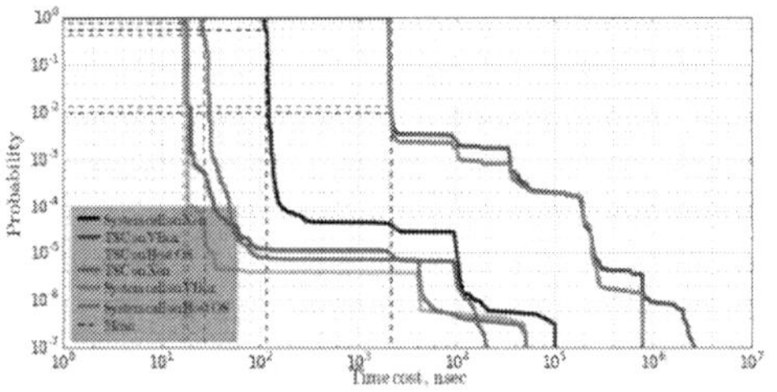

Fig. 2 CCDF representation of measurements of timer cost for the Idle testbed

3.2 CPU – bound testbed

Work lateness can be expected due to limited hardware resources in case of CPU background workload. The median value does not extremely differ from the median values in idle mode. The same can be concluded for QEMU and VMware ESXi [1]. However, this type of load causes long tails on CCDF plot (figure 4), which shows significant influence on mean value. The probability of occurrences of value less then median value and is still 99%. As expected, the best performance achieved on Host OS and the worst one is on Virtual Box virtualized OS. As for CCDF of QEMU and VMware ESXi, their results are mostly repeated results for Host OS, and not greater than System Call for Xen virtualized OS [1]. Virtual Box still shows worst results.

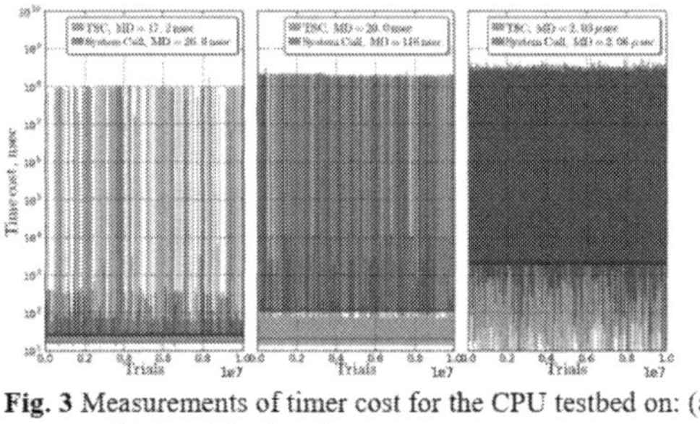

Fig. 3 Measurements of timer cost for the CPU testbed on: (a) Host OS (b) Xen (c) VirtualBox

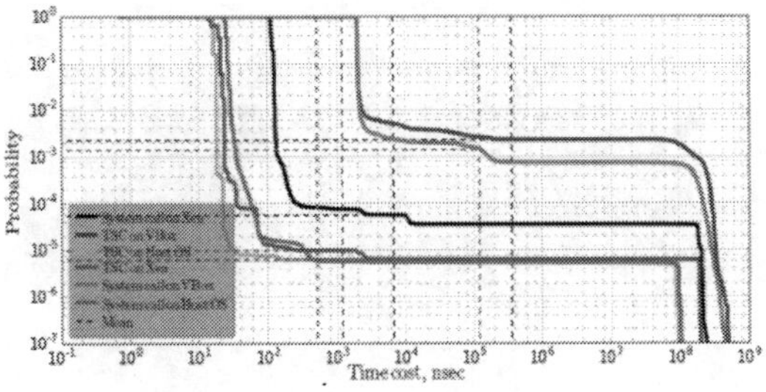

Fig. 4 CCDF representation of measurements of timer cost for the CPU testbed

3.3 I/O – bound testbed

Under a load of interrupt handling caused by IO-bound tasks, the median values are still staying without significant changes.

The range of values is slightly higher than in idle mode. At the same time, median values for TSC on QEMU and on VMware ESXi are 15.8 nsec and 21.5 nsec respectively. For the case of System call on QEMU, the median is 60 nsec, on VMware ESXi it is 34 nsec.

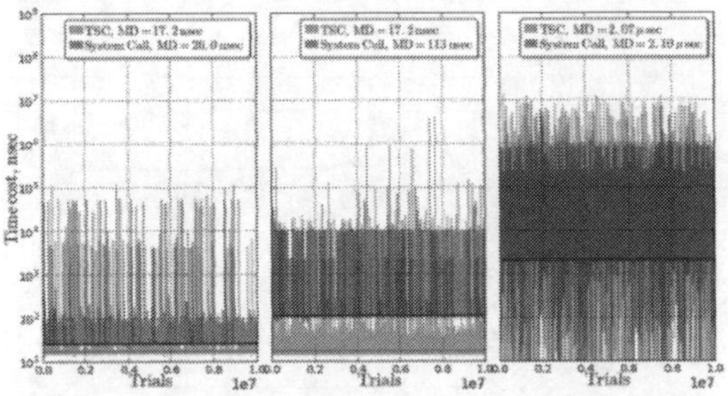

Fig. 5 Measurements of timer cost for the I/O testbed on: (a) Host OS (b) Xen (c) VirtualBox

CCDF plots show longer tails, but the mean value is almost the same as in idle mode, which tells that the probability of obtaining results is greater than median value is extremely small.

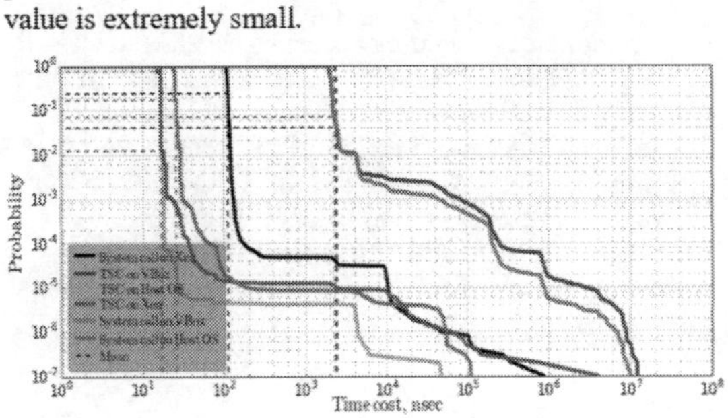

Fig. 6 CCDF representation of measurements of timer cost for the I/O testbed

3.5 Network – bound testbed

Median values for this type of load keeps the same. But the values are changing frequently during the test. That is caused by workflow of load type.

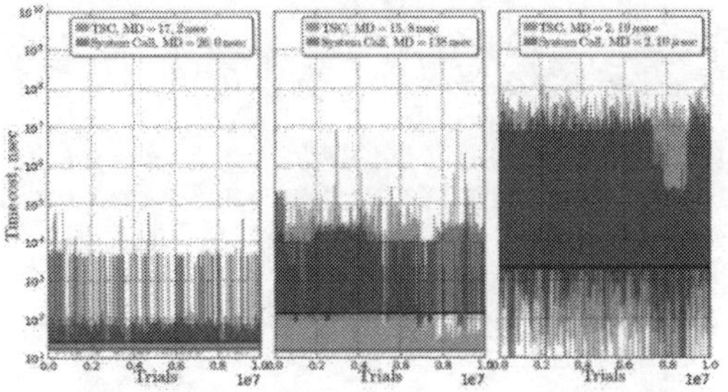

Fig. 7 Measurements of timer cost for the Network (RMDT download) testbed on: (a) Host OS (b) Xen (c) VirtualBox

On CCDF plot, represented on figure 10, the tail length remains almost the same as in idle case only for Host OS. For Xen and Virtual Box tails are much longer. At the same time, probability of obtaining time cost value greater than median value is greater for all cases but still median value occurs with probability around 99%.

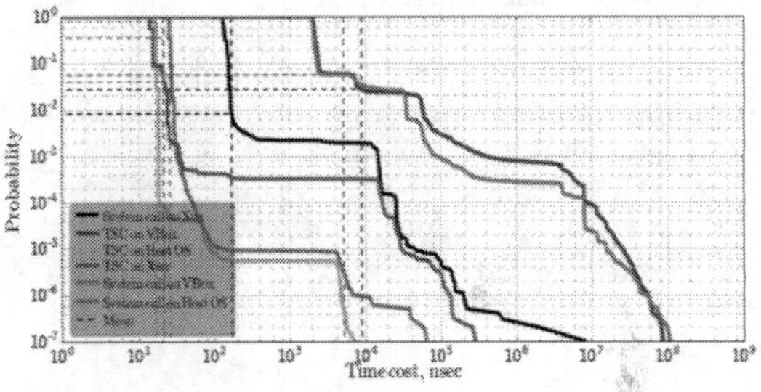

Fig. 8 CCDF representation of measurements of timer cost for the Network (RMDT download) testbed

4 Experimental results for Sleep Measurements

Sleep measurements are based on the comparison of the sleep function of HighPerTimer library [5] and the usleep() function from the standard C library in three scenarios: Host OS, Xen and Virtual Box. Further experiments with the sleep operation from HighPerTimer library are called HPTSleep and experiments with C library are called uSleep.

4.1 Idle testbed

The results for the idle mode with sleep operations are shown on figure 9. Median values for Host OS and Xen virtualized OS for HPTSleep are almost the same, while median of HPTSleep for Virtual Box is significantly greater. In the same time median values for uSleep for all cases are in the same range – between 62 µsec and 131 µsec. The difference of median magnitude between uSleep and HPTSleep is the greatest for Xen: 94.5 nsec for HPTSleep and 109 µsec for uSleep. For Host OS difference do not significantly differ from Xen: 82 nsec for HPTSleep against 62 µsec for uSleep. For Virtual Box the difference between HPTSleep and uSleep is the least: 5.19 µsec for HPTSleep and 131 µsec for uSleep. The values within the same testbed for QEMU are 97 µsec for uSleep and 116 nsec for HPTSleep, for VMware ESXi – 63.1 µsec for uSleep and 95 nsec for HPTSleep. So Xen results are almost the same as for QEMU. System call for Xen is almost twice greater than same value for VMware ESXi while HPTSleep remains almost the same [1].

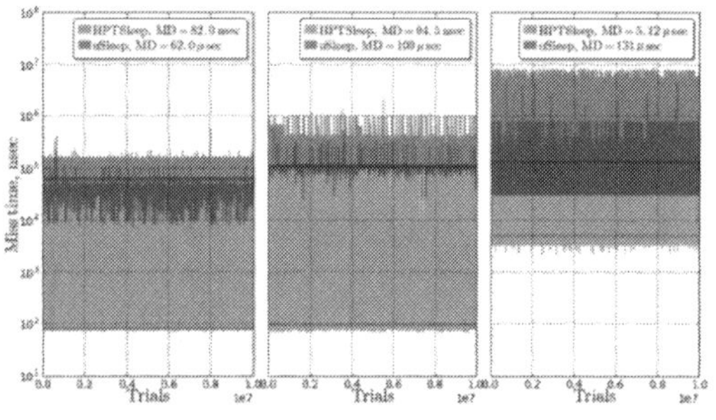

Fig. 9 Measurements of miss time for the idle testbed on: (a) Host OS (b) Xen (c) VirtualBox

Considering the CCDF plots on figure 10, measurements performed with uSleep method are grouped in the right part of the plot. However, HPTSleep for Xen and Host OS are situated in the left part of the plot. HPTSleep for Virtual Box keeps to the right, because of the higher order of values in comparison to Host OS and Xen. QEMU and VMware ESXi behave almost the same as Host OS and Xen.

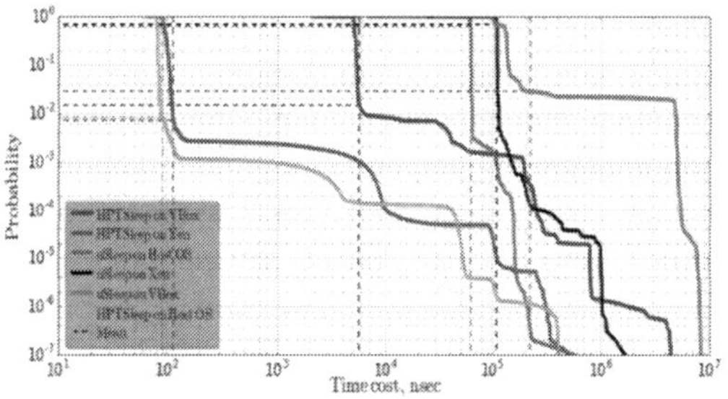

Fig. 10 CCDF representation of measurements of miss time for the idle testbed

4.2 CPU – bound testbed

In the case of CPU load, median (fig. 11) value of uSleep for Virtual Box is greater then in Ideal case: 7.6 msec for CPU load and 131 µsec for Idle case. Median values for all other cases remain the same as for Idle case. Considering QEMU and ESXi [7], System call values are 61.3 µsec and 56 µsec and HPTSleep values are 104 nsec and 114 nsec correspondingly. So, uSleep for Xen shows almost twice greater overhead than for QEMU and VMware ESXi, while HPTSleep value is almost the same.

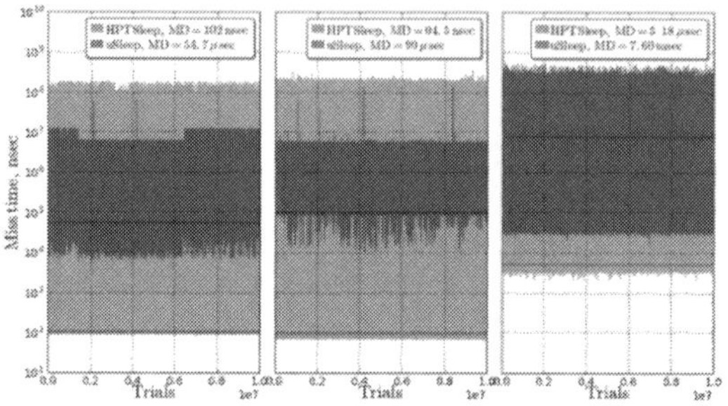

Fig. 11 Measurements of miss time for the CPU-bound testbed on: (a) Host OS (b) Xen (c) VirtualBox

CCDF plot (fig. 12) shows that the tails are significantly increased in comparison to Idle mode. The mean values for HPTSleep of Host OS and Xen are extremely different from mean values of Idle case. Other mean values are increased as well, but not so significantly. System call for Virtual Box differs in comparison to other plots: the plot has a slight structure representation, but not the stepwise. In the same time the probability of obtaining value greater then mean value is much greater then for Ideal case. But still the probability of obtaining value less then median value is about 99%. The behavior of QEMU and VMware ESXi almost repeat the behavior of Host OS [7].

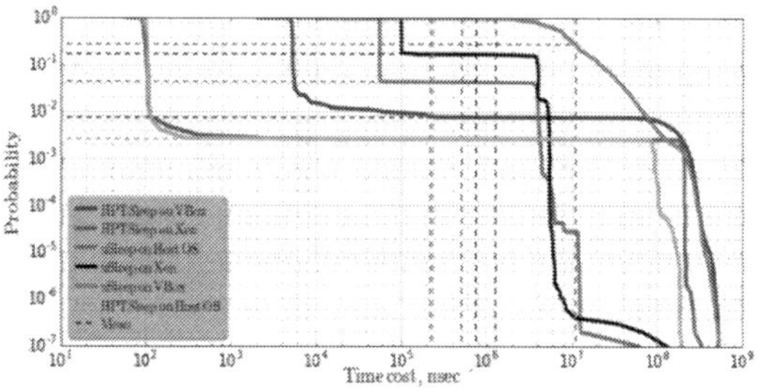

Fig. 12 CCDF representation of measurements of miss time for the CPU-bound testbed

4.3 I/O – bound testbed

In comparison to idle mode, median values remain the same while the range of values is greater. The values of QEMU for HPTSleep and uSleep are 113 nsec and 71.1 µsec and for VMware ESXi are 117 nsec and 62.2 µsec correspondingly. Accordingly, Xen introduces extra time overhead in comparison to QEMU and ESXi.

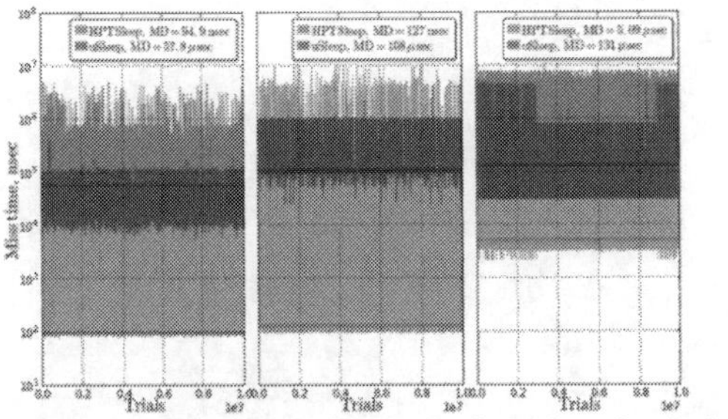

Fig. 13 Measurements of miss time for the IO-bound testbed on: (a) Host OS (b) Xen (c) VirtualBox

CCDF plot shows (fig. 14) that HPTSleep for Host OS is almost the same but probability of getting value greater then minimal is greater then in Ideal case but still not significantly. The mean value a bit greater. The tail of HPTSleep for Xen is much greater than in the idle state. Mean value is greater with less probability of getting mean value then in Ideal case. For all the other cases, mean value remains the same as for Ideal case with not significantly changed tails.

4.4 Network – bound testbed

Observing the influence of network load on sleep operations on figure 16, the median values for this type of load remains approximately the same as for the previous scenario.

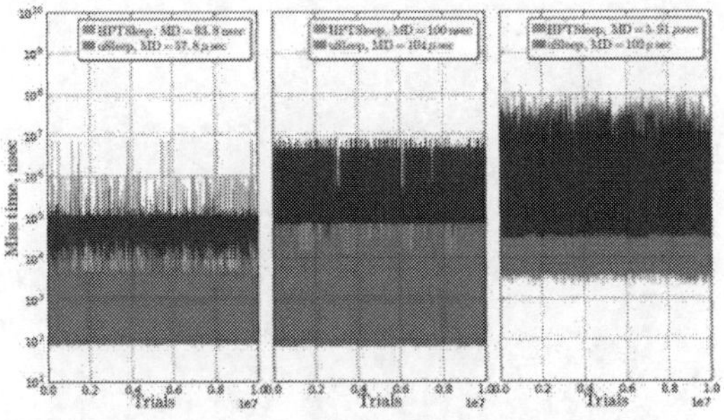

Fig. 14 CCDF representation of measurements of miss time for the Network-bound (RMDT download) testbed

Taking into account CCDF plot (figure 16), for HPTSleep on Xen and Virtual Box, tails are much longer and mean values become significantly greater in comparison to Idle testbed. Probability of obtaining value is greater then mean value is much greater then in Ideal case, but still for values less then median value is around 99%. HPTSleep for Host OS keeps without significant changes. Mean value of uSleep for all cases remains in the same order but tails become significantly longer. The probability for obtaining mean value is greater then in the Idle case. The probability for getting value is less then median value and remains around 99%.

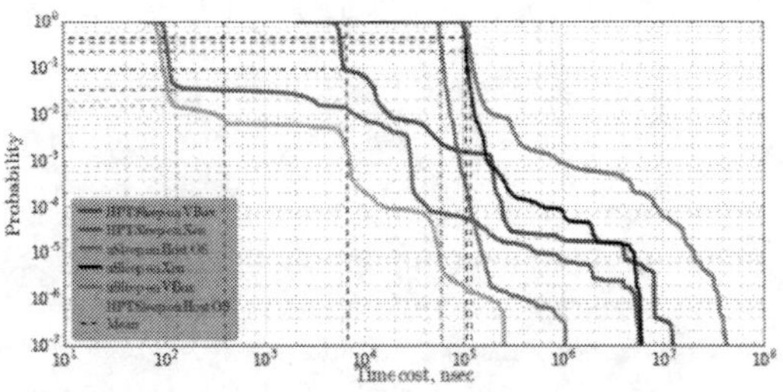

Fig. 15 CCDF representation of measurements of miss time for the Network-bound (RMDT download) testbed

5 Conclusion

Various measurement results for the case, when one virtual instance is located on hardware, allow to conclude the following: (i) While Xen has almost no basis overhead, Virtual Box's access to the hardware adds the highest overhead among all virtualization platforms (QEMU, Xen, VMware ESXi). (ii) In the case of sleep accuracy, the means of sleeps methods prevail over the virtualization environment effects, except Virtual Box, when miss time median values are 100 times higher then other virtualization in case of HighPerTimer sleep. (iii) Considering the worst-case execution time, the used virtualization plays almost neglecting role, the majority of values are measurably better on host machine then (at least) on QEMU; (iv) hypervisor-based approach brings clearly measurable benefits against OS-based one. (v) Virtual Box has the highest median time overheads and the highest deviation. Also based on minimal overhead values it's make sense to assume that this platform does not provide direct access to hardware. Since that this platform is not recommended for time-critical applications.

Acknowledgement

This work has been funded by Volkswagen Foundation for trilateral partnership between scholars and scientists from Ukraine, Russia and Germany within the project CloudBDT: Algorithms and Methods for Big Data Transport in Cloud Environments.

References

1. K. Karpov, I. Fedotova and E. Siemens, *"Impact of Machine Virtualization on Timing Precision for Performance-critical Tasks"* J. of Physics: Conference Series, **870** (2017)
2. K. Karpov, I. Fedotova, D. Kachan, V. Kirova and E. Siemens, *"Impact of virtualization on timing precision under stressful network conditions,"* Smart Technologies, IEEE EUROCON 2017-17th International Conference on. IEEE, (2017)

3. S. G. Langer, T. French, "*Virtual Machine performance benchmarking*", Journal of digital imaging, **24**, 883 (2011)
4. P. Barham et al. *"Xen and the art of virtualization"*, ACM SIGOPS operating systems review, **37**, 164 (2003)
5. I. Fedotova, E. Siemens, and H. Hu, "*A high-precision time handling library*", Journal of Communication and Computer, **10**, 1076 (2013)
6. I. Fedotova and E. Siemens, *"High-precision process sleeping aspects in the time handling library HighPerTimer for Linux OS"*, The Forth Scientific and Practical Conference "Information and Measuring Equipment and Technology", (2013)

Composing Internet of Things Platforms in Smart Grid

Ervin Varga, *Bojan* Blagojević, and *Dejan* Mijić

Faculty of Technical Sciences, University of Novi Sad, Trg Dositeja Obradovića 6, Novi Sad, Serbia

Abstract. The abundance of smart appliances and concomitant data in Internet of Things (IoT) poses new industrial challenges. Devices must be managed, and data efficiently harvested. IoT platforms are the rescue; they are key architectural components in combatting complexity and scalability issues. This paper proposes a pragmatic composition method of these platforms, hence helps solving IoT interoperability conundrums. The described approach results in a higher-level abstraction, that shields user applications from the underlying turmoil. Our article presents an approach of wrapping such a mesh under a high-level domain API. To verify the soundness of our proposal we have implemented a fully functional proof of concept implementation. The source code is freely available at request. The outcome decisively demonstrates the proposed method's feasibility, power, and usefulness.

1 Introduction

The vast amount of data, produced by sensors and smart appliances, hides many opportunities to boost both business and operational intelligence [1]. For example, in manufacturing there is an urgent need to predict machine failures, and schedule maintenance work accordingly. The return on investment in applying IoT is obvious; less downtime results in huge savings. Moreover, by creating historical reports, the management may better organize supply chains, and make sound business plans [2]. Before the IoT era, the industry relied on traditional SCADA systems, or similar centralized solutions, for factory automation. Such approaches had serious limitations stemming from data compartmentalization. IoT tries to dynamically bundle disparate data channels, as multiple data sources may provide new insights (something not possible by looking at each of them separately). Nonetheless, IoT also introduces novel challenges. The major obstacles toward making IoT a commodity can be segregated into the following categories: device management, real-time event handling, data analytics (embraces Big Data problems) [3], end user applications, and methods to develop IoT systems [4]. Security is a cross-cutting concern that is exaggerated by a higher exposure of devices to all sorts of attacks i.e., the attack surface is volatile and voluminous [5, 6]. However, the enormous gain from harvesting IoT's capabilities mandates us to find resolutions to the previously enrolled issues.

An IoT platform is a kind of an enabler such as a powerful software framework and integrated development environment for software development (nobody is writing million lines of code in assembly, nor is it a conceivable feast). At one point, it is impossible to manually tackle the task of controlling myriad of devices that swamp the target system with data. This is where an IoT platform plays a central role. It consists of components to manage devices, incorporates an event handling engine, and has facilities to perform data analytics. The large number of both commercial and open-source offerings (like Amazon Web Services IoT, Microsoft Azure IoT, Mainflux, Eclipse IoT, Kaa) is an excellent proxy for showcasing the importance of an IoT platform.

End users rarely embark on extensive software development; they prefer appropriate applications and services. Although, these could be delivered in many shapes, their common goal is to implement the required features. The same is true in the case of IoT. Users wouldn't use IoT platforms directly, but search for custom applications and services that will expose suitable, high-level Application Programming Interfaces (API) or Graphical User Interfaces (GUI). Figure 1 shows the major use cases pertaining to customers. These are expected to be delivered in the form of UI dashboards, Web service APIs (SOAP or REST), and

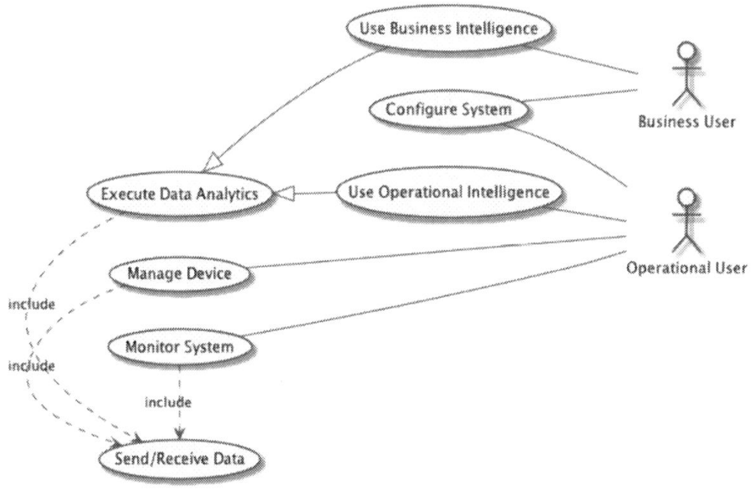

Fig. 1 The major IoT use cases, and their relations to the two groups of users: business and operations.

straightforward configuration mechanisms. In other words, the inner details of an IoT platform should be properly hidden.

The large number of data sources cannot be efficiently encompassed by a single IoT platform (either on premise or in the cloud). Consequently, each IoT platform is responsible only for a subset of devices; IoT platforms are frequently organized into a mesh network, having devices as leafs. The wisdom is based on the main tenet of edge computing, i.e., bring processing near data. This has multiple benefits, as listed below:

- Latency is reduced when handling critical events (geographical scalability).
- Irrelevant events are filtered out, thus only crucial ones are passed further (load scalability)
- Data handling logic is distributed; hence transformations happen early (functional scalability)

Evidently, IoT platforms must interoperate to form such networks, and make application development easier. This problem has been partially tackled by the EdgeX Foundry open-source umbrella project (see https://www.edgexfoundry.org), which aims to standardize what happens between devices and IoT platforms. Such a standardization prevents vendor lock-in at device level, thus enables consumption of data from disparate sources (many times combining data sources happen in an ad-hoc fashion). The authors are part of the core team of the Mainflux open-source IoT platform (https://www.mainflux.com), and members of the EdgeX Foundry project.

This paper describes a backend application/service construction method, that provides a domain-specific high-level API, and allows single point of control (administrative scalability) for users. The work is accompanied by the proof of concept (POC) implementation, in the domain of smart grids (a principal territory of IoT [7]).

2 Method

Our IoT application/service construction method embodies the next three characteristics, which are elaborated in separate sub-sections:

- API-driven development [8]
- Micro-services architectural style [9, 10]
- Stratified (layered) architectural pattern [10]

2.1 API-Driven Development

Figure 2 shows the UML activity diagram depicting the metaprocess of creating an API. A real (concrete) process would be an instance of this metaprocess configured with a set of principles, patterns, and techniques applied at each activity [8]. For example, in case of a RESTful API, there are additional constraints and specifics how to properly craft such an API [8, 11, 12], and in what ways is the concrete API process different than in the case of a SOAP based API [13].

Here are the most salient benefits of starting with an underpinning use case [8]:

- It encompasses behavioral requirements.
- It has a clear goal from the primary actor's perspective.
- It has a designated level helping to identify the target software layer, where this use case belongs.
- It lists all the stakeholders interested in the use case.
- It describes the main scenario, possibly with alternate flows.

The use case must be validated to properly reflect the intention of the primary actor. The use case later serves as a reference to verify the API. A use case is the generator of abstractions, which will become part of an API. Note that a single use case could entail multiple APIs if someone decides, for example, to apply the interface segregation principle to enhance maintainability,

 https://doi.org/10.1051/matecconf/201820802007

usability and reusability (this is reflected in the grouping abstractions meta-activity on Figure 2) [8].

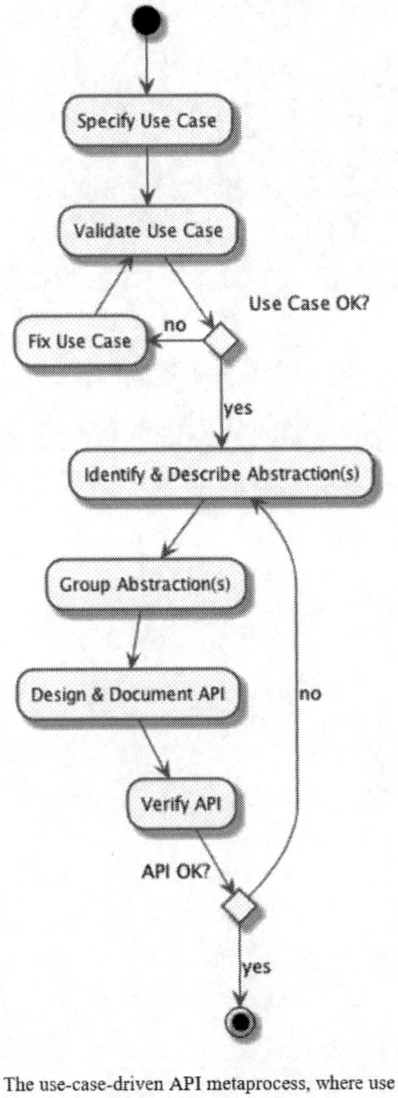

Fig. 2. The use-case-driven API metaprocess, where use cases play a central role in shaping the seams of the system's architecture (adapted with permission from [8]).

Figure 3 shows a typical build-up of an IoT platform (in this case Mainflux), that would serve as a model to explain the importance of use cases, and how they impact APIs. On the left side, we see the gateway facing devices, while on the right, we observe an integration point with custom applications. Devices and applications are on a completely different scale and abstraction level. Consequently, the associated APIs would serve a different purpose. Device APIs reflect specific low-level communication protocols (like, Level 1 REST, MQTT, CoAP, OSLP, Modbus, etc. [2, 7]) and goals; custom applications demand higher-level resource APIs (usually level 2 REST according to the Richardson's maturity model [12]), that hide device handling complexities, and provide aggregated views on the system.

On the other hand, custom applications would realize even more domain specific APIs for end users. These would be truly RESTful (level 3 REST), completely shielding users from the underlying IoT platforms. This approach is compliant with the tenet of the layered design, where as we move up on the layers' ladder we encounter more business oriented abstractions.

APIs also help in organizing work across distributed development teams, that are often geographically spread out with concomitant communication issues, which is crucial at the scale of IoT. A mobile application should be developed independently from an extension to the core IoT platform, or a new firmware sitting on devices. Finally, domain specific vertical APIs boost productivity of custom mobile applications, since working against IoT platform (or devices) would be cumbersome, and require a steep learning curve.

2.2 Micro-Services Architectural Style and the Layered Architectural Pattern

The micro-services architectural style is inherently based upon the layered design, and revolves around the following principles [9]:

- Domain-driven design [14]
- High automation with autonomous deployment model (aligned with the continuous delivery paradigm)
- Information hiding and encapsulation
- Decentralization of functional aspects of a system
- Dependability via failure isolation and powerful observability

Figure 4 depicts the module decomposition view of a generic vertical micro-service exposing a RESTful API toward clients (these may be mobile applications providing a decent user interface for users). The service communicates via proxies with the corresponding IoT platforms. Nonetheless, the existence of these platforms can be either completely hidden, or published in a very abstract manner. For example, the platform may be mentioned as a data source to be grouped with other similar units to form a composite. At any rate, none of the intricacies of dealing with platform specific details should be revealed in the service's API.

It is easy to notice layers inside the micro-service: API provider, engine, and database (frequently a relational database management system for the sophisticated query mechanism). The HTTP interface is common for all REST levels (from 1 to 3). The IoT platforms are treated as black boxes (they belong to a lower layer). Nevertheless, it is possible to peak into one of them to unravel a separate world of micro-services. This is illustrated on Figure 3 in the case of the Mainflux platform. The micro-services there are used

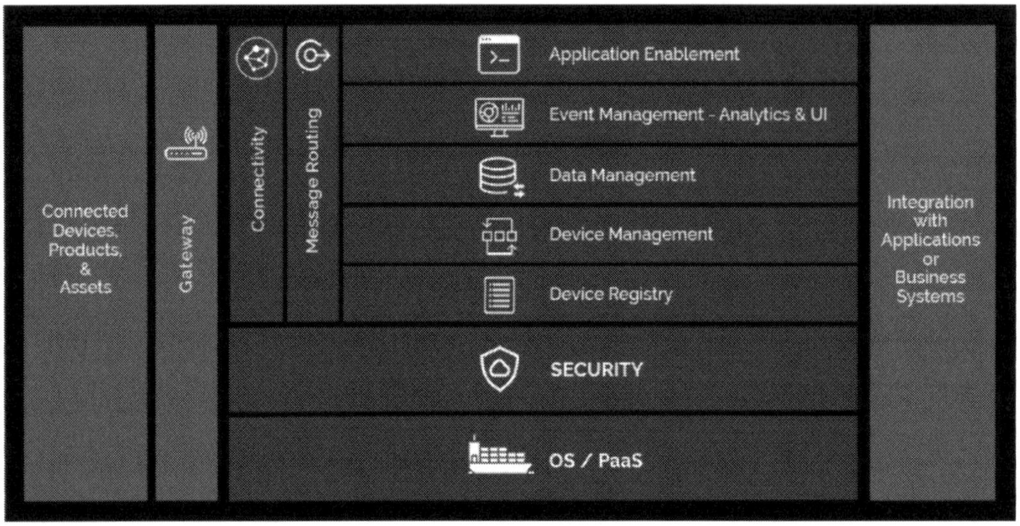

Fig. 3. The Mainflux software infrastructure stack, that contains all necessary components and micro-services for implementing fully functional IoT solutions (adapted with permission from http://mainflux.com). PaaS stands for Platform as a Service.

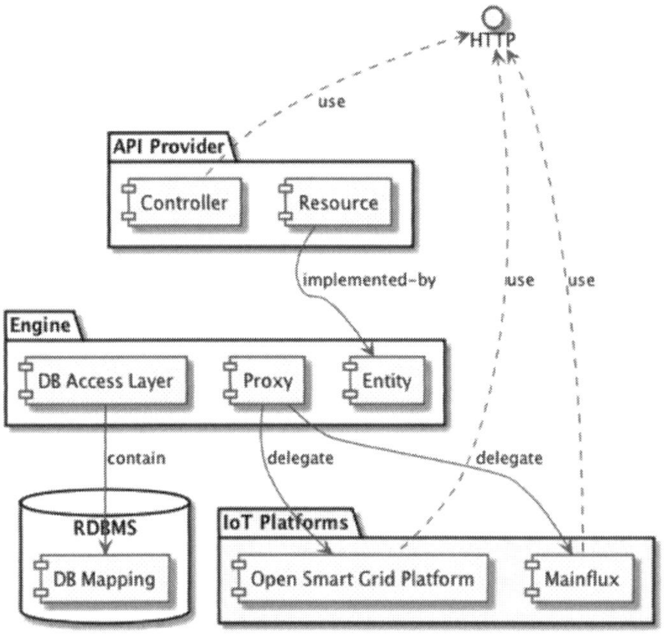

Fig. 4. The universal structure of a vertical IoT micro-service.

to implement the IoT platform in a scalable and fault-tolerant fashion. This is the principal power behind a layered design; it is always possible to zoom in/out as necessary (abstractions are major design artifacts to tame complexity, and layers group them in the most natural manner).

3 Result

The POC illustrates the method from the previous section. It showcases the management of microgrids as major building blocks of a smart grid. Each microgrid is controlled by a separate IoT platform, which is the Open Smart Grid Platform (OSGP). The POC uses the OSGP's microgrids domain for controlling and monitoring microgrids. Our custom application partially realizes the vision of OSGP's load management (https://opensmartgridplatform.org/load-management/). The POC offers a new OSGP domain to control the electricity load.

Besides getting loads from microgrids, the POC expands the load management use case with microgrids administration API. It allows an end user mobile application to create and delete microgrids in the system, group them together, and subscribe for reports about aggregated loads. The load management API is specified using on Open API Specification v3.0 (https://www.openapis.org/) leveraging JSON API v1.0 in the background (http://jsonapi.org/). The service is implemented as a micro-service in the Scala programming language using similar layering as depicted on Figure 4. Deployment is done via Docker Compose (see https://www.docker.com).

The end-points of the POC's Resource API are classified into three groups: microgrids, groups, and subscribers. The first denotes operations to manage microgrids, the second represents operations to handle groups, and the last contains end-points for dealing with subscriptions. This delineation of the API is the application of the interface segregation principle [15] in the realm of RESTful services.

The management interface would be driven by a separate client application. By using a callback notification mechanism (via the subscription facility), the communication patterns between the POC and each IoT platform may vary independently of subscribed clients; they are completely shielded from these details. Also, the POC's API is on a considerably higher level, than those used for application integration purposes in IoT platforms. An elevated domain specific API is crucial to enable massive adoption of IoT, and make the benefits available to end users.

4 Discussion

To expand the number of use cases related to IoT we need to attain the following objectives:

- Use IoT platforms as basic building blocks.
- Standardize the edge to allow interoperability among devices and IoT platforms.
- Combine IoT platforms under a domain specific umbrella API.

This paper shows a generic method to accomplish the last goal from the above list. The method is testified by a full POC system, that is freely available on-line. Once there is a high-level API it may be used by various user friendly mobile applications. Moreover, the API itself may be treated as a key asset, and published via API management platforms (like, Apigee, 3scale, etc.). According to [16] around 45-80% of market spending in IoT will be targeted for applications and value-added services. Having sophisticated vertical APIs is the precondition to achieve this forethought.

Our future work is related to incorporate security mechanism into the POC, and integrate many more different IoT platforms. Also, we plan to extend the load management related functions to support the establishment of dynamic pricing model at the level of microgrids. Such attempts are buttressed by the advancements in blockchain technologies for local electricity markets [17].

References

1. K. Lakhani and M. Iansiti, The Internet of Things, Harvard Business Review, 2014.
2. R. Barton, G. Salgueiro, and D. Hanes, IoT Fundamentals: Networking Technologies, Protocols, and Use Cases for the Internet of Things, Cisco Press, 2017.
3. H. Geng, Internet of Things and Data Analytics Handbook, John Wiley & Sons, 2017.
4. I. Jacobson, I. Spence, and P.W. Ng, "Is There a Single Method for the Internet of Things?", Communications of the ACM, Vol. 60, No. 11, Pages 46-53, 2017.
5. N. Dhanjani, Abusing the Internet of Things, O'Reilly, 2015.
6. S. Smith, The Internet of Risky Things, O'Reilly, 2017.
7. O. Elloumi, D. Boswarthick, and O. Hersent, The Internet of Things: Key Applications and Protocols, John Wiley & Sons, 2012.
8. E. Varga, Creating Maintainable APIs - A Practical, Case-Study Approach, Apress, 2016.
9. S. Newman, Building Microservices, O'Reilly, 2015.
10. M. Richards, Software Architecture Patterns, O'Reilly, 2015.
11. R. Daigneau, Service Design Patterns: Fundamental Design Solutions for SOAP/WSDL and RESTful Web Services, Pearson Education, Inc., 2012.
12. L. Richardson, Sam Ruby, Mike Amundsen, RESTful Web APIs, O'Reilly, 2013.
13. F. Belqasmi, J. Singh, S.Y.B. Melhem, and R.H. Glitho, "SOAP-Based vs. RESTful Web Services: A Case Study for Multimedia Conferencing", IEEE Internet Computing, Vol. 16, No. 4, Pages 54-63, 2012.
14. V. Vernon, Domain-Driven Design Distilled, Addison-Wesley Professional, 2016.
15. E.J. Braude and M.E. Bernstein, Software Engineering: Modern Approaches (2nd Edition), Waveland Press, Inc., 2011.
16. I. Hughes and D. Immerman, "Ericsson shifts gears, introduces its IoT Accelerator and a new Experience Center", 451 Research, LLC, 2017.
17. M.E. Peck and D. Wagman, "Blockchains Will Allow Rooftop Solar Energy Trading for Fun and Profit", IEEE Spectrum, October 2017.

Study on the spatial roll angle measurement based on magneto-optic modulation

Wei Cai[1, 2], *Zhiyong* Yang[1], *Junhui* Xing[1] and *Youan* Xu[1]

[1]Xi'an High-Tech Institute, 710025 Xi'an, China
[2]School of Telecommunication Engineering, Xidian University, 710071 Xi'an, China

Abstract. Faraday rotation angle formula for polarized light propagating in magneto-optic medium is investigated theoretically. Based on the Biot-Savart Law and Maxwell's equations, a calculation model for electromagnetic field in solenoid paraxial region is deduced. Taking advantage of this model, together with the Bessel functions, the axial magnetic field in a typical medium is calculated and analyzed. In addition, to thoroughly analyze the influence mechanisms of the incident wavelength and temperature on the Verdet constant and then uncover its essence, a hypothesis which suggests the Faraday effect results from the combination of various factors is proposed. Furthermore, a theory of wave-transition contribution to the Verdet constant is deduced by adopting the theory of wave-particle duality. Taking the typical diamagnetic material ZF1 and paramagnetic material TGG as examples, the deduced theory together with the corresponding models is tested and verified by analyzing the relevant parameters and the test data. At last, a testing system for magneto-optical rotation together with its finite element model is introduced.

1 Introduction

Spatial roll angle measurement based on magneto-optical modulation refers to measuring the roll angle between non-connected instruments in different horizontal planes, with polarized light and Faraday effect [1]. With the advantage of non-contact and high precision, this angle measurement technology has been gradually used in industrial, medical, biological, chemical detection and other fields, and it also has a broad application prospects in spacecraft docking.

Many works have been devoted to magneto-optical modulator signal generation, processing, and measurement system design, to improve the accuracy of roll angle measurement [2-5]. On the other hand, some researchers have studied the physical origin of the Faraday effect, and theories of the interactions among different physical fields in magneto-optical effect. Freiser derived the Faraday effect from Maxwell's equations and discussed heuristically its microscopic origin [6]. The research on its microscopic origin developed progressively and formed a sophisticated quantum theory [7, 8]. However, the uncertain probabilities of electrons' distribution in different energy levels and of dipole transitions cause the usage of the quantum theory inconvenient. In order to describing the Faraday effect precisely and integrally, some efforts have been devoted to combine these two theories together in recent years [9, 10]. Besides the electromagnetic fields induced by the incident light and the applied magnetic field, the temperature around the magneto-optical material is another important factor which interacts with other physical fields and complicates the polarized light rotation. Liu researched the temperature dependence of Faraday effect in typical magneto-optical medium [11], and some practically applicable methods of compensating the temperature drift caused by the linear birefringence in Faraday material were developed [12].

The purpose of this paper is to analyze the factors which influence magneto-optical rotation. Following with a brief introduction of roll angle measurement principle in section 2, two important variables in Faraday rotation angle formula, magnetic field strength and Verdet constant, are studied and modelled theoretically in section 3 and section 4, respectively. In addition, a testing system for magneto-optical rotation together with its finite element model is presented in section 5.

2 Principle of roll angle measurement

Fig. 1 schematically shows the principle of roll angle measurement based on magneto-optical modulation. A linear polarized light from the polarizer in upper instrument passes through the magneto-optical rotation glass in modulator, and its optical polarization plane will turn some angle because of the Faraday effect. The modulated signal received by analyzer in lower instrument is processed with lens, opto-electric conversion, amplification and certain algorithm. At the same time, the synchronous signal from upper instrument is transmitted to lower instrument, after that the roll angle is acquired. Various measurement models and corresponding algorithms have been developed [1, 5, 13].

 https://doi.org/10.1051/matecconf/201820803001

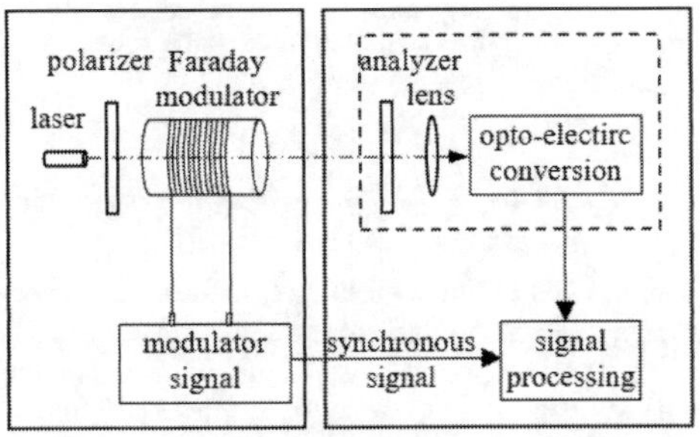

Fig 1. Principle of spatial roll angle measurement system based on magneto-optical modulation.

It can be proven that any linearly polarized wave can be seen as a superposition of two circularly polarized components, namely a left-circularly polarized (LCP) and a right-circularly polarized (RCP) one. The rotation of the light polarization plane, known as the Faraday rotation angle θ_F, is given

$$\theta_F = \pi L(n_+ - n_-)/\lambda = VLB \tag{1}$$

where L is the length of magneto-optical medium, n_+ and n_- are the refractive indices for RCP light and LCP light, respectively, λ is the light wavelength, B is the strength of applied magnetic field, V is the medium's Verdet constant. To acquire θ_F accurately, the magnetic field in modulator, together with the medium's Verdet constant, which is dependant on light wavelength and temperature, should be modelled and calculated precisely.

3 Magnetic field modelling in a solenoid with embedded magneto-optical medium

Although different waveforms and multi-frequency wave compound are used in magneto-optic modulation [13], sine wave magneto-optical modulation is still the broadest used form, which is the base for studying other methods [14].

Based on the Biot-Savart Law, the magnetic flux density around a solenoid, together with its three components (B_z, B_r, B_φ) in cylindrical coordinate, can be expressed as

$$\boldsymbol{B}(r,\varphi,z) = \frac{\mu_0}{4\pi}\int_s \frac{\boldsymbol{J}_s \times \boldsymbol{R}}{D^3}\mathrm{d}S \tag{2}$$

$$B_z = \frac{\mu_0 naI}{4\pi}\int_{-\pi}^{\pi}\int_{-b}^{b}\frac{a-\cos\varphi}{D^3}\mathrm{d}z d\varphi \tag{3}$$

$$B_r = \frac{\mu_0 naI}{4\pi}\int_{-\pi}^{\pi}\int_{-b}^{b}\frac{(b-z)\cos\varphi}{D^3}\mathrm{d}z\mathrm{d}\varphi \tag{4}$$

$$B_\varphi = -\frac{\mu_0 nI}{4\pi}\int_{-b}^{b}(b-z)\left(\int_{-\pi}^{\pi}\frac{\sin\varphi}{D^3}\mathrm{d}\varphi\right)\mathrm{d}z = 0 \tag{5}$$

where $\boldsymbol{J}_s$ and $\boldsymbol{R}$ are the current density vector and radius vector, respectively, a and b are the radius and half length of the solenoid. Considering the solenoid coil stimulated by $I(t) = I_m e^{-\mathrm{i}\omega t}$, we get the conclusion that B_φ, E_z and E_r are equal to zero approximately. Furthermore, Maxwell's equations are adopted to study the relationship between the electric and magnetic fields, then we have

$$\frac{\partial E_\varphi}{\partial z} = -\mathrm{i}\omega B_r \tag{6}$$

$$\frac{1}{r}\frac{\partial(rE_\varphi)}{\partial r} = \mathrm{i}\omega B_z \tag{7}$$

$$\frac{\partial B_r}{\partial z} - \frac{\partial B_z}{\partial r} = -\mathrm{i}\omega\mu_0\varepsilon_0 E_\varphi \tag{8}$$

Through Eqs. (6) to (8), both the electric field and magnetic field around the solenoid can be work out.

Furthermore, the linear polarized light passes through the magneto-optical medium along the solenoid axis and forms a slender light beam, whose scope can be regarded as a paraxial region in which the interaction between the magnetic field and polarized light takes place. As well, we can get the conclusion that B_r is approximately equal to zero in the paraxial region. Hence, by combining Eqs. (6) to (8) , we can acquire B_z and E_φ as follow

$$r^2\frac{\partial^2 B_z}{\partial r^2} + r\frac{\partial B_z}{\partial r} + k_0^2 r^2 B_z = 0 \tag{9}$$

$$r^2\frac{\partial^2 E_\varphi}{\partial r^2} + r\frac{\partial E_\varphi}{\partial r} + (k_0^2 r^2 - 1)E_\varphi = 0 \tag{10}$$

Eqs. (9) and (10) are the basic calculation models for the electromagnetic field in paraxial region of the solenoid stimulated by alternating current, in which $k_0 = \omega\sqrt{\mu_0\varepsilon_0}$ is the wave vector in vacuum. For example, taking a typical medium without polarization relaxation into consideration, the axial magnetic component in (9) can be calculated by adopting the Bessel functions,

$$B_z(r,t) = B_0 \mathrm{J}_0(k_0 r)\cdot e^{-\mathrm{i}\omega t} \tag{11}$$

where B_0 is the magnetic amplitude stimulated by direct current I_m, and $\mathrm{J}_0(k_0 r)$ is the zero order Bessel function.

Moreover, adopting (11) and keeping the solenoid length as a constant, the influence of solenoid radius on the amplitude of axial magnetic component B_{zm} can be calculated, and its effect is shown in Fig. 2, in which the the solenoid length is 10cm, and the amplitude of the alternating current is 2A with the coil density being 1000 turns/m.

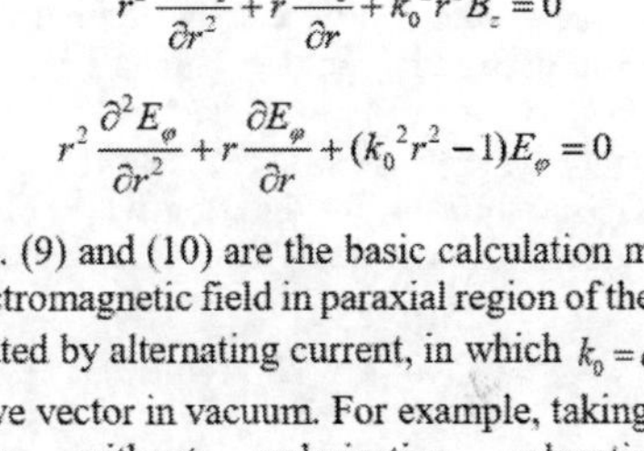

Fig 2. B_{zm} in paraxial region with different radius a

4 Verdet constant of magneto-optical materials

In order to explore the influence mechanisms of the incident wavelength and temperature on the Verdet constant and then uncover its essence, two different theories are developed, including the classical electronic dynamics theory and quantum theory.

The former, based on the Lorentz dispersion theory, focuses on the oscillations of the electrons in magneto-optical materials under the combined influence of the polarized light and the applied magnetic field. Based on the classical electronic dynamics theory, Verdet constants for both the diamagnetic medium V_{dia} and paramagnetic medium V_{par} are obtained,

$$V_{dia}=\frac{e\lambda}{2mc}\frac{\mathrm{d}n}{\mathrm{d}\lambda} \quad (12)$$

$$V_{par}=\frac{e\lambda}{2mc}\frac{\mathrm{d}n}{\mathrm{d}\lambda}(1+\upsilon\chi) \quad (13)$$

in which e and m are the charge and mass of the electron, respectively, c is the velocity of light in vacumm, $\mathrm{d}n/\mathrm{d}\lambda$ is the dispersion, υ is a coefficient in connection with Weiss field, χ is the magnetic susceptibility.

The latter, quantum theory, takes the differences of the probability and separation between ground and excited states, caused by applied magnetic field or medium magnetization, as the microcosmic essence of the Faraday rotation. Typical formulas for the paramagnetic medium V_{dia} and diamagnetic medium V_{par} can be described as

$$V_{par}=\frac{4\pi^2 v^2\chi}{g\mu_B ch}\cdot\sum_{ij}\left[C_{ij}/(v^2-v_{ij}^2)\right] \quad (14)$$

$$V_{dia}=4\pi N v^2\cdot\sum_{ij}\left[A_{ij}/(v^2-v_{ij}^2)^2\right] \quad (15)$$

where v is the frequency of incident light, C_{ij} denotes the oscillator strengths of transitions with v_{ij}, g is the Lande splitting factor, μ_B is the Bohr magneton number, N is particles per volume, A_{ij} denotes the transition moments.

However, neither of the two theories can be separately used to fully explain the Verdet constant and the correlative test data. Therefore, in order to thoroughly analyze the influence mechanisms of the incident wavelength and temperature on the Verdet constant and then uncover its essence, based on a hypothesis which suggests the Faraday effect results from the combination of various factors, we proposed a theory of wave-transition contribution to the Verdet constant by adopting the theory of wave–particle duality [10]. That is, the Faraday effect is caused by two different contributions simultaneously. One is the wave contribution, denoted as Vl, which is the interaction between the wave aspect of light and the magneto-optical medium, and the other refers to the transition contribution, denoted as Vh, which comes from the electronic transition. When it comes to the deflection angle, the wave contribution is positive while the transition contribution is negative.

Based on the theory of wave-transition contribution, together with an approximation method, simplified formulas of the Verdet constant can be expressed as

$$V_{dia}=\frac{q_e\lambda}{2mc}\frac{\mathrm{d}n}{\mathrm{d}\lambda}-\frac{A\lambda_t^4\lambda^2}{(\lambda^2-\lambda_t^2)^2} \quad (16)$$

$$V_{par}=\frac{q_e\lambda}{2mc}\frac{\mathrm{d}n}{\mathrm{d}\lambda}\cdot\left(1+\upsilon\frac{T_c}{T-T_p}\right)-\frac{\lambda_t^2}{\lambda^2-\lambda_t^2}\frac{E}{T-T_p} \quad (17)$$

Where T_c is the Curie constant, T_p is the paramagnetic Curie point, T is the temperature, λ_t and C_t are the dominant transition wavelength and its corresponding probability, respectively, q_e=-e<0, $E=4\pi^2C_tT_c/(g\mu_B ch)$ and $A=-4\pi NA_t/c^2$ are coefficients independent with the wavelength λ.

Taking the typical diamagnetic material ZF1 and the typical paramagnetic material TGG as examples, the influences of the incident wavelength and the temperature on the Verdet constant are analyzed, and the deduced theory together with the corresponding models is tested and verified by analyzing the relevant parameters and the test data.

Fig. 3 and Fig. 4 show that the theoretical results correspond to the real values, which proves the rationality of the hypothesis and the authenticity of the deduced theory.

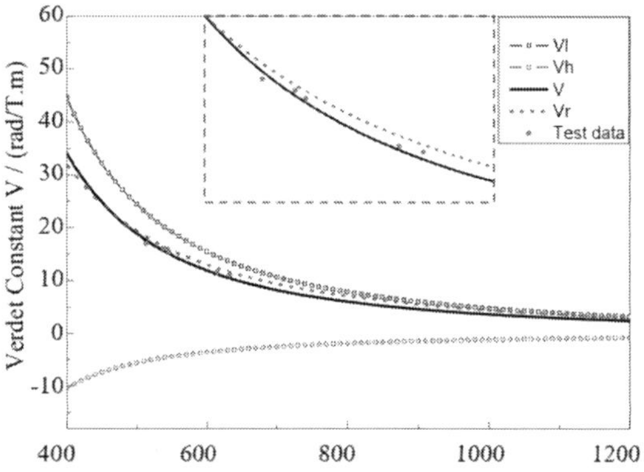

Fig 3. Wavelength dependence of Verdet constant of ZF1.

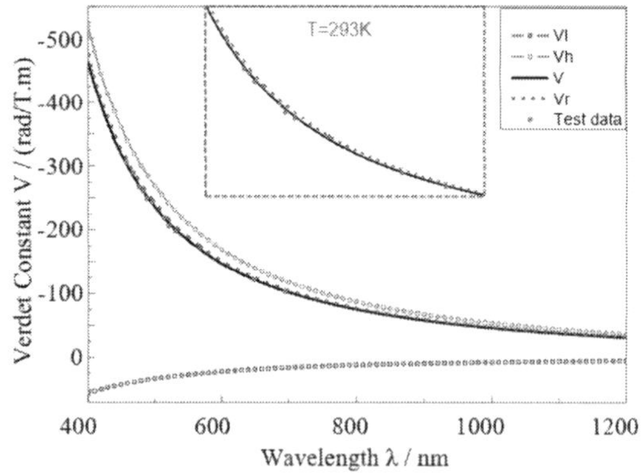

Fig 4. Wavelength dependence of Verdet constant of TGG (T=293K).

5 Testing system for magneto-optical rotation

To measure the magneto-optical rotation angle's dependencies on both incident light wavelength and magnetic field strength, a testing system was designed as Fig. 5, in which both the incident light source and stimulating signal can be adjusted, and the magneto-optical glass in the modulator can be replaced with different magneto-optical materials as well.

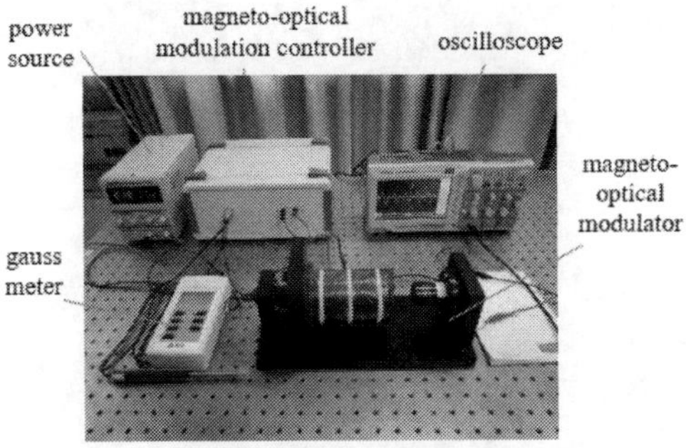

Fig 5. Testing system for magneto-optical rotation

The interaction between the polarized light and the applied magnetic field takes place in the magneto-optical glass surrounded by the solenoid coil, which is one of the key component in the magneto-optical modulator. In order to analyze the three-dimensional magnetic field stimulated by the solenoid coil visually, a finite element model of the modulator was built by using COMSOL software [14]. The geometric model and its mesh are shown in Fig. 6, and the solved magnetic field is given in Fig. 7.

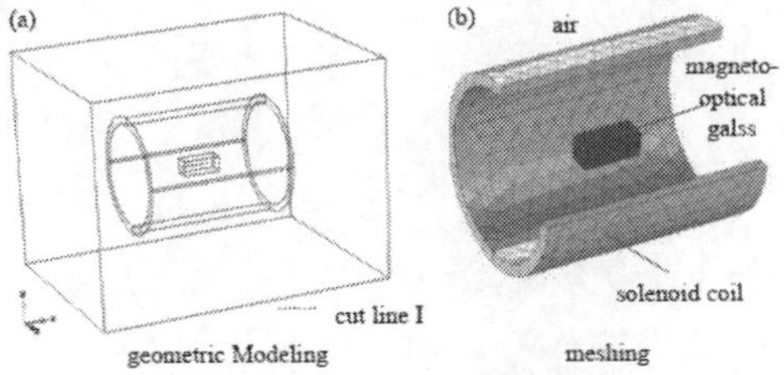

Fig 6. (a) Geometric model ; (b) mesh generation

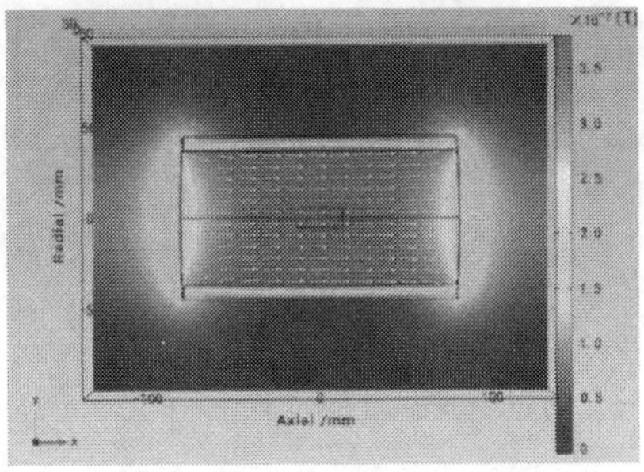

Fig 7. Magnetic field distribution arround the solenoid

6 Conclusions

The paper deals with two variables, B and V, in the Faraday rotation angle formula. To the former, by adopting the Biot-Savart Law and Maxwell's equations, the electric field and magnetic field around the solenoid are analysed. Then, focusing on the solenoid axis in which the magneto-optical glass is set and the magneto-optical effect takes place, the calculation models for the electromagnetic field in paraxial region of the solenoid stimulated by alternating current are built, and the influence of solenoid radius on B_{zm} is simulated numerically. To the latter, based on the essential property of the magneto-optical effect, a theory of wave-transition contribution to the Verdet constant is deduced by adopting the theory of wave-particle duality, and the deduced theory together with the corresponding models is tested and verified by analyzing the relevant parameters and the test data. The obtained results show that the simplified formulas of the Verdet constant for both diamagnetic and paramagnetic materials match well with the real values, and the proposed finite element model together with the magnetic field model gives a convenient approach to analysing physical fields in magneto-optical effect.

It should be noticed that either B or V is dependent on the temperature in solenoid, which is influenced by various of factors, and more serious, is time-varying. Therefore, both the temperature field in the working solenoid and its influence on B and V should be studied deeply in the future research, and multi-physical fields coupling analysis may be an effective method to address this challenge.

ACKNOWLEDGEMENT

The work presented in this paper was partially funded by the National Natural Sciences Foundation of China (grant No. 61505254) and the Aviation Science Foundation of China (grant No. 201551U8088).

References

1. X.N. Dong, L.M. Gao, X.J. Shen, L.Y. Chen, Acta Photonica Sinica **11**, 1389-1391 (2001)
2. S.G. Li, C.X. Yang, E.Y. Zhang, G.F. Jin, Opt. Lett. **30** (3), 242-244 (2005)
3. P. Mihailovic, S. Petricevic, J. Radunovic, JINST **1** (12), 1-11 (2006)
4. S. Egami, H. Watarai, Rev. Sci. Instrum. **80** (9), 093705 (2009)
5. Z.Y. Yang, W.Cai, X.X. Huang, J.H. Xing, Optik **134**, 194-202 (2017)
6. M. Freiser, IEEE Trans. on Mag. **Mag-4** (2), 152-161 (1968)
7. F. Zhang, Y. Xu, J.H. Yang, M. Guillot, Eur. Phys. J. B **20**, 165-175 (2001)
8. W. Wang, M. Gu, G. Liu, Phys. Rev. B **72**, 014415 (2005)
9. T. Haider, Int. J. Electrom. & App. **7** (1), 17-24 (2017)

10. W. Cai, J.H. Xing, Z.Y. Yang, Acta Phys. Sinica **66** (18), 187801 (2017)
11. G.Q. Liu, W.K. Zhang, X. Zhang, Phys. Rev. B **48**, 16091 (1993)
12. P. Menke, T. Bosselmann, J. Lightwave Technol. **13** (7), 1362-1370 (1995)
13. Z.Y. Yang, X.X. Huang, Z.F. Zhou, Z.L. Zhang, Optik **125** (3), 1042-1048 (2014)
14. Z.Y. Yang, X.X. Huang, Z.F. Zhou, Z.L. Zhang, Lasers in Eng. **26**, 115-125 (2013)

Comparison of estimates calculated on the energy flow problem basis and using testing equations method

Elena Kochneva[1] and *Aleksandar* Sukalo[2]

[1]Automated Electrical Systems Department, Ural Federal University named after the first President of Russia B.N. Yeltsin, 19, Mira St., Yekaterinburg, Russia
[2]Elektroperenos – Elektroprijenos BiH, Banja Luka, Bosna and Hercegovina

Abstract. The article is devoted to the electrical energy measurements estimates calculation. Automatic Meter Reading systems are the source of measurements. Two methods are compared. The first one is based on the energy flow problem, which allows to receive the estimates of the electrical energy flows and losses on all elements of the electrical network. Calculations using the first method were made using the software product "Balance". The second technique is simplified and uses a system of testing equations to calculate the measurements estimates. Comparison showed the use of a simplified procedure gives adequate accuracy on condition of sufficient redundancy level.

1 Introduction

The current state of the electric power industry is characterized by the following problems [1]:

- a high level of commercial losses,
- the difficulty of determining the commercial measurement information reliability degree,
- the complexity of the element-wise calculation of electrical energy technical and commercial losses in confined networks.

Many enterprises, grid and generating companies are equipped by Automatic Meter Reading systems. These systems include the modern microprocessor devices that provide high accuracy measurements and time synchronization. The main purpose of these systems is to provide financial settlements for electricity with reliable and operative measuring information. At the same time, the level of commercial power losses remains high. Despite the technical complexity of modern accounting systems, mathematical processing of the data is practically not done.

Existing approaches to measurement data processing allow to solve a wide range of problems: identification of unreliable data, localization of sources of commercial power losses, assessment of the characteristics of electricity metering complexes, and others [2].

2 Energy flow calculation on the network elements using energy flow problem

The task of calculating energy flows and losses on all elements of the electrical network using measurement information was defined as the energy flow problem. Initially, attempts to model power distribution modes were associated with the use of steady-state equations [3-5]. However, it was found out that averaging the modes over a time interval leads to the appearance of unrecoverable unbalances in steady-state equations. The main reason is topological changes in the network [6]. The energy flow problem state system differ from the traditional state estimation one [7-11]. The electric energy balances equations in the nodes and branches of the network are to be used as the state equations describing the process of energy distribution. These equations are identities for any topological and mode changes.

Since the measured values of electrical energy contain metrological errors, the balance energy ratios in the nodes and branches of the network diagram, which include these measurements, may be nonzero. The energy flow problem allows to get the estimated energy flows and the balance ratios for these calculated flows will be correct. Calculations are made on the basis of the energy flow problem mathematical model. It is important to note that when using measured energy flows, the balance equations will include technical and commercial energy losses, and when compiling balances for the calculation model, the equations will include only technical losses.

The objective function of the energy distribution problem is the weighted squares sum of relative measurement errors:

$$F = \sum_{l=1}^{K} a_l \left(\frac{W_l^{meas} - W_l^{calc}(X)}{W_l^{meas*}} \right)^2 \rightarrow \min \qquad (1)$$

where K is the number of energy measurements for the power system fragment under consideration.

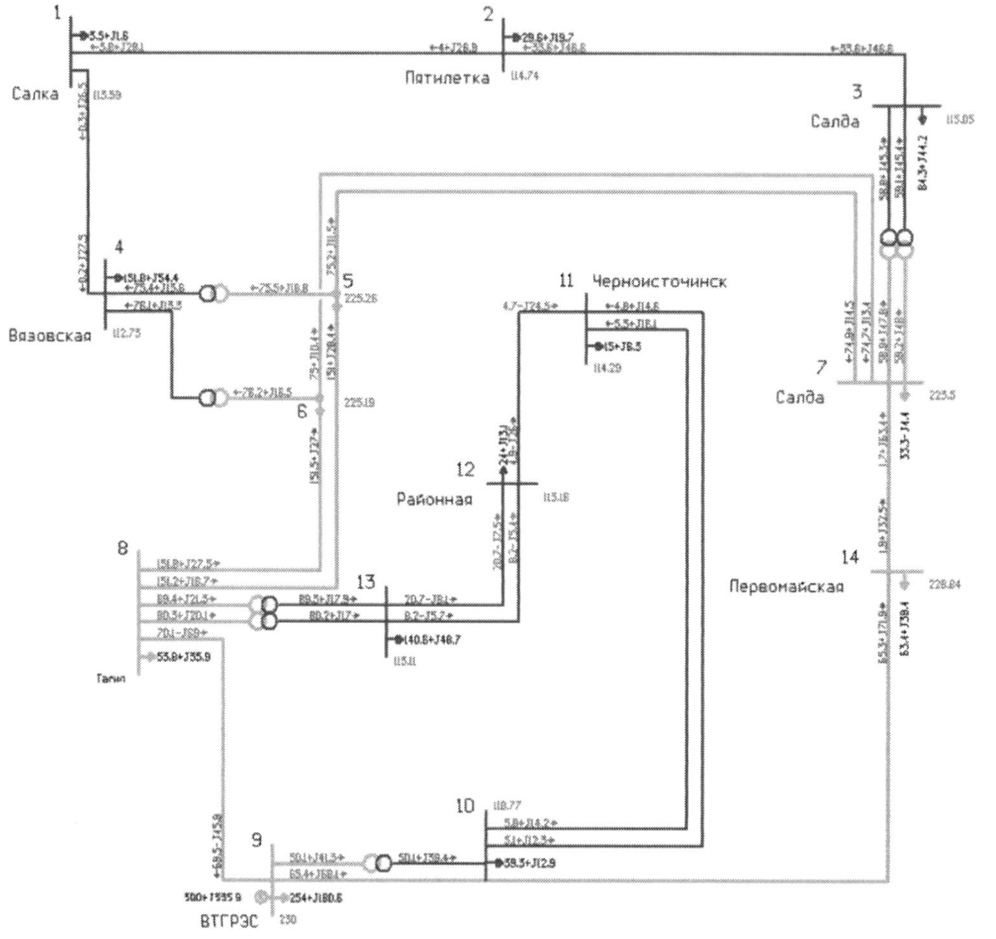

Fig 1. Test scheme

The weight coefficients a_l from (1) are necessary to take into account the information on the relative errors of automatic meter reading complexes [12-13].

Thus, the network model for the solution of the energy flow problem is the traditional equivalent electric circuit with lumped parameters. It should be noted that the equivalent circuit parameters determine only the amount of active and reactive energy losses on the longitudinal and transverse elements of the network. The magnitude and direction of energy flows are determined by the available nodal and linear measurements, which are located in the branches and nodes of the equivalent circuit of the network under consideration.

3 Testing equations method for energy flows calculated analogues estimation

The algorithm for estimation the energy flows calculated analogues can be based on the testing equations system.

The method is based on the small coherence of the energy flow problem state equations. This method is an a priori one. According to the terminology of the state estimation theory, equations that include only the measured variables are called testing ones. The unbalance (mismatch) of such an equation will be zero if all the measurements in its composition contain no errors, and the unbalance is different from zero if the measurement includes an error. The presence of gross errors leads to the appearance of significant unbalances. Thus, the method of testing equations is close to the method of comparing actual and permissible imbalances, but it allows to analyze a connected balance equations system.

In real power systems, there is almost never a complete coverage of connections by energy measurements, in this case a testing equations system can be obtained from the energy flow problem state equations system by mathematical exclusion of all unmeasured variables.

The topological method of forming the testing equations system is visual and is based on the topology of the network and the placement of measuring complexes. The formalized procedure for obtaining control equations (the algebraic method) is based on the elimination of all non-measured variables from the energy flow problem state equations system by algebraic means.

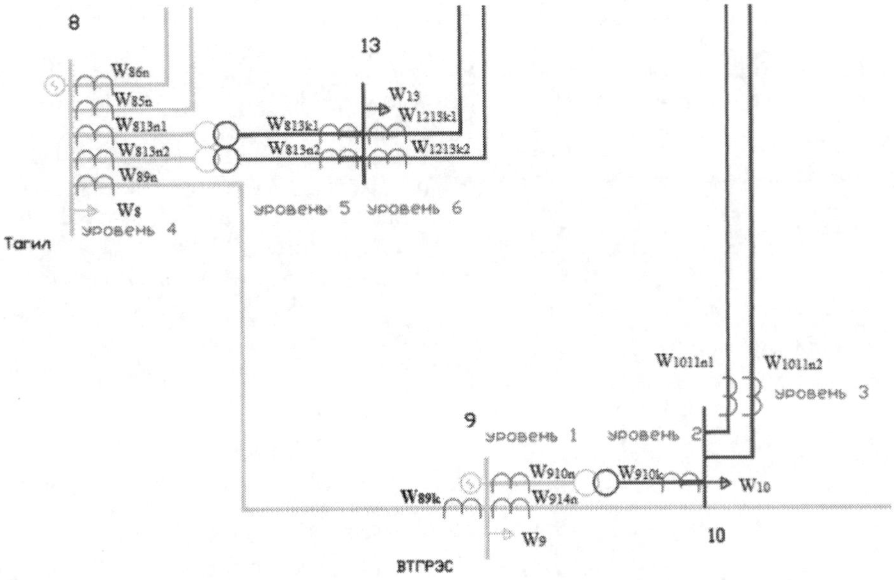

Fig 2. Line 8-9k measurement levels

Having transferred all the nodal and linear streams of energy to the left side, a system of linear equations with variables can be obtained. Considering nodal and linear energy flows as equal variables, all unmetered variables can be excluded by the Gauss method.

The presence of redundant measurements and the testing equations system construction makes the calculated measurement analogue computation possible. Each testing equation allows to compute the calculated measurement analogue. The final calculated measurement estimate can be obtained based on these calculated analogues. This process is described in [] in detail.

4 Comparison of estimates calculated on the energy flow problem basis and using testing equations method

To compare the proposed methods, a scheme of the real power system fragment was compiled. Figure 1 shows the circuit configuration under consideration.

The software complex "Balance" is intended for solving tasks on calculation, analysis and optimization of electrical networks modes, as well as systems of various configurations and arbitrary dimension. The energy flow problem for the network in question was solved using the software package "Balance". The results obtained during the calculation in the software package will be compared with the results of the manual calculation performed using the testing equation system.

Equivalent circuit including the load capacities of the nodes, as well as measuring of the electrical energy at all points where the Automatic Meter Reading complexes are installed are introduced into the program complex. The accuracy classes of the measuring complexes are also entered in the "Balance". The calculation result is the computed estimates of all electricity flows on network elements, as well as technical losses.

The lineal measurement 8-9k is selected as an investigated one. Figure 2 shows the levels of its measurement. That is, the nodal energy is measured not only by means of a specific Automatic Meter Reading complex installed on the outgoing line, but also it is reserved by a number of other measuring complexes located at a distance. In general, the composition of the measurements is redundant, which makes obtaining a testing equation system and computing a number of calculated analogues of measurement possible. The testing equations system is:

$$\begin{cases} W_{12n} + W_{23k} - W_2 + dW_{23} + dW_{12} = 0 \\ W_{14k} - W_{12n} - W_1 + dW_{14} = 0 \\ W_{45n} + W_{45k} + dW_{45} = 0 \\ W_{46n} + W_{46k} + dW_{46} = 0 \\ W_{86n} - W_{46k} - W_6 + dW_{86} + W_{67k} + dW_{67} = 0 \\ \ldots. \end{cases} \tag{2}$$

There are 27 redundant measurements in the network under consideration. The number of testing equations is the same. Testing equations system construction allows to solve a number of problems: estimate observability, to allocate the metering complexes [13-14], to reveal bad data [15].

In previous studies [15] it was shown that the use of more than 6 levels to calculate the estimated analogue does not lead to an increase in the accuracy of calculations. In this case, 6 measurement levels were accepted for calculations. First verifying equation is:

$$W_{89k}{}^{1} = -W_{910n} - W_{914n} - W_9 \tag{3}$$

The accuracy of the calculated estimate is 0.66% at a measurement complex accuracy of 1%. Calculated value is 34,6679 kWh. The same variable calculated using Energy Flow Problem solving in «Balance» is 34,658 kWh.

The final calculated analog is computed using the results of verifying equations taking into account the accuracy of each. The verifying equation accuracy is calculated through the accuracy classes of all Metering complexes whose indications are included in the verifying equation.

Accuracy calculation for the first verifying equation [16]:

$$\delta_1 = \sqrt{(\delta W_{910n})^2 + (\delta W_{914n})^2 + (\delta W_9)^2} \quad (4)$$

Similar comparisons were made for a number of measurements. In all cases, the difference of the calculated estimates was less than 1%, which is within the limits of the permissible errors of the Automatic Meter Reading complexes.

5 Conclusion

Obtaining the calculated measurement estimate allows to solve the problem of Automatic Meter Reading systems bad data validation. The ways of obtaining such estimates can be different. Studies and test calculations have shown that, if it is necessary to obtain a calculation estimate for single measurements, it is possible to use simplified methods based on a testing equations system. The magnitude of the calculated estimate is practically the same as the value obtained by solving the complete energy flow problem.

References

1. A. Pazderin, V. Samoylenko, Localization of non-technical energy losses based on the energy flow problem solution. *Proceedings of the 6th IASTED Asian Conference on Power and Energy Systems, AsiaPES 2013*, pp.100-103, (2013).
2. A. Egorov, E. Kochneva, A. Pazderin, Detection of systematical errors of AMR system complexes, Advanced *Materials Research* 960-961, pp 1342-46, (2014).
3. A. Monticelli, F. F. Wu, M. Yen, Multiple Bad Data Identification for State Estimation by Combinatorial Optimization. *Proceesings of the PICA Conference*, pp. 452-460, (1985)
4. X. Nian-De, W. Shi-Ying, Y. Er-Keng, A new approach for Detection and Identification of Multiple Bad Data in Power System State Estimation. *IEEE Transactions on Power Systems*, PAS-101, no. 2, pp.454-462, (1982).
5. F. Schweppe, J. Douglas, D. Rom, Power System State Estimation, Part I - Exact Model, *IEEE Transactions on Power Apparatus and Systems*, PAS-89, pp. 120-125, (1970).
6. A. Pazderin, S. Kokin, A. Egorov, E. Kochneva, Solution of energy flow problem using state estimation technique, *ECON Proceedings (Industrial Electronics Conference)*, pp. 1736-41.
7. Abur A, Exposito A G, *Power System State Estimation*. Marcel Dekker inc., New York, (2004)
8. A. Gamm, I. Kolosok, *Bad data detection in measurements in electric power system*. Nauka, Novosibirsk, Sib. Enterpr. RAS, (in Russian), (2000).
9. F. Schweppe, J. Douglas, D Rom, Power System State Estimation, Part II - Approximate Model, *IEEE Transactions on Power Apparatus and Systems*, PAS-89, pp. 125-130, (1970).
10. A. Monticelli, Electric Power Syaytem State Estimation. *Proceedings of the IEEE*, vol. 88, no.2, pp. 262-282, (February 2000).
11. P. Chusovitin, I. Polyakov, A. Pazderin, Three-phase state estimation model for distribution grids. *Proceedings of International Conference on the Science of Electrical Engineering ICSEE*, (2016).
12. E. Plesniaev, A. Pazderin, Data acquisition system faults detection. *Proceedings of IEEE Conference on Contril Applications*, pp.1390-94, (2003).
13. E. Plesniaev, A. Pazderin, Analysis of the computation techniques for energy flow problem solving. *Proceedings of the International Conference on* Computer *as a Tool II EUROPCON*, pp.1469-1472, (2005).
14. A. Pazderin, A. Egorov, S. Eroshenko. The energy meters allocation in electric systems on the basis of observability theory, *Proceedings of 9th Conference on Environment and Electrical Engineering, EEEIC*, pp. 167-170, (2010).
15. A. Pazderin, E. Kochneva, Bad data validation on the basis of a posteriori analysis. *Proceedings of the IEEE International Energy Conference ENERGYCON*, pp. 386-391, (2014).
16. RD 34.09.101-94. *Typical instruction for* electricity *metering in its generation, transmission and distribution*. M. ORGRES, S. (in Russian) (1995).

Acknowledgments

The reported study was funded by the Ministry of Education and Science of the Russian Federation under Federal Targeted Programme according to the agreement № 14.578.21.0226 (Project identifier: RFMEFI57817X0226).

Defining the nodal equivalent parameters using synchrophasor measurements

Alexander Berdin, *Dmitriy* Bliznyuk and *Pavel* Kovalenko

Automated Electrical Systems Department, Ural Federal University named after the first President of Russia B.N. Yeltsin. 19 Mira St., Yekaterinburg, Russia

Abstract. The method for defining the nodal equivalent parameters of grid elements based on synchrophasor measure-ments during transients in a power system is proposed. It allows to define the topology and parameters of nodal equivalents and makes it possible to implement adaptive models for problems of power system control.

1 introduction

The rapid development and steadily widening introduction of synchrophasor measurements technology (or just synchrophasors) in power systems all over the world enable new possibilities to improve the quality of primary equipment models involved in power system control applications. This is much more important taking into account the growing complexity of power systems and the impossibility to replace them in all fields of humanity existence and sustainable socioeconomic development.

Conventional approach to emergency control implies utilizing detailed models, characterized by a large amount of parameters [1]. At that, the problem of their definition is complicated due to various factors, including the absence of factory reference values, the complexity of obtaining the parameters of equipment supplied from other countries, change of the parameters over time etc. Hence, utilizing those models can potentially lead to obtaining the results significantly different from the real object in operation and applying the improper control actions to the system.

To date there are numerous projects carried out on developing the adaptive models for power systems primary equipment with the parameters being defined based on the synchrophasor measurements. Specifically, the methodology for defining the instantaneous power system performance parameters was developed [2], allowing to significantly improve the accuracy of the performance parameters calculation during electromechanical transients, including the changes over periods shorter than the industrial frequency; the possibility of defining the inertia constant of a power generating unit [3] and its damping capabilities [4], constituting the most valuable part of a synchronous machine adaptive model were investigated in detail. The next necessary stage of this research is evaluating the mutual influence of synchronous machines in the same power system and rectifying the parameters of their models by taking this influence into account. In order to address this challenge the technique for defining the parameters of power transmission network nodal equivalents based on synchrophasor measurements obtained from WAMS.

The concept of forming and utilizing the nodal equivalents also known as REI-models was proposed by a Romanian scientist P. Dimo back in 1970s [5, 6]. Till present days their parameters have been being defined on the basis of conventional network equivalent transformations and several original techniques. The models were suggested to be employed for solving a wide range of power system problems. Modern progress in measurement technology allows to adapt some of those investigation results for the purpose of building the adaptive models of power system components.

2 Defining the equivalent topology

REI-equivalent is a substitution circuit, completely equivalent to the initial network and independent from the conditions of the rest of the system. REI-equivalent includes the nodes, voltage of which are assumed independent (usually generation nodes) as well as the nodes essential for analysis purposes. Then a network fragment can be selected from the resulting equivalent circuit, containing the node under consideration, all nodes, connected to it, and the connection lines. This fragment is referred to as the nodal model, its distinctive feature is that it is radial and allows to analyze the influence on the node under consideration caused by the other network nodes [5].

The conventional approach to the equivalent topology and parameters definition implies that all nodes, not included into the resulting equivalent circuit, are eliminated by means of Gauss technique. This methodology's drawback is that it can lead to significant errors [7, 8], since the reference initial data values might substantially differ from the actual ones. The issue is made even worse by the consequential essence of Gaussian elimination procedure, that results in errors accumulation.

It is proposed to preserve the Gaussian elimination procedure for defining the electrical network equivalent topology. However, the problem of defining the equivalent model parameters is proposed to be addressed by utilizing the synchrophasor measurements – associated with the nodes which are not eliminated – collected during transients resulting from random disturbances (the methodology for defining the parameters is presented in the following section).

The widely known Kundur model [9] (Figure 1) is investigated as an example for the proposed methodology validation. The model comprises 4 generating nodes, 2 load nodes and connecting lines. The initial equivalent circuit is shown in Figure 2. The nodal equivalent implies eliminating all nodes except for the generating ones. Hence, the nodes 1, 2, 3, and 4 are preserved while the nodes 7, 8, and 9 are eliminated. The network reduction results are presented in Figure 3.

REI-equivalent, intended for the analysis of a part of the network influence on the node under examination, can be obtained from the complete network equivalent by eliminating the nodes having no connection with the considered one. Equivalents for each of the network nodes are shown in Figure 4.

The equivalent impedances were defined as well during the equivalent calculations. They are to be used as the reference values in order to compare the experiment results.

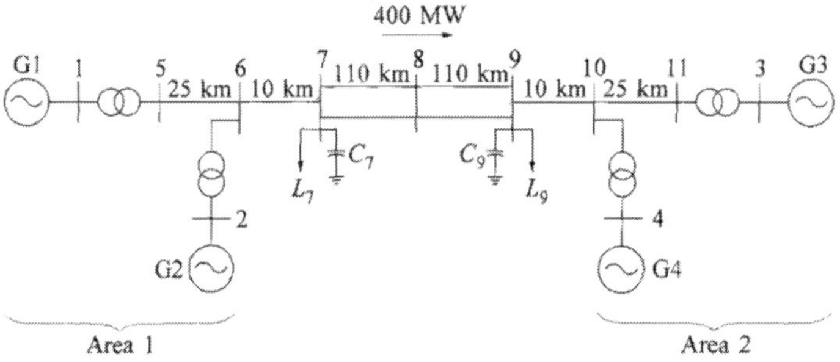

Fig 1. The power system under consideration

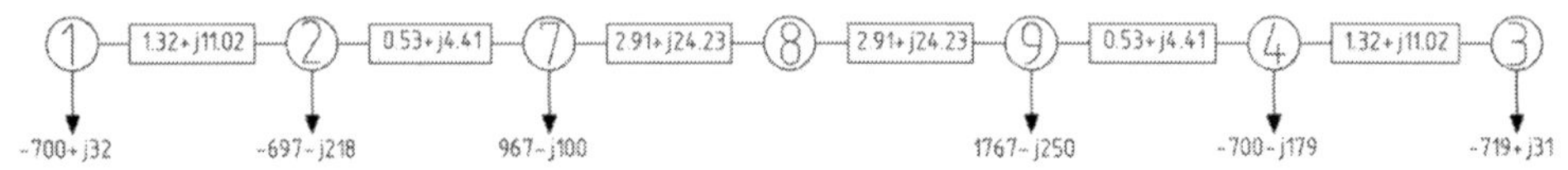

Fig 2. The initial equivalent circuit

Fig 3. The network equivalent

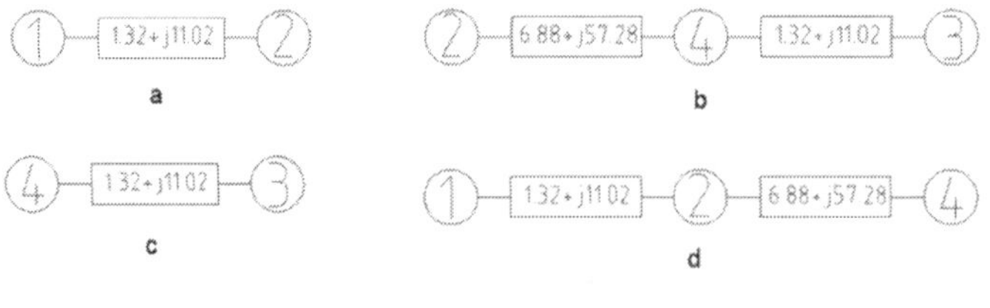

Fig 4. The equivalents associated to each network node (a – for node 1, b – for node 4, c – for node 3, d – for node 2)

3 Defining the nodal equivalent parameters

Nodal equivalent parameters definition is based on the commonly known electrical engineering laws. Apparent power in a line connecting the i and j nodes is defined according to the following expression:

$$\dot{S}_{ij} = \sqrt{3}\dot{U}_i \bar{I}_{ij} \quad (1)$$

where $\dot{U}_i$ is the i node voltage vector,

$\bar{I}_{ij}$ is the line connecting the i and j nodes current complex-conjugate vector.

Hence, the apparent power complex-conjugate corresponding to the same line after substituting the current expression and algebraic transformation can be rewritten as follows:

$$\bar{S}_{ij} = Y_{ij} U_i (U_j (\cos\delta_{ij} + j\sin\delta_{ij}) - U_i) \quad (2)$$

where Y_{ij} is the conductance of the line connecting the i and j nodes,

U is the corresponding node voltage,

δ_{ij} is the angle between the i and j nodes.

In 220 kV (and higher) networks, which are primarily the relevant field for current research, the resistance is significantly smaller, then the reactance, and hence can be neglected. Taking this into account, the real part of (2) can be expressed as follows:

$$I_{Pij} = b_{ij} U_j \sin\delta_{ij} \quad (3)$$

where $I_{Pij} = \frac{P_{ij}}{U_i}$, and P_{ij} is the active power flow over the line connecting the i and j nodes;

$b_{ij} = \frac{1}{x_{ij}}$ is the reactive conductance of the line connecting the i and j nodes, and x_{ij} is its impedance.

By totalizing the corresponding currents for every line connecting the i node to the n j nodes connected to it, one can obtain:

$$I_{Pi} = \sum_{j=1}^{n} b_{ij} U_j \sin\delta_{ij} \quad (4)$$

where $I_{Pi} = \frac{P_i}{U_i}$, and P_i is the equivalent active power for the i node.

The expression (4) is used for the purpose of defining the parameters of a nodal model for the i node. The unknown is the reactive conductance of the lines, while the rest of the parameters are WAMS-measured. The unknown can be defined by utilizing the multi-parameter model [10] or the least square technique.

The same expression can be obtained for the imaginary part of (2) as well. Therefore, the expression containing reactive power instead of active power is derived. The combined usage of (4) and the same expression with the reactive power for defining the equivalent parameters is to improve the calculations accuracy.

Equivalent nodal active power P_i is a vector sum of the considered node generation, its load and part of a network load distributed into the node. For example, the equivalent nodal active power of 332 MW (see Figure 3) for the node 2 of the considered network shown in Figure 2 is the aggregate of the node generation (which is –700 MW) and the part of the nodes 7 and 9 loads. At that, the equivalent active power for the node 1 is only its own generation, since the other nodes load is not distributed into the node.

Therefore, each stage of Gaussian elimination must be accompanied by preserving the coefficients reflecting the share of the load distributed from the eliminated nodes to the remaining ones. Thus, when subtracting the equation x, corresponding the eliminated node, from the equation y, corresponding to another node, the coefficient of this subtraction must be preserved. These coefficients are referred to as the load distribution coefficients k_x^y. They reflect the share of the node x load transferred to the node y. the coefficients for the investigated example are the following:

$k_2^7 = 0.923$; $k_2^9 = 0.077$;

$k_4^7 = 0.077$; $k_4^9 = 0.923$.

Then the equivalent nodal active power can be derived from the node generation measurements by taking into account all loads, distributed into the node through the load distribution coefficients.

4 Method validation

4.1 Steady state simulation

The general possibility of utilizing the proposed methodology for the nodal equivalents parameters was validated using Kundur's digital model shown in Figure 1. At the first stage a set of steady-state conditions were simulated using RastrWin software.

10 steady-state conditions were simulated with nodes 1, 3, and 4 generation (node 2 is the swing-bus) being varied from 650 MW to 720 MW. Based on the obtained results the impedances of the nodal equivalent networks were calculated for each of the 4 generating nodes. There were two calculations performed – with taking resistance into account and without. The results are summarized in Tables 1 and 2.

Table 1. The results of evaluating the equivalent parameters based on steady-state simulation measurements (with considering resistance)

Considered node	X_{12}	X_{24}	X_{34}	R_{12}	R_{24}	R_{34}
1	10.96			1.62		
2	11.40	56.17		-0.48	0.75	

3			11.02			1.36
4		48.90	6.00		15.05	-6.50
Average	11.18	52.54	8.51	0.57	7.90	-2.57
Reference	**11.02**	**57.28**	**11.02**	**1.32**	**6.88**	**1.32**
Error	1.45 %	-8.28 %	-22.78 %	-56.86 %	14.84 %	-294.6 %

Table 1 indicates that taking resistance into account renders the calculation results deviation from the reference values quite significant. The reference was obtained by means of Gaussian elimination (see Figure 3). At that, Table 2 shows that in case calculation is performed without taking resistance into account, the results are close to the reference values. Low accuracy of the calculation with taking resistance into account is due to its small value compared to the reactance. Excluding the resistance from the analysis (Table 2) significantly improves the results accuracy. Apparently, in-deep research of this phenomenon requires the detailed investigation of the utilized computational technique.

Table 2. The results of evaluating the equivalent parameters based on steady-state simulation measurements (without considering resistance)

Considered node	X_{12}	X_{24}	X_{34}
1	11.08		
2	11.37	55.82	
3			11.08
4		62.92	11.47
Average	11.22	59.37	11.28
Reference	**11.02**	**57.28**	**11.02**
Error	1.86 %	3.64 %	2.35 %

However, the combination of calculation results for different nodes allows to obtain the calculated values close to the actual ones. The maximum average values deviation of the calculated values from the reference ones amounted for as low as 3.64 % (without considering resistance).

The experiment results prove the applicability of the proposed method for defining the nodal equivalents parameters based on the steady-state measurements. However, the steady-state measurements change insignificantly over short periods of operation, which leads to the results accuracy decrease. Although longer periods can be considered, the power system topology and performance parameters might change. Therefore, applying the proposed method to the analysis based on steady-state measurements requires that one ensures the limits of the equipment parameters deviations and the fact that the power system topology stays the same.

4.2 Transient simulation

Actual practice highlights it is preferable to define the equivalents parameters based on transient measurements since it is transients when the measured parameters deviate in a wide range over short period of time. During transients the network topology generally remains the same, while the significant deviations of the measured parameters decrease the influence of measurement errors, which improves the calculation accuracy. The following experiment was conducted in order to validate the proposed method application to the WAMS emergency recordings.

A disturbance was simulated in Kundur model using MATLAB Simulink. The excitation winding of the generator in node 3 was subjected to a +5 % 20 ms voltage impulse. At the same tome the excitation winding of the generator in node 4 was subjected to a – 5 % voltage impulse of the same duration. As a result the low-frequency oscillations emerged. Figures 5 and 6 show the simulated WAMS measurements of the active power and voltage in the generation nodes. The nodal equivalent parameters calculation results for all generation nodes are summarized in Table 3.

The results clearly indicate that the simulated WAMS measurements during transients the accuracy of the nodal equivalents parameters calculation proved to be high. The maximum deviation from the value assumed to be the reference, amounted for 4.02 %.

Table 3. The results of evaluating the equivalent parameters based on transient (low-frequency oscillations) simulation measurements

Considered node	X_{12}	X_{24}	X_{34}
1	10.93		
2	11.57	58.29	
3			10.92
4		60.88	11.51
Average	11.25	59.58	11.22
Reference	**11.02**	**57.28**	**11.02**
Error	2.08 %	4.02 %	1.86 %

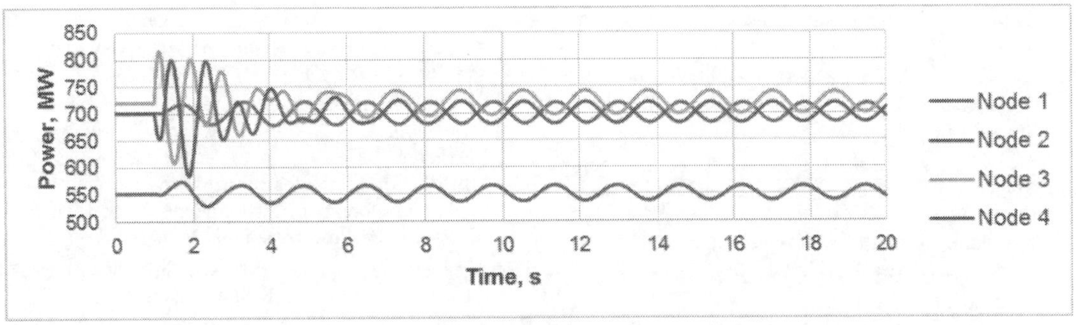

Fig 5. Active power measurements in all nodes of the considered system

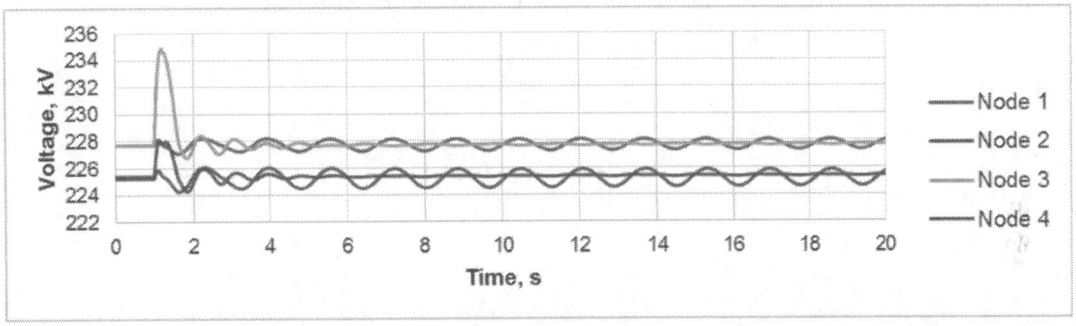

Fig 6. Voltage measurements in all nodes of the considered system

5 Conclusion

Current research proposes the method allowing to define the parameters of the network nodal equivalents based on the synchrophasor measurements performed during power system transients. Accumulating the corresponding measurements archive allows to rectify the parameters and define them for different network topologies and operating conditions [11, 12].

The method was validated and proved its efficiency using the digital simulation of both steady-state conditions and transients. The defined equivalents parameters values differ from the reference ones by less than 5 %, which indicates the proposed method can be applied to defining the nodal equivalents parameters based on the transient measurements.

The developed method plays an important role in assessing the mutual influence of the generators in a power system. The nodal equivalents clearly outline the nodes influencing the considered one and evaluate this influence with the electrical distance between them being calculated.

The next stage of the method development implies its validation using the physical simulation results and actual data collected from the power system objects. The WAMS systems development and improving the synchrophasor observability of the power system suggests a wide range of potential opportunities for implementing the proposed method and employing it in power system dispatch and automatic emergency control.

Acknowledgement

The work was supported by Act 211 Government of the Russian Federation, contract № 02.A03.21.0006 and the Ministry of Education and Science of Russian Federation (in the framework of state assignment, № 13.1928.2014/K (project № 1928)).

References

1. G. Shabalin, A. Pazderin, P. Bannykh, E. Balakh. Voltage stability analysis using quadratic objective function taking into account equality constraint // 2016 IEEE International Conference on the Science of Electrical Engineering Proceedings.
2. Berdin A.S., Bliznyuk D.I., Kovalenko P.Y., Egorov A.O., Cherepov A.S. Otsenka mgnovennyh znacheniy parametrov elektricheskogo rezhima v seti peremennogo toka // Elektricheskie stantsii, #8 – 2015.
3. Berdin A.S., Bliznyuk D.I., Gerasimov A.S. Opredelenie ehkvivalentnoj inercionnoj postoyannoj po dannym izmerenij ehlektromehanicheskogo perekhodnogo processa // Izvestiya NTC Edinoj ehnergeticheskoj sistemy #1 – 2016.
4. Berdin A.S., Bliznyuk D.I., Gerasimov A.S. Ocenka dempfernyh svojstv ehnergobloka s ispolzovaniem ehksperimentalnyh dannyh // Izvestiya NTC Edinoj ehnergeticheskoj sistemy #2 – 2016.

5. Dimo P. Uzlovoj analiz ehlektricheskih system. Moscow, 1973.
6. Dimo P. Modeli REI i parametry rezhima. Obyedinennye ehnergosistemy. Moscow, Energoatomizdat – 1987.
7. Kochneva, E.S., Pazderin, A.V., Sukalo, A. Testing equations method for electrical energy measurements systematic errors detection and measurements results adjustment. Journal of Physics: Conference Series 870(1), 2017.
8. Kochneva, E.S., Pazderin, A.V. Improving the method of test equations for the validation of electricity measurements. Power Technology and Engineering 50(6), 2017.
9. Kundur P. Power system stability and control. – New York: McGraw-hill, 1994.
10. Bliznyuk D.I., Romanov I.B. Ocenka dostovernosti opredeleniya harakteristik nagruzki po dannym vektornyh izmerenij // Izvestiya NTC Edinoj ehnergeticheskoj sistemy #2 – 2016.
11. Mukhlynin, N.D., Pazderin, A.V., Sukalo, A. Estimating operating mode parameters in distribution grids with regard to the peculiarities of their development. International Symposium on Industrial Electronics, INDEL 2016.
12. Pazderin, A.V., Pazderin, A.A., Mukhlynin, N.D. Technical and economic model of energy transmission and distribution based on the smart metering technologies. IEEE International Symposium on Industrial Electronics, 2017.

Influence of measuring force on CMM accuracy

WANG Bin

Research Institute of Physical and Chemical Engineering of Nuclear Industry, Tianjin, China

Abstract. The measuring accuracy of CMM (Coordinate Measuring Machine) is influenced by many factors, such as temperature, humidity, measuring force and method of signal acquisition. For thin parts, the influence of measuring force is especially obvious. In this paper, the relationship between measuring force and measurement accuracy is studied for a thin part with a U-shaped cross-section. By analyzing the structure of the probe and establishing the force model, the influencing factors of the accuracy of CMM are obtained, and the influence of the contact deformation and the bending deformation on the measurement accuracy is analyzed from the point of view of material mechanics. At the same time, the measurement accuracy of different measuring cross-sections is analyzed. Through the research of this paper, the relationship between measuring force and CMM is established, and an effective method to improve the accuracy of CMM is also found.

1 Structural analysis of trigger probe and error principle

The probe is an important part of CMM, consisting of a measuring rod and a ruby head, which is mainly used to measure the working surface, move through the mechanical position of the measuring end, trigger the signal and collect a measurement data. The internal structure of the RDS XXT (Carl Zeiss' products) probe is shown in Fig. 1, the rod is fitted on the ring and the outer wall of the ring has a uniform contact terminals. When the rod is not measured, the outer wall of the ring is kept in contact with the terminals. One contact detachment is caused at least when the spherical end of the measuring rod is contacted with the workpiece, thus causing the circuit to disconnect, produce step signal, through the sampling circuit, will be along three axes direction of the coordinate data storage processing. Fig. 1-measuring rod, 2-ring cage.

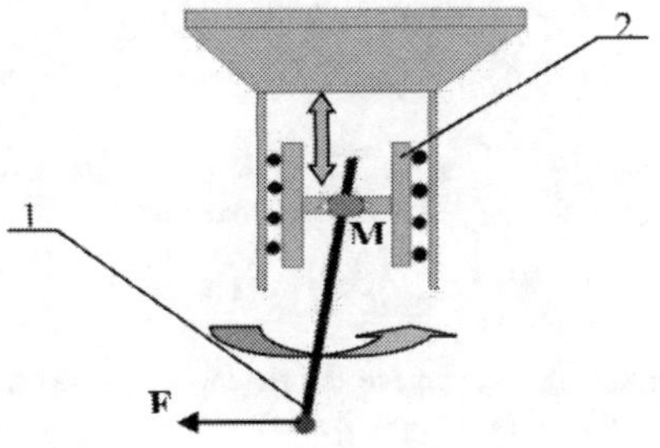

Fig. 1 Schematic diagram of the head probe

The process between the probe end and the measured part contacting to the probe trigger is "Pre-range" during the CMM is working. The Pre-range is not an error source, because a standard sphere with known radius is used to determine a reasonable probe radius to eliminate the effect of the average Pre-range during the probe certification process. The different triggering forces, that is the production of the Pre-range changes, as shown in Figure 2, the figure can be seen visible three convex angle, reflecting the amount of the change in the Pre-range "δ". This mode shows symmetry of the probe, which greatly reduces the ability of CMM to perform high-precision measurement.

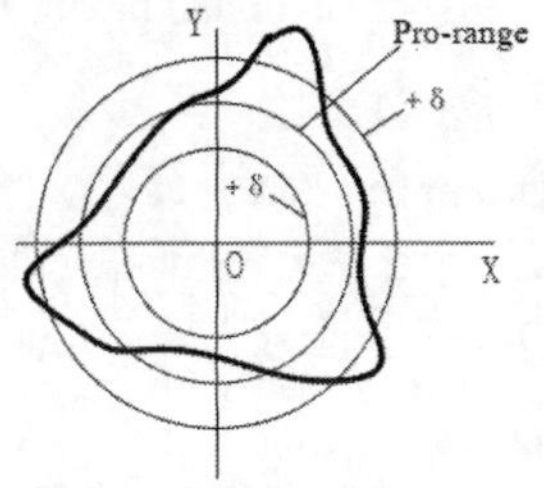

Fig. 2 The sketch of the convex angle of probe

2 Stress analysis of the probe model

The probe system of CMM is to issue the trigger pulse when the probe ball and workpiece collide, to send out the sampling signal to CMM, and the three coordinate values of the probe are stored at this time to determine its space position. The space coordinate equation can be obtained by more than three contact measurements on a plane. The center coordinate and radius can be obtained by touching the plane circle more than three times, and the shape of any complex space object may be determined [1].

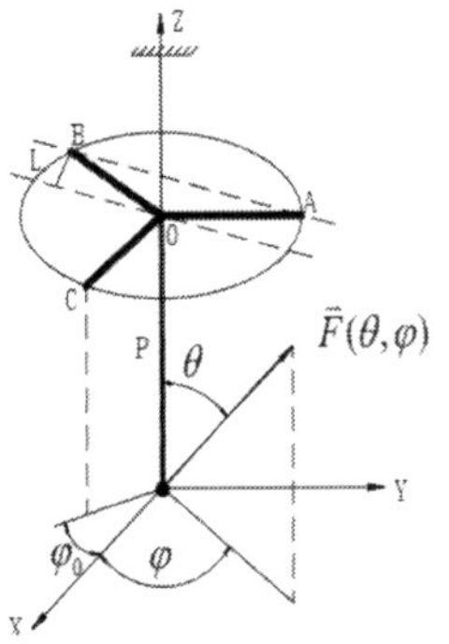

Fig. 3 Force model of dynamic trigger probe

As shown in Figure 3, the probe is positioned in the Cartesian coordinate system, and the probe is coincident with the Z-axis. The probe length is P, the projection of the cylindrical and the X-axis angle areφ_0, the Force F (θ,φ) is triggered by the end and the component surface. The direction of force F is determined by the polar angle θ and the azimuth φ. assuming that the direction of the probe is in line with the normal direction of the part surface, the probe is moved along the direction of the force.

$$\vec{n} = \vec{F}/F = \vec{i}\sin\theta\cos\varphi + \vec{j}\sin\theta\sin\varphi + \vec{k}\cos\theta \qquad (1)$$

If the polar angle "θ" and azimuth "φ" are in a suitable range, "F" will cause the probe to rotate around the *A-B* axis, and the signal will trigger when the cylinder *C* leaves the contact terminals. According to the different F-direction of the force, the other two possible axes of rotation are *B-C* and *A-C*, and the corresponding angle is $\varphi_0 \pm 120°$. In addition to rotation, Force "F " also causes the probe to bend, resulting in bending displacemen.

2.1 Contact deformation

In order to ensure the reliable contact between CMM and the workpiece surface, it is necessary to impose a certain force acting on the workpiece, but the force will cause contact deformation of the workpiece. The size of contact deformation is related to the material and measuring force of the workpiece.

$$\delta_{contact} = 1.231\left((F/E)^2(R_1+R_2)/R_1R_2\right)^{1/2} \qquad (2)$$

"F" is the measuring force, "E" is the elastic modulus of the workpiece, the "R_1, R_2" are the workpiece and the radius of the probe ball.

2.2 Bending deformation

The measuring force can also cause the bending deformation, and the bending deformation is mainly affected by the measuring position and the measuring force value. According to the knowledge of material mechanics, the bending deformation of workpiece under the action of measuring force is:

$$\delta_{bending} = \frac{FL^3}{3EI} \qquad (3)$$

"F" is measuring force, "E" is the elastic modulus of workpiece, "I" is the workpiece section inertia moment, "L" is the length of probe.

3 Modal analysis of force measurement

There are two kinds of CMM probes, one is the dynamic probe, the other is the static probe. The RDS XXT probe system is a dual-mode signal acquisition system which can realize dynamic (scanning) and static (triggering) measurement, under which the influence factors of the measuring force are analyzed respectively.

3.1 Dynamic measurement mode

It is possible for the probe to leave the surface of the workpiece and cause the measurement error. Because of the relative movement of the probe and the workpiece, when the relative motion speed is higher and the measuring force is small. In order to analyze the dynamic measuring force, the probe and the rod are simplified into the rod, which mass is "m", length is "L", and the static measuring force on the workpiece is "F_0". When measuring the probe relative to the workpiece surface, the measuring rod element revolves around the fulcrum, and the pendulum angle is "θ", the dynamic equation of the measuring rod element is:

$$J\ddot{\theta} = -\frac{M_{gl}}{2}\sin\theta + (F-F_0)L \qquad (4)$$

Because the measuring rod pendulum angle is very small, "θ"can be approximated as:

$$\theta = \frac{h_m}{L}\sin\omega t \qquad (5)$$

the influence of the gravity moment of the measuring rod element can be ignored, the formula (4) can be reduced:

$$F = F_0 - \frac{M}{2}h_m\omega^2\sin\omega t \qquad (6)$$

In order to ensure reliable contact during measurement, the force value is $f \geq 0$, calculation:

$$\omega_{max} = (2F_0/Mh_m)^{1/2} \qquad (7)$$

Therefore, the workpiece is related to the maximum relative rotational speed and the static measuring force of the measuring ball during the scanning process. The probe and workpiece are separated when workpiece speed is greater than "Ω_{max}".

Therefore, in order to maintain good contact between the probe and the workpiece and to prevent the probe ball from being detached from the workpiece with lower

https://doi.org/10.1051/matecconf/201820803004

rotational speed, which is equivalent to the static measurement, so that the influence of the measuring force change on the measurement result can be neglected [2]

3.2 Static measurement mode

The three-coordinate probe should maintain a constant measuring force in its contact with the static mode. The measuring force is mainly able to overcome the dust and oil film on the surface of workpiece, so as to ensure the stability of the measured value. However, in ensuring its stability, it should also control the value of the measuring force, to avoid deformation and scratches on the workpiece surface due to the excessive force. In the process of measuring the workpiece, there are two kinds of errors in the measurement result of geometrical dimension, one is the contact deformation of the surface, the other is the error due to the longitudinal deformation and bending [3].

When two objects are in contact with each other, a certain deformation is produced by squeezing. At the same time, because of the influence of the stress on the value of the deformation, the Gelz theory is used to analyze and calculate the deformation of the stress relation and stress effect. The spherical probe is the most commonly used in the contact measuring method, in which the contact deformation affects the measurement precision of the geometrical dimension, in which case, the effect of measuring force on the measurement result can be reduced by increasing the radius of the sphere.

According to the above analysis, the main influencing factors in the static measurement process are contact deformation and bending deformation. The RDS XXT probe system is configured with three triggering measuring force modes, which are sensitive trigger, standard trigger and strong trigger. The sensitive mode is mainly used to measure the smooth surface of the workpiece, such as plastic and glass. The standard mode is used for most metal materials. The strong mode is used for surface roughness testing of composite materials. Therefore, the standard trigger mode is more suitable for the detection of metal parts, and the relationship between the measuring force and length of the bar is shown in Fig. 4.

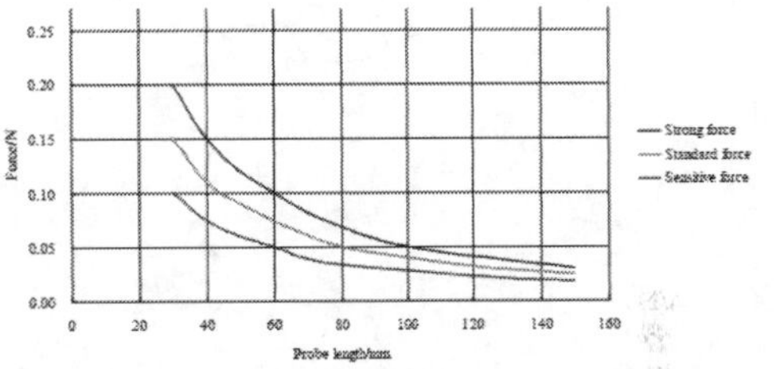

Fig. 4 Relationship between measuring force and probe length

3.2.1 Calculation of contact deformation

The probe ball radius is 1.5mm, substituting formula (2), and the deformation is shown in table 1.

Table 1. Calculation value of measuring force and contact deformation

Probe length/mm	Measuring force/mN	Contact deformation /μm
30	150	1.8
40	110	1.5
50	90	1.3
75	55	0.9
100	40	0.7
125	30	0.6
150	25	0.5

The contact deformation of workpiece is only 1.8μm, which corresponded to the minimum probe length. Because the CMM measurement error is 1.9μm, so the deformation can be neglected.

3.2.2 Calculation of bending deformation [4]

The bending deformation of workpiece is not only related to the material of workpiece, but also has great influence on the fixed height. The structure of the U-shaped piece is shown in Figure 5.

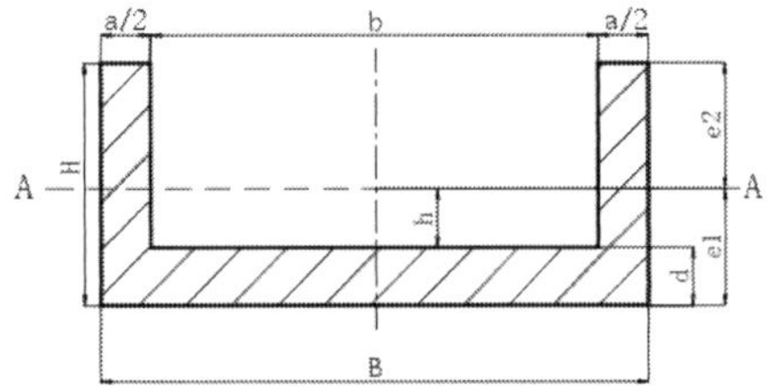

Fig. 5 Simplified model diagram of inertial moment

The inertia moment is calculated by the formula (8).

$$I = \frac{Be_1^3 - bh^3 + ae_2^3}{3} \tag{8}$$

The relationship between the bending deformation val ue and the different fixed height H can be calculated [5]. Height variation △ H is 1mm. Substituting formula (3), t he relationship between the measuring force and the bend ing deformation of the U-shaped parts in different measu ring sections can be calculated as shown in table 2[6].

Table 2 .Calculation of measuring force and bending deformation

Measuring force /mN Fixing height ΔH/mm	150	90	55	40	25
2	0.5	0.6	0.8	0.9	1.2
4	0.6	0.9	1.1	1.4	1.7
6	1.2	1.3	1.6	1.9	2.2
8	2.2	2.4	2.5	2.8	3.1
10	4.0	4.2	4.5	5.2	6.2

4 Conclusion

With the increase of measuring force, the bending deformation is also increasing. And with the increase of the fixed height, the bending deformation reaches 6.2μm, so the influence of bending deformation is the key factor of the measurement error.

By analyzing the principle of three coordinate measuring machine and the deformation mechanism caused by measuring force, it is shown that the influence of the U-shaped parts is mainly reflected in the bending deformation. The instability of the measured results caused by deformation can be improved by reducing the force and determining the reasonable measuring section.

References

1. Wang Yunqing, Li Qingxiang, Zhou Chiying. Analysis of stylus measurement force in contact Contour Measurement [j]. Modern metrology Test, 1996, (1); 18-21.
2. De Guangyun, Jianhua, Li Ping. Dynamic performance analysis of contact micrometer for probe [j]. Xi ' an Industrial college journal, 2004, 24 (1): 11-14
3. Statham A. Coordinate measurement Centre for Metrology, The University of Birmingham, Birmingham, 1991
4. Hou Yu, Lu Jin, CMM theory and method of shape error evaluation [J], Journal of Instrumentation, 1996, 17(6): 618-623
5. Fei Y. T. et al., Symmetrizable connection and combined calibration method for accuracy measurement of CMM. Measurement Tech. And Intelligent Instruments. 1993; 1462-1464

Simulation Studies of a Trench MOS Device Structure with Small Figures of Merit

Junhong Li

State Key Laboratory of Electronic Thin Films and Integrated Devices, University of Electronic Science and Technology of China

Abstract. We proposed a vertical high permittivity trench power MOS (HKTMOS) device with alternating N&P drift region and high permittivity (HK) trench sandwiched in between. The unique structure guarantees uniform potential distribution for wide voltage range at block state owing to both HK potential modulation effect and superjunction (SJ) charge balance. The specific on-resistance (Rons) of HKTMOS is in orders of magnitude lower than the SJ counterparts at the on state because of the strong accumulation effect brought by HK trench. Although the gate charge also significantly rises because of the accumulation, the figures of merit (FOM) of HKTMOS still reduces considerably than the SJ under same BV. An expression for FOM is derived demonstrating that the FOM of HKTMOS is proportional to the square of HK trench depth, which agrees on with simulation results well. The simulation results indicate that within the BV range of 500~2000V, the Rons and FOM of HKTMOS are in 1~2 orders of magnitude lower and 17.4%~44.1% of SJ, respectively under the same BV condition. Furthermore, HKTMOS also demonstrates better charge imbalance tolerance than SJ.

1 Introduction

The performance of silicon power devices are essential for the energy conversion system. The introduction of high permittivity (HK) material into power MOS device allows better potential distribution in silicon and therefore improves device breakdown voltage (BV) [1]. The effect is called the HK potential modulation (PM) effect [1], which had been demonstrated both theoretically [2] and experimentally[3]. On other hand, because HK material exhibits large permittivity, it is potential to be used to activate the carrier accumulation (CA) effect in silicon to significantly reduce the device on-resistance (Ron). In this letter, we propose a novel HK trench MOS (HKTMOS), which utilizes both PM and CA effects from HK, resulting in specific-Ron(R_{ons}) decrease in orders of magnitude compared with conventional superjunction device (SJ). Although such R_{ons} decrease is at the cost of larger switch loss due to the extra accumulation charge on the gate, the relationship between R_{ons} and gate charge is linear, and the total figures of merit (FOM) of HKTMOS still experiences a significant improvement than the SJ. Furthermore, HKTMOS also indicated strong doping-imbalance tolerance in N and P drift region.

2 Device structure and mechanism

The cross section view of HKTMOS is showed in Figure 1a. Similar with SJ, HKTMOS also features alternating N and P drift region pillar, whereas a layer of HK material is sandwiched in between. Together with the buried oxide in the N+ drain region, the P drift region is completely isolated from the N drift region and the drain, but directly contacts with the gate.

At the device blocking state, with the same mechanism as [4], although there is HK isolation between N and P drift region, their electric field still depleted with each other because of the large permittivity of HK, so that the charge balance effect still exists. Moreover, the total permittivity of drift region is also boosted with the introduction of HK trench, according to [1], the slope of electrical field in drift region is given by (1),

$$\frac{\partial^2 \varphi}{\partial^2 x} = -\frac{\partial E}{\partial x} = -\frac{qN_d}{\varepsilon_{total}}, \qquad (1)$$

Where N_d and ε_{total} are the doping concentration and total permittivity of the drift region, respectively. With large ε_{total}, the slope of the electric field is small, giving better potential distribution and higher BV on the basis on charge balance. The effect is attributing to the HK PM effect. As showed in Figure 1b, the potential distribution of HKTMOS is almost ideal at the blocking state. Besides, better doping imbalance tolerance between N and P pillar can be also realized with effect of HK, which will be verified later. Above all, at the device blocking state, both effect of HK PM and charge balance co-exists to guarantee the high BV not only for the short drift region, but also for long drift region devices.

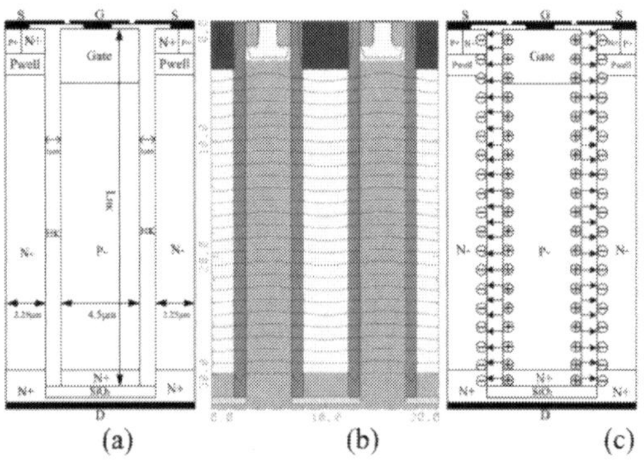

Fig 1. (a) HKTMOS cross section view with N- and P- doping of 10^{-15}cm^{-3}. (b) Potential distribution of HKTMOS at block state. (c) Illustration of the accumulation effect.

Although the reduction of device R_{ons} by mean of CA effect has been reported in [5], the CA is generated by SiO_2 layer, with the low permittivity of SiO_2, its effect is limited. Moreover, small ε of the SiO_2 also brings negative effect for BV according to (1). On contrast, HK material is capable of overcoming the above problems. When the device in Figure 1a is in on state, the gate voltage is high, the P pillar share the same high potential with gate as their directly contact. Whereas both drain and source voltage is much lower than the gate at the on state, consequently, CA effect happens at the interface of silicon and HK trench as showed in Figure 1c. The large permittivity of HK amplified the CA effect and forms a high carrier density path between drain and source, the device R_{ons} is thereby significantly reduced.

CA effect allows significant R_{ons} reduction with no additional drift region doping. However, large trench capacitance is necessary to accumulate enough charge at the device on-state to reduce the R_{ons}. And the trench capacitor will be discharged at off-state to guarantee high BV, such charge-discharge cycle with large capacitance results in high switch loss. Rarely had any literatures qualitatively investigated such effect and its impact to FOM. For the HKTMOS device, as a low resistance path exists between drain and source generated by HK CA effect, the device R_{ons} is major determined by the accumulation resistance (R_A). According to [6], $R_A \propto 1/C_g$, where C_g is the unit area trench capacitance, we can derive that R_{ons} of HKTMOS is given by

$$R_{ons} = \frac{K_{HKT} L_{HK}}{V_g C_g W} \tag{2}$$

Where the K_{HKT}, L_{HK}, V_g, and W are the HKTMOS constant, depth of HK trench, gate voltage and width of the device, respectively. Although the N+ source junction and the drain not contributing to the total resistance are covered in the L_{HK} in (2), both of their depth are overwhelming smaller than the drift region length and thereby negligible. We see that the R_{ons} is proportional to L_{HK} for HKTMOS from (2).

On the other hand, the gate charges are contributed by both HK trenches and SiO_2 buries. The contribution from SiO_2 buries is also overwhelming smaller than the HK trenches, for both permittivity and area. Therefore, it is neglected and gate charge is given by

$$Q_g = W L_{HK} V_g C_g \tag{3}$$

According to (2) and (3), we see that the R_{ons} is inverse proportional to the gate charge for a given device dimension. The definition[6] for FOM is $FOM=R_{ons}*Q_g$, which gives

$$FOM_{HKT} = K_{HKT} L_{HK}^{\ 2} \tag{4}$$

An interesting result is observable from (4) that FOM of the HKTMOS is irrespective of the unit area trench capacitance and device width. It is only proportional to the square of the HK trench depth with ratio of HKTMOS constant K_{HKT}. According to analysis and simulation results, the analytical value of K_{HKT} is $7.812\times10^{-4}\Omega$*nC/μm^2.

3 Simulation verification and discussion

We use TMA-MEDICI to make 2D simulation and verify above analysis. Figure 2 shows the BV dependence for R_{ons} and switch delay for both HKTMOS with different HK permittivity ε_{HK} (a) and SJ device (b).

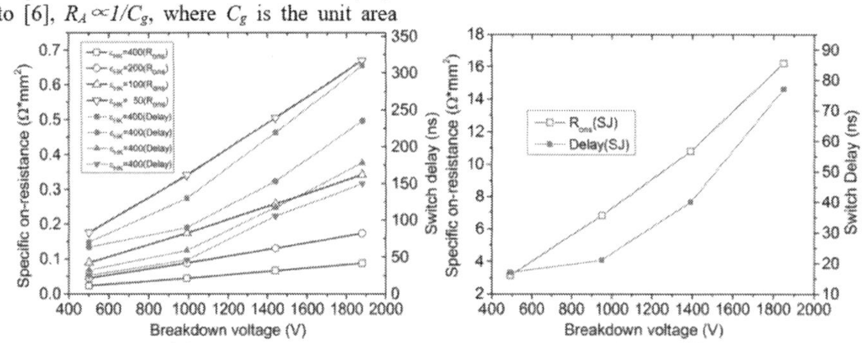

(a) (b)

Fig 2. Under the 15 V gate voltage, the R_{ons} vs. BV and switch delay (sum of the transient rise and fall time for the drain voltage) vs. BV with a drain resistance of 4×10^6 Ω for HKTMOS (a) and 4×10^7 Ω for SJ (b) under the dimension of 20 μm×1 μm for both devices. The R_{ons} and the switching delay are simulated under a drain voltage of 0.1 V and 100 V, respectively.

As revealed in Figure 2b, the R_{ons} of the SJ rises rapidly with increase of BV at the exponential of 1.33[4]. On contrast, under the effect of strong CA effect brought by HK, the R_{ons} of HKTMOS is in the orders of magnitude lower than the SJ as showed in Figure 2a. Higher ε_{HK} always brings smaller R_{ons}, owing to the reason that higher permittivity always provides larger trench capacitance, thereby stronger CA effect and smaller R_{ons}. Moreover, expression (2) suggests that the R_{ons} of HKTMOS is proportional of the drift region length. Because the BV is also proportional to the drift region length, the R_{ons} of HKTMOS rises linearly with the increase of BV as Figure 2a reveals, which agrees on with (2).

Although the CA effect of HK reduces R_{ons} in orders of magnitude, the large trench capacitance also slow down the switching speed. As showed in Figure 2, the switch delay of the HKTMOS is much larger than the SJ device, in the orders of magnitude as well. Higher ε_{HK} value always causes worse switch delay. To investigate the comprehensive performance of the HKTMOS, the device FOMs are simulated and calculated as shown in Figure 3a. R_{ons} is simulated under the condition of a 15 V gate voltage and 0.1 V drain voltage. Q_g is simulated schematically as shown in Figure 3a, where a 15 V square wave voltage is applied on the gate. Q_g is determined by the current integral at the time interval when the drain drops from 100 V to the voltage of the fully on-state. As Figure 3a shows, both R_{ons} and Q_g increase linearly with the rise of L_{HK}, which is consistent with (3) and (5). (3) and (5) also indicated that R_{ons} and Q_g are proportional and inversely proportional to ε_{HK}, respectively. The relation is also verified by simulation, as Figure 3a shows; a larger ε_{HK} always provides a smaller R_{ons} but a larger Q_g and vice versa.

Figure 3b shows the FOM and BV vs. L_{HK} for both HKTMOS and SJ, where the FOM of HKTMOS with changing ε_{HK} are plotted in different line styles. As mentioned above, R_{ons} and Q_g are inversely proportional and proportional to L_{HK}, respectively, which cancel each other out when the multiplication for FOM occurs. It can be observed that all of the lines almost overlap with each other for different ε_{HK} when L_{HK} ranges from 30 μm to 120 μm (Figure 3b) according to simulation, which agrees with (6) very well. The proportional coefficient between the FOM and L_{HK} square may be different depending on the cell geometry; however, for a device with a given cell geometry, the FOM is only determined by L_{HK} regardless of the HK permittivity and trench thickness.

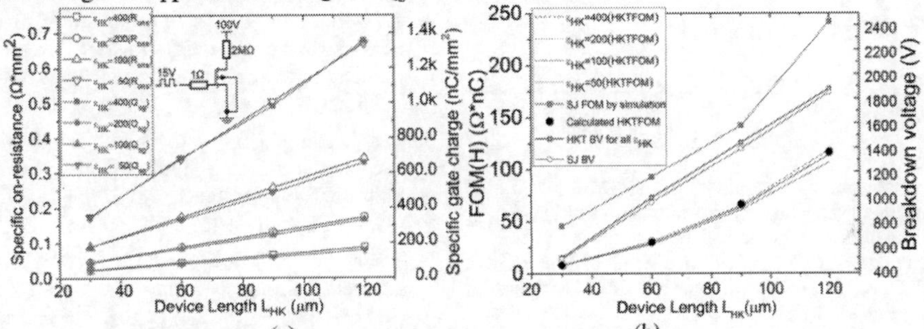

(a) (b)

Fig 3. The R_{ons} (simulated under 15 V gate voltage and 0.1 V drain voltage) and Q_g (simulated by the integral of the gate current as the schematics show in the figure) vs. L_{HK} for HKTMOS (a) The FOM(H) vs. L_{HK} for HKTMOS by both simulation and calculation, together with the FOM(H) of SJ and BV by simulation (b).

For HKTMOS with the size displayed in Figure 2a, the analytical value of the proportional coefficient is derived as 7.812×10^{-3} Ω*nC/μm². Utilizing such a coefficient, we are able to calculate the FOM using (6), which is shown by the round dots in Figure 3b. It is clear that the calculation results agree well with the simulation results. Figure 3b also reveals that for the HKTMOS with smaller ε_{HK} (ε_{HK}=50), the FOM by simulation is slightly smaller than the calculated value, especially at larger L_{HK}. This is because Q_g is determined by the sum of Q_{HK} and Q_{DC}, then, Q_g will drop linearly with the linear decrease of ε_{HK} as long as Q_{HK} is still larger than Q_{DC} according to (4). However, as R_A parallels with R_{drift}, with smaller ε_{HK}, R_A in (2) is less dominant, and the contribution to R_{ons} from the drift region R_{drift} becomes more significant although R_A is still much lower than R_{drift}. Consequently, the total R_{ons} will rise sub-linearly with the decrease of the small ε_{HK} due to the contribution from the drift region bypass, which has been neglected in previous analysis. Moreover, according to (6), the FOM of HKTMOS is proportional to the square of L_{HK}; the square amplifies the model inaccuracy if L_{HK} is large, which causes a larger FOM difference between the calculated and simulated values under the condition of a large L_{HK} and a small ε_{HK} as shown in Figure 3b.

Figure 3b also shows the relationship between BV and L_{HK} for both HKTMOS and SJ. The BVs of HKTMOS with different ε_{HK} completely overlapped with each other so that only one curve is shown for HKTMOS. As the BV of SJ relies on CB only, while HKTMOS takes both effects of PM and CB to achieve high BV, the BV of HKTMOS will be slightly better than that of SJ, as shown in Figure 3b. Last but not least, HKTMOS always demonstrates significant FOM improvement over that of SJ under all L_{HK} in the figure; the FOM of HKTMOS is 48% of that of SJ at the L_{HK} of 120 μm, and only 17% at the L_{HK} of 30 μm. Although the FOM takes the total gate charges into consideration instead of the gate-to-drain charges only, HKTMOS still exhibits a significant FOM improvement over that of SJ.

https://doi.org/10.1051/matecconf/201820803005

Charge imbalance between N&P drift region in actual fabrication process is another issue of the SJ, even slight charge imbalance caused by doping difference can severely impact the BV of SJ device. Adversely, HKTMOS exhibit excellent doping difference tolerance because their BVs not only depend on charge balance but also PM effect from HK trench. As Figure 4 shows, with the rise of the N&P drift region doping difference, the BV of HKTMOS with ε_{HK} =400 almost remains unchanged. However, smaller ε_{HK} value always causes higher BV sensitivity to doping difference, this is because with smaller ε_{HK}, the PM capability of HK trench is impaired, consequently, the BV drops more rapidly with the rise of doping difference. As the BV of SJ solo relies on the charge balance and free of PM effect, SJ gives the worst performance. The BV of SJ at the doping difference of 4 ± 10^{-16}cm^{-3} drops to 30% of the BV under the charge-balance condition.

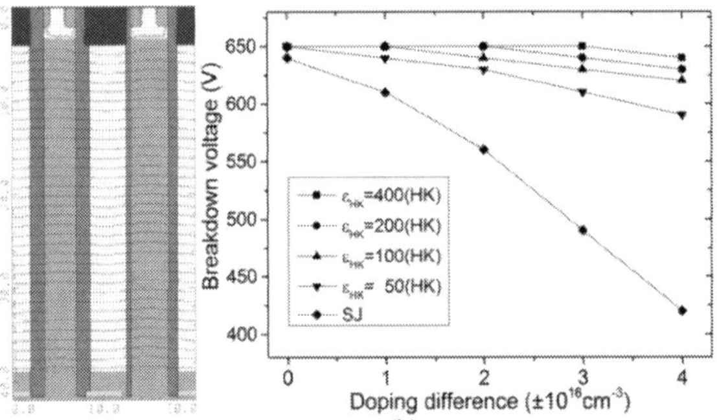

(a) (b)

Fig 4. (a) The potential distribution for a HKTMOS with L_{HK} of 39 μm and HK trench randomly shifted in both width and position. (b) The impact of charge-imbalance to BV for SJ and HKTMOS with different ε_{HK} for the device length of 39 μm under the baseline doping of 10^{15} cm^{-3} for both N and P pillar.

4 Conclusion

The HKTMOS we proposed in this letter indicates excellent performance for wide BV range according to investigation. Its FOM is predictable using the HKTMOS constant and shows significant improvement than that of the SJ. Besides, HKTMOS also demonstrates its good charge-imbalance tolerance than its SJ counterparts. Above all, the HKTMOS is potential for the application on the energy conversion system for better efficiency and further promotion of the silicon limit.

References

1. X. Chen, "Super-junction voltage sustaining layers with alternating semiconductor and high-K dielectric regions," U.S. Patent, 7230310B2, Jun. 12, 2007
2. X. Luo, Y. Jiang, K. Zhou, P. Wang, X. Wang, Q. Wang, G. Yao, B. Zhang, and Z. Li, "Ultralow specific on-resistance superjunction vertical DMOS with high-K dielectric pillar," IEEE Electron Device Lett., vol. 33, no. 7, pp. 1042–1044, Jul. 2012.
3. J. Li P. Li, W. Huo, G. Zhang, Y. Zhai, and X. Chen, "Analysis and fabrication of an LDMOS with high-permittivity dielectric," IEEE Electron Device Lett., vol. 32, no. 9, pp. 1266–1268, Sep. 2011.
4. X. Chen and J. Sin. "Optimization of the specific on-resistance of the COOLMOS (TM)," IEEE Trans. Electron Devices, vol. 48, no. 2, pp. 344-348, Feb. 2001.
5. B. J. Baliga, T. Syau, and P. Venkatraman, "The accumulation-mode field effect transistor: a new ultralow on-resistance MOSFET" IEEE Electron Device Lett., vol. 13, no. 8, pp. 427–429, Aug. 1992.
6. B. J. Baliga, "Fundamentals of power semiconductor devices," Springer-Verlag, 2008, pp. 362.
7. B. J. Baliga, "Advanced power MOSFET concepts," Springer-Verlag, 2010, pp. 55.

A Simple Integrated Geometrical Error Model for Laser Tracer

Huixu Song , *Hongfang* Chen and *Zhaoyao* Shi

Beijing Engineering Research Center of Precision Measurement Technology and Instruments, Beijing University of Technology, Beijing, 100124, China

Abstract. Geometrical errors of 2D gimbal mount axes are studied for laser tracer for the sake of realizing rapid detection for numerical control equipment. The high precision reference sphere, which serves as the reflection unit and is set on the base, avoids the influence on measurement accuracy caused by the movements of rotatory axes, and ensures that laser tracer owns large tracing angle. The nonlinear coupling relationship between the geometrical errors and the measurement accuracy of laser tracer is analyzed; the integrated geometrical error model of 2D gimbal mount axes and laser interferometry system is established; the error map of the measuring system is brought. The simulation results demonstrate that the measurement error of the laser tracer is certain and unique after the value and direction of all the geometrical errors determined. (Key words: Laser tracer; gimbal mount axes; geometrical error model; error map.)

1 Introduction

The laser tracer holds a high precision reference sphere as the reflection unit whose form error is less than 50 nm, and it can ensure the measurement accuracy by using high precision and high stability laser interferometry system [1-3]. Laser tracer owns the advantage of high precision and high efficiency, however, the geometrical errors of 2D gimbal mount axes are the major factors for the measurement accuracy of laser tracer, therefore it is necessary to detect and calibrate these geometrical errors. Traditional commercial laser tracker and laser tracer were developed on the basis of theodolite and total station. Because of similar shafting structures, the analytical methods for error models of instruments were also analogous. Deumlich analyzed the geometrical errors' sources which could affect the measurement accuracy of theodolite, and gave the relationship between the single geometrical error of rotatory axes and the measurement error of theodolite [4]. K. Lau et al. were the first to give a model of the squareness error for the laser tracker, a model of dead path error for laser interferometry and a model of alignment error for laser beam [5]. B. Muralikrishnan et al. intensively studied the influence of the angular measurement error for the laser tracker on the laser ranging measurement accuracy, and provided a geometrical model for calibrating the measurement error of the laser tracker [6-9]. Hitherto, there was not any research on the geometrical error model for laser tracer.

This paper deeply analyzes the geometrical errors' sources that influence the laser ranging measurement accuracy of laser tracer. In establishing a unified geometrical error model, the nonlinear coupling relationship between the geometrical errors and the laser ranging measurement error is presented, and the error map is drawn then [10-11].

2 Geometrical error sources of rotatory axes in Laser Tracer

The schematic diagram of the laser tracer is shown in Fig. 1 [12].

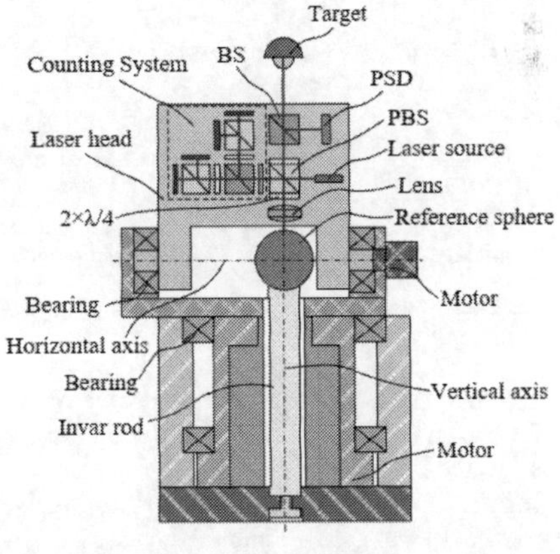

Fig 1. Schematic diagram of laser tracer

In order to clearly identify the characteristics of all the geometrical errors, the geometrical errors are divided into two categories according to the principle of generation - originally geometrical errors and derivatively geometrical errors. The originally geometrical errors include the coaxiality error of horizontal axis (left section and right section), the squareness error of two rotatory axes, the different surface error of two rotatory axes and the cone angle error of vertical axis. The derivatively geometrical errors include the eccentricity error of reference sphere and the collimation error of the laser beam.

2.1 Originally geometrical errors of rotatory axes in Laser Tracer

As shown in Fig. 2, the coordinate system *O*-XYZ is established, with the center of reference sphere *O* as the origin of coordinate system, the left section of horizontal axis as X, and the vertical axis as Z. α represents the angle between the actual axis and the theoretical axis of horizontal axis; L_C represents the offset from the base of laser tracer to horizontal axis; θ represents the rotation angle of the vertical axis; φ represents the rotation angle of the horizontal axis. The geometrical errors' sources of 2D gimbal mount axes can be expressed as follows: β - squareness error of two rotatory axes; γ - cone angle error of vertical axis; e_Z – swing distance of horizontal axis caused by cone angle error of vertical axis; e_T - coaxiality error of horizontal axis; e_Y - different surface distance of two rotatory axes, e_{Y1} - distance between horizontal axis and the center of reference sphere, e_{Y2} - distance between vertical axis and the center of reference sphere.

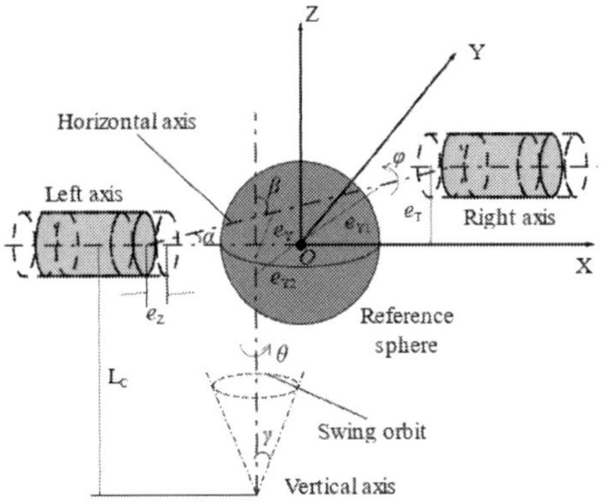

Fig 2. Originally geometrical errors of 2D gimbal mount axes

2.1.1 Coaxiality error of horizontal axis

It is considered that there is only the coaxiality error of horizontal axis - e_T, and the intersection of two rotatory axes is located on the axis Z. When the vertical axis and the horizontal axis are rotated respectively, the measurement errors, Δ_{TV} and Δ_{TH}, caused by the coaxiality error of horizontal axis are shown as (1) and (2):

$$\Delta_{TV} = 0 \tag{1}$$

$$\Delta_{TH} = \frac{e_T \cdot \cos\alpha}{4} \cdot \left[\sin\left(\varphi - \frac{\pi}{2} \right) + 1 \right] \tag{2}$$

2.1.2 Squareness error of two rotatory axes

It is considered that there is only squareness error of two rotatory axes - β, and this error does not produce any eccentricity between the two rotatory axes and the reference sphere. Laser tracer is based on the principle of multi-lateration and does not need the angles of two rotatory axes to determine the location of the target. Therefore, the measurement error, Δ_C, caused by the squareness error of two rotatory axes is zero.

2.1.3 Different surface error of two rotatory axes

It is considered that there is only the different surface error of two rotatory axes - e_Y, and e_{Y1} and e_{Y2} are both located on the axis Y. When the vertical axis and the horizontal axis are rotated respectively, the measurement errors, Δ_{YV} and Δ_{YH}, caused by the different surface error of two rotatory axes are shown as (3) and (4):

$$\Delta_{YV} = \frac{e_{Y2}}{2} \cdot \left[\cos(\theta - \pi) - 1 \right] \tag{3}$$

$$\Delta_{YH} = \frac{e_{Y1}}{2} \cdot \left[\sin\left(\varphi - \frac{\pi}{2} \right) + 1 \right] \tag{4}$$

2.1.4 Cone angle error of vertical axis

It is considered that there is only the cone angle error of the vertical axis - γ, and this error produces a certain eccentricity between the two rotatory axes and the reference sphere, namely $e_Z = L_C \times \tan\gamma$. Under the action of this eccentricity, the laser head is still rotated around the reference sphere at a constant radius. Therefore, measurement errors, Δ_{ZV} and Δ_{ZH}, caused by the cone angle error of vertical axis are both zero, when the vertical axis and the horizontal axis are rotated respectively.

2.2 Derivatively geometrical errors of rotatory axes in Laser Tracer

The analysis above shows that the geometrical errors of 2D gimbal mount axes for laser tracer not only can affect the intersection state of two rotatory axes but also can affect the relationship between two rotatory axes and reference sphere. Therefore, the originally geometrical errors will result in the eccentricity error of reference sphere and the collimation error of laser beam, shown as Fig. 3.

When there are only the eccentricity error of reference sphere and the collimation error of the laser beam in laser tracer, the center of reference sphere does not overlap the origin of coordinate system, and the laser beam does not pass through the center of reference sphere. The coordinate system *O*-XYZ is established, with the horizontal axis as Y, the vertical axis as Z, and the intersection *O* of two rotatory axes as the origin of coordinate system. Point b_1 represents the original position of the target; point s_1 represents the position of laser spot on the surface of reference sphere when the target is at the original position; point b_0 represents the intersection between laser beam and plane YOZ; point b_1' represents the final position of the target after being rotated around vertical axis as θ and horizontal axis as φ with the radius of L; point s_1' represents the position of laser spot on the surface of reference sphere when the target is at the final position; point b_0' represents the final position of b_0 after being rotated around the two rotatory axes. The laser beam reflected by reference sphere does not backtrack into counting system, due to the deviation between the laser beam and the center of reference sphere. Change in the direction of reflected beam results in the change in optical path of measuring beam. Influenced by the refraction of optical elements, it is relatively complex to calculate the optical path difference between measuring beam and reference beam. Therefore, we study the characteristics of the collimation error of the laser beam first.

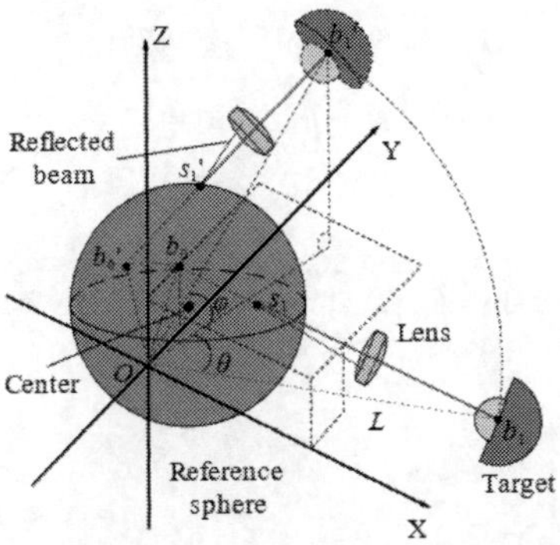

Fig 3. Derivatively geometrical errors of 2D gimbal mount axes

The analysis of the collimation error of the laser beam is shown in Fig. 4. The position of red dashed line represents the theoretical positions of the laser beam and optical elements without the collimation error of the laser beam. The position of blue dashed line represents the theoretical positions of the laser beam and optical elements with the collimation error of the laser beam. The position of purple line represents the actual positions of the laser beam and optical elements when the rotatory axes are controlled by the signals of PSD. l_1 represents the distance between the target and the center of PBS; l_2 represents the distance between the center of lens and the center of PBS; l_3 represents the distance between the center of counting system and the center of PBS; f represents the focal length of lens; r represents the radius of reference sphere; δ represents the collimation error of laser beam; ω represents the angle of incident beam and reflected beam; γ represents the angle of transmitted beam and the optical axis of lens; point P_A to point P_I represent the intersections between laser beam and optical elements.

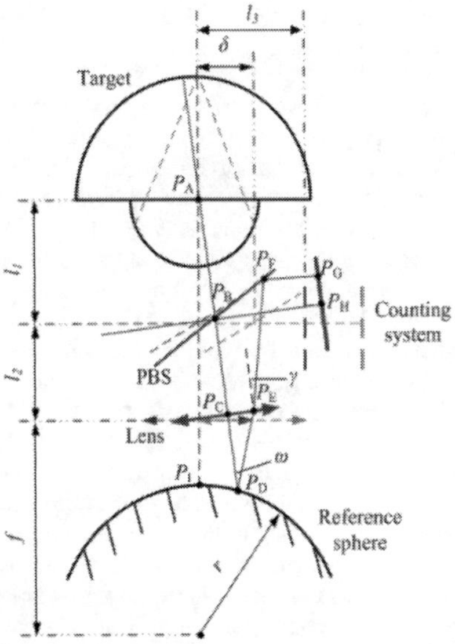

Fig 4. Collimation error of laser beam

When there is the collimation error of the laser beam, the optical path difference between measuring beam and reference beam is shown (see (5)). However, when there is no collimation error of the laser beam, the optical path difference between measuring beam and reference beam is shown (see (6)).

$$\Delta_1 = l_{P_BP_A} + l_{P_AP_D} + l_{P_DP_E} + l_{P_EP_F} + l_{P_FP_G} - l_{P_BP_H} \tag{5}$$

$$\Delta_2 = 2\left(l_1 + l_2 + f - r\right) \tag{6}$$

Therefore, the measurement error of the laser ranging caused by the collimation error of the laser beam is given as (7):

$$\begin{aligned}\Delta_D = {} & 2\sqrt{\left(l_1 + l_2 + f\right)^2 - \delta^2} - \sqrt{r^2 - \delta^2} - \left(l_2 + f\right) + \\ & \frac{f - \sqrt{r^2 - \delta^2}}{\cos\omega} + \frac{l_2 + l_3}{\cos\gamma} - l_3 - 2\left(l_1 + l_2 + f - r\right)\end{aligned} \tag{7}$$

Laser ranging measurement error caused by the collimation error of the laser beam is less than 0.1 μm if the collimation error is within ±10 μm [13-14]. In order to simplify the analysis process, the situation that the incident beam does not overlap the reflected beam can be ignored.

Actually, during the rotation of two rotatory axes, the laser ranging measurement error is coupled nonlinearly by all the geometrical errors of 2D gimbal mount axes in laser tracer. Therefore, it is necessary to make a comprehensive analysis about all the geometrical errors and calculate the laser ranging measurement error in a unified coordinate system, in order to realize error compensation.

3 Integrated model of geometrical errors of axes

Integrated model of geometrical errors of axes is set as Fig. 5. According to the effect of each geometrical error, the coaxiality error of the horizontal axis is equivalent to the eccentricity error of reference sphere and the squareness error of two rotatory axes. The cone angle error of the vertical axis is equivalent to the collimation error of the laser beam. The coordinate system *O*-XYZ is established, with the vertical axis (V) as Z. Axis X is perpendicular to both the vertical axis and horizontal axis (H) simultaneously. O_1 (c_x, c_y, c_z) represents the center of reference sphere, namely the eccentricity error of reference sphere. b_0 $(0, b_y, b_z)$ represents the intersection between the laser beam and plane YOZ, and ζ represents the angle between the laser beam and plane XOY, which both belong to the collimation error of the laser beam. ε represents the angle between the horizontal axis and plane XOY, namely the squareness error of two rotatory axes. e_x represents the distance between the horizontal axis and plane YOZ, namely the different surface error of two rotatory axes. L is the initial distance between the target and the origin of coordinate system *O*-XYZ, namely constant radius. Point b is the projection of point b_1 and is on the same plane with point b_0. R is the distance between the origin O and point b. r represents the radius of reference sphere. Point h_0 represents the initial intersection between the horizontal axis and axis X.

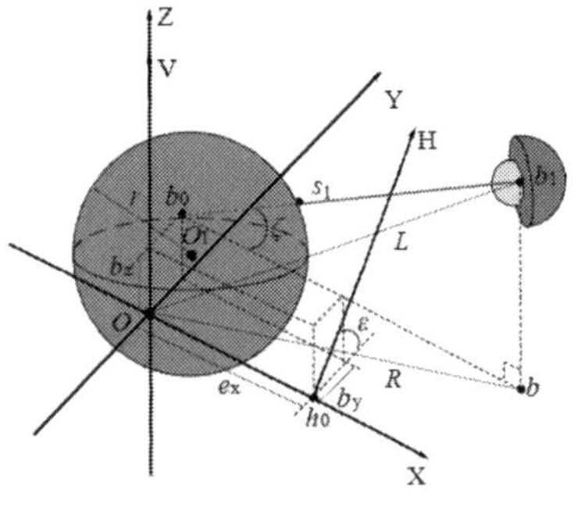

Fig 5. Integrated geometrical error model

The initial intersection h_0 and direction vector $\mathbf{n_{H0}}$ are given as (8) and (9):

$$\mathbf{h_0} = \begin{bmatrix} e_x & 0 & 0 \end{bmatrix}^T \tag{8}$$

$$\mathbf{n_{H0}} = \begin{bmatrix} 1 & 0 & 0 \\ 0 & \cos\varepsilon & -\sin\varepsilon \\ 0 & \sin\varepsilon & \cos\varepsilon \end{bmatrix} \begin{bmatrix} 0 \\ 1 \\ 0 \end{bmatrix} \approx \begin{bmatrix} 0 \\ 1 \\ \varepsilon \end{bmatrix} \tag{9}$$

Point b_0, b_1 and direction vector $\mathbf{n_{B0}}$ are given as (10) - (12):

$$\mathbf{b_0} = \begin{bmatrix} 0 & b_y & b_z \end{bmatrix}^T \tag{10}$$

$$\mathbf{b_1} = \begin{bmatrix} \sqrt{R^2 - b_y^2 - b_z^2} - e_x \\ b_y \\ b_z - \xi\sqrt{R^2 - b_y^2 - b_z^2} \end{bmatrix} \tag{11}$$

$$\mathbf{n_{B0}} = \begin{bmatrix} \cos\xi & 0 & \sin\xi \\ 0 & 1 & 0 \\ -\sin\xi & 0 & \cos\xi \end{bmatrix} \begin{bmatrix} 1 \\ 0 \\ 0 \end{bmatrix} \approx \begin{bmatrix} 1 \\ 0 \\ -\xi \end{bmatrix} \tag{12}$$

Point $b_0(\theta, \varphi)$, $b_1(\theta, \varphi)$ and direction vector $\mathbf{n_B}(\theta, \varphi)$ has changed after rotation, and are given as (13) - (15):

$$\begin{aligned} \mathbf{b_0}(\theta,\varphi) &= \begin{bmatrix} x_{b0} & y_{b0} & z_{b0} \end{bmatrix} = R_z(\theta)\mathbf{h_0} \\ &+ R_z(\theta)R_x(\varepsilon)R_y(\varphi)R_x(-\varepsilon)[\mathbf{b_0} - \mathbf{h_0}] \end{aligned} \tag{13}$$

$$\begin{aligned} \mathbf{b_1}(\theta,\varphi) &= \begin{bmatrix} x_{b1} & y_{b1} & z_{b1} \end{bmatrix} = R_z(\theta)\mathbf{h_0} \\ &+ R_z(\theta)R_x(\varepsilon)R_y(\varphi)R_x(-\varepsilon)[\mathbf{b_1} - \mathbf{h_0}] \end{aligned} \tag{14}$$

$$\begin{aligned} \mathbf{n_B}(\theta,\varphi) &= \begin{bmatrix} m & n & p \end{bmatrix} = \\ & R_z(\theta)R_x(\varepsilon)R_y(\varphi)R_x(-\varepsilon)\mathbf{n_{B0}} \end{aligned} \tag{15}$$

Laser spot $s_1(\theta, \varphi)$=(x_{s1}, y_{s1}, z_{s1}) can be calculated by the equation of spatial line b_0b_1 and the equation of reference sphere, and is given as (16):

$$\begin{cases} \dfrac{x - x_{b0}}{m} = \dfrac{y - y_{b0}}{n} = \dfrac{z - z_{b0}}{p} \\ (x - c_x)^2 + (y - c_y)^2 + (z - c_z)^2 = r^2 \end{cases} \tag{16}$$

The distance of $l_{s1b1}(\theta, \varphi)$ between s_1 and b_1 is given as (17):

$$\begin{aligned} & l_{s1b1}(\theta,\varphi) = \\ & \sqrt{(x_{s1} - x_{b1})^2 + (y_{s1} - y_{b1})^2 + (z_{s1} - z_{b1})^2} \end{aligned} \tag{17}$$

The dead path of laser interferometry system is set as $l_{s1b1}(0, 0)$. Therefore, Δl, the measurement error of the laser tracer, is given as (18):

$$\Delta l = \frac{l_{s1b1}(\theta,\varphi) - l_{s1b1}(0,0)}{2} \tag{18}$$

4 Simulation and analysis

Firstly, all the geometrical errors of 2D gimbal mount axes are measured. Secondly, the measurement error of laser tracer is computed by a unified coordinate system. Finally, the error map of the measuring system for laser tracer is drawn. The integrated geometrical error model is established in *Mathematica*, and the artificial data of all the geometrical errors are shown in table 1.

Table 1. Simulation value for geometrical error.

Types of geometrical error	Linear error μm	Angular error μrad

Eccentricity error of reference sphere	$c_x = 5.00$, $c_y = 5.00$, $c_z = 5.00$	-
Collimation error of laser beam	$b_y = 5\sqrt{2}$, $b_z = 5\sqrt{2}$	$\zeta = -0.10$
Squareness error between two rotatory axes	-	$\varepsilon = 0.10$
Skew lines error between two rotatory axes	$e_x = 15.00$	-

Through simulation and analysis, the measurement error of laser tracer caused by the geometrical errors of 2D gimbal mount axes is obtained, shown as Fig. 6.

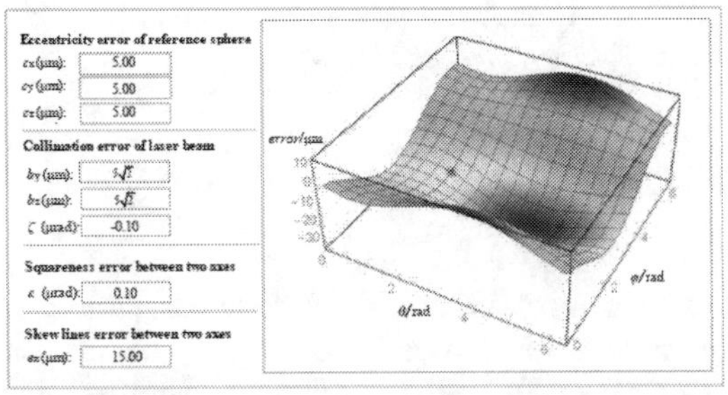

Fig 6. Simulation results for measurement error

Depending on the error map in Fig. 6, the measurement error of laser tracer is the positive maximum +12.64 μm, when θ = 3927.54 μrad and φ = 6059.66 μrad. The measurement error of laser tracer is the negative maximum -32.61 μm, when θ = 3927.54 μrad and φ = 2886.82 μrad. The variation of the measurement error is 45 μm, in which the contributors for the eccentricity error of reference sphere and the different surface error of two rotatory axes are 17.3 μm and 30 μm respectively.

For laser tracer, all the geometrical errors of 2D gimbal mount axes must be measured first, and the error map of the measuring system for laser tracer is uploaded to a computer after being drawn by *Mathematica*. In the process of tracing, the laser ranging measurement value and the compensated data are summed automatically, namely $l+\Delta l$, in which l represents the laser ranging measurement value.

In summary, the different surface error of two rotatory axes and the eccentricity error of reference sphere are major geometrical errors, which need to be strictly controlled. The squareness error of two rotatory axes and the collimation error of the laser beam are minor geometrical errors. Therefore, research on geometrical errors of 2D gimbal mount axes in laser tracer is conducive to improve the measurement accuracy of laser tracer and reduce the processing cost.

5 Conclusion

The geometrical errors of 2D gimbal mount axes in laser tracer are studied, and the nonlinear coupling relationship between these geometrical errors and the measurement accuracy is determined. The originally geometrical errors not only can affect the intersection state of two rotatory axes but also can affect the relationship between two rotatory axes and reference sphere. These geometrical errors above will result in the eccentricity error of reference sphere and the collimation error of the laser beam. The integrated geometrical error model is established, and the error map is drawn depending on the relation of spatial geometric transformation. The simulation results demonstrate that the measurement error of laser tracer is a two-variable function of rotation angles of two rotatory axes. This function will not have anything to do with the distance to the target. In other words, the measurement error owns the characteristic of certainty and uniqueness.

References

1. H. Schwenke, M. Franke, J. Hannaford, H. Kunzmann, CIRP ANN-MANUF TECHN, **54**, 475-478 (2005)
2. H. Schwenke, R. Schmitt, P. Jatzkowski, C. Warmann, CIRP ANN-MANUF TECHN, **58**, 477-480 (2008)
3. H.W. Lee, J.R. Chen, S.P. Pan, H.C Liou, P.E. Hsu, APPL SCI, **6**, 105 (2016)
4. F. Deumlich, *Surveying instruments*, 126-146 (1982)
5. K. Lau, R.J. Hocken, W.C. Haight, PRECIS ENG, **8**, 3–8 (1986)
6. B. Muralikrishnan, D. Sawyer, C. Blackburn, S. Phillips, B. Borchardt, W.T. Estler, J RES NATL INST STAN, **114**, 21-35 (2009)
7. B. Muralikrishnan, S. Phillips, D. Sawyer, PRECIS ENG, **44**, 13-28 (2016)
8. B. Muralikrishnan, C. Blackburn, D. Sawyer, S. Phillips, J RES NATL INST STAN, **115**, 291-301 (2010)
9. B. Muralikrishnan, M. Shilling, P. Rachakonda, W. Ren, V. Lee, D. Sawyer, J MANUF SYST, **37**, 550-557 (2015)
10. B. Hughes, A.B. Forbes, A. Lewis, W.J Sun, MEAS SCI TECHNOL, **22**, 045103 (2011)
11. B. Hughes, A.B. Forbes, A. Lewis, W.J Sun, Macroscale, (2011)
12. Z.Y. Shi, H.X. Song, H.F. Chen, Y.Q. Sun, MEAS SCI REV, **18**, 13-19 (2018)
13. H.X. Song, Z.Y. Shi, H.F. Chen, Y.Q. Sun, *Proc. of SPIE*, **10621**,0C1-0C8 (2018)
14. H.X. Song, Z.Y. Shi, H.F. Chen, Y.Q. Sun, Chinese Journal of Lasers, **45**, 0504001 (2018)

A Fuzzy Adaptive Control of Doubly Fed Induction Machine (DFIM)

*A.*Saidi, *Lamia* Youb, *Farid* Naceri, *Sebti* Belkacem

Uinversite Batna-2, Departement of Electrical Engineering, Batna, Algeria

Abstract. In this paper, we are interested in the adaptive fuzzy control a technique has been studied and applied, namely adaptive fuzzy control based on theory of Lyapunov. The system based on the stability theory is used to approximate the gains Ke and kdce to ensure the stability of the control in real time .the simulations results obtained by using Matlab environment gives that the fuzzy adaptive control more robust, also it has superior dynamics performances. The results and test of robustness will be presented.

1 Introduction

The asynchronous machine with double feed (DFIM) is very popular since it profit from certain advantages compared to all the other types at variable speed, sound use in the chains of electromechanical conversion as an aero generator or engine knew a spectacular growth during last years. Indeed, it converter of energy used in order to rectify-undulate the alternating currents of the rotor has a fractional nominal output nominal of that of the generator, which reduces its cost by report/ratio with concurrent topologies(Saidi and Naceri, 2016).

The DFIM is essentially non linear, due to the coupling between the flux and the electromagnetic torque. The vector control or field orientation control allows a decoupling between the torque and the flux (Payam, et al., 2010; Chafaa, et al., 2014).

The system of fuzzy logic aims to model human reasoning and thinking ,the process with linguistic variables.They are very useful when the process to control with some uncertainties or unknown variables(Sundarapandian, et al., 2016; Souli, et al., 2015; Rabie, et al., 2015;Calcev, et al., 2014).

However, in order to maintain consistent execution in the presence of true uncertainties, the use of adaptive control is,in most cases, inevitable.Fuzzy adaptive control has been the subject of intensive research during the last decade (Minth, et al., 2009; Annis, et al., 2004; Chen, et al., 1999;Chih, et al., 2004).

Therefore, the combination of the study of adaptability and uncertainty has allowed researchers to derive adaptive fuzzy controllers (Chekkouri, et al., 2003; Rabie, et al., 2015).

Several adaptive fuzzy control schemes have been proposed for complex systems In these adaptive schemes,the study of stability is based on the Lyapunov approach(Fischle, et al., 1999; Yusupbekov, et al., 2016). An adaptive fuzzy control approach can be used: the direct approach is based on the adaptation mechanism.In the direct approach,the adaptation system is used to estimate the gains of the PI-fuzzy and the control law is obtained from this PI-fuzzy controller.

Despite their dynamic capabilities and their practical implementation, the first PI-fuzzy suffered from lack of stability analysis, That is to say the stability of the closed loop is not guaranteed (Sarala, et al., 2016;Furuhashi, et al., 1998; Faquir, et al., 2016). Recently, a large class of adaptive fuzzy systems, which use the fuzzy systems of Mamdani or TS was developed to control non-linear processes, and stability has been guaranteed in the sense of the Lyapunov theory.

Treatment in this paper adaptive fuzzy control based on stability theory to recalculate real-time PI-fuzzy gains.we will develop a new order strategy which is: adaptive fuzzy control based on the theory of Lyapunov.

2 Adaptive fuzzy control based on Lyapunov theory

Today, there are a number of studies on the stability of fuzzy systems. However,restrictive because of nonexistence tools appropriate. To do this, the fuzzy controller being non-linear,We have to do use of non-linear methods . such as the Lyapunov method, the theory of hyper stability or the criterion of Popov.Most of these methods are quite limiting,providing only sufficient conditions for stability in a domain restricted.

For adaptive fuzzy control based on the Lyapunov theory,we follow two steps. Lapremière is devoted to the design of a PI-fuzzy controller. The second is to define the methodology for determining the gains of a fuzzy regulator based on the Lyapunov theory. The control applied to the Dual Power Asynchronous Machine is provided by a PI-fuzzy controller.

We proposed a new adaptive fuzzy control strategy based on the theory of Lyapunov to determine the gains of K_e and K_{dce} (normalization) , which can be applied for a large class of non-linear systems. It combines the advantages of two robust techniques and which are the control by the fuzzy logic and the control adaptive(Laamayad, et al., 2012; Chen, et al., 1999).

2.1 Control law of a PI-fuzzy controller

We consider a nonlinear system whose dynamic equation of the command is described in the following form:

$$T^*_{em}(t) = T^*_{em}(t-1) + K_{dce} dT_{nem} \tag{1}$$

2.2 Fuzzification

The error between the reference signal and the process serves to act directly (the gains of the mad regulator are recalculated in real time as a function of this error and the response of the regulator itself and gains and coefficients of adaptation).The speed error denoted e (t) is defined by:

$$e(t) = \omega_{ref} - \omega_r(t) \tag{2}$$

The derivative of the speed error noted $\frac{de(t)}{dt}$ is approximated by:

$$\frac{de(t)}{dt} = \dot{e}(t) \approx \frac{de(t+1) - e(t)}{T_e} \tag{3}$$

T_e : Being the sampling period.

The error and the derivative of the error are adapted as follows:

$$e(t) = K_e e_n(t) \tag{4}$$

$$\frac{de(t)}{dt} = K_{de} \frac{de_n(t)}{dt} \tag{5}$$

2.3 Inference table

Once the fuzzy presentation stage is complete,the gains estimated by the adjustment mechanism which uses the Lyapunov theory are sent to the fuzzy regulator to construct the output dT_{nem} (t) Therefore, the fuzzy regulator rule takes the form next :

Table 1. Inference matrix of fuzzy rules.

dT_{nem}		e_n				
		NB	NM	Z	PM	PB
$\frac{de_n}{dt}$	NB	NB	NB	NM	NVS	Z
	NM	NB	NM	NVS	Z	PS
	Z	NM	NVS	Z	PS	PM
	PM	NVS	Z	FS	PM	PB
	PB	Z	PS	PM	PB	PB

2.4 Deffuzification

For defuzzification, the method of the center of gravity presented above, we obtain:

$$dT_{nem}(t) = \frac{\sum_{j=1}^{25} = 1\mu_{A1J}(e_n(t))\mu_{A2J}(\frac{de_n(t)}{dt})C_J S_J}{\sum_{j=1}^{25} \mu_{A1J}(e_n(t))\mu_{A2J}(\frac{de_n(t)}{dt})C_J S_J} \tag{6}$$

3. Schematic diagram

To minimize the instantaneous error between the actual speed of the machine DFIM and that of the reference speed we recalculate in real time the gains k_e and k_{dce} of the premises and the consequences by the application of an adaptation algorithm of study type stability of control in the sense of Lyapunov (Laamayad, et al., 2012; Chen, et al., 1999).

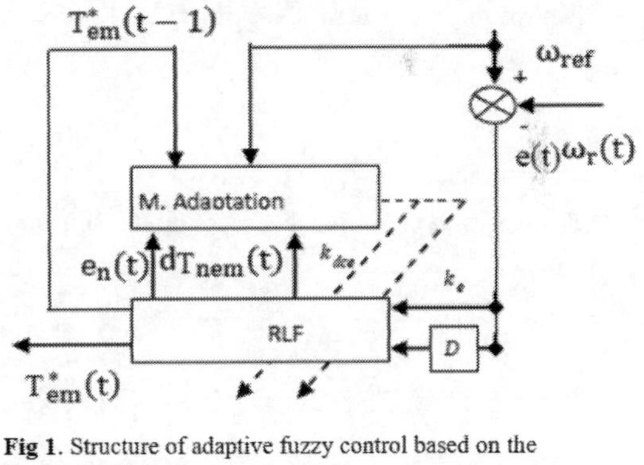

Fig 1. Structure of adaptive fuzzy control based on the Lyapunov theory.

4. Analysis of the stability of the proposed control

4.1 Function of Lyapunov candidate

Function of Lyapunov can be considered as a function of error and the variation of the gains. This is a decreasing function along the trajectories of the system. We consider a Lyapunov candidate function:

$$V = \frac{1}{2}(e^2 + \frac{1}{\gamma_1} k_e^2 + \frac{1}{\gamma_2} k_{dce}^2) \tag{7}$$

$\gamma_1 \rangle 0$ and $\gamma_2 \rangle 0$

The derivative of the Lyapunov function with respect to real time:

$$\frac{dV}{dt} = e\frac{de}{dt} + \frac{k_e}{\gamma_1}\frac{dk_e}{dt} + \frac{k_{dce}}{\gamma_2}\frac{dk_{dce}}{dt} \quad (8)$$

The following expression can be obtained:

$$\frac{dV}{dt} = e(a_p\omega_r(t) - b_pT^*_{em}(t-1) - b_pk_{dce}dT_{nem}(t) + d_pT_r) + \frac{k_e}{\gamma_1}\frac{dk_e}{dt} + \frac{k_{dce}}{\gamma_2}\frac{dk_{dce}}{dt} \quad (9)$$

We note that we can replace the necessary expression of error e (t) by the form next error $e(t) = k_e e_n(t)$ in the derivative of the Lyapunov function.

$$A = |(a_p\omega_{ref} - b_pT^*_{em}(t-1)| \quad (10)$$

The gains of a fuzzy regulator are to be defined from a single necessary and sufficient condition for existence of the stability regime in the sense from Lyapunov.

$$\frac{dV}{dt}\langle 0 \quad (11)$$

The inequality is verified by imposing the following adaptation law:

$$\frac{dk_e}{dt} = -\gamma_1 e_n A \quad (12)$$

$$\frac{dk_{dce}}{dt} = \gamma_2 b_p k_e e_n dT_{nem} \quad (13)$$

The form of the Lyapunov function is proposed to determine the PI-fuzzy gains which ensure a convergence of the actual speed of the DFIM towards its reference value. Adaptation gains are recalculated in real time in the purpose of imposing desired stability on the controlled system that is to say the speed of adjustment at the level of the assembly

PI-fuzzy controller and the DFIM machine in the transitional regime. So , for the speed loop, the parameters of the Lyapunov function are chosen to ensure: the derivative of the Lyapunov candidate always negative (eq14), as well as an accuracy of speed control.

5. Application for doubly fed induction machine

For the adjustment of the speed of a DFIM by the adaptive fuzzy control based on the theory of lyapunov, the following method is followed:

From the reference speed and the actual speed, the adaptive fuzzy controller provides

the torque $T^*_{em}(t)$. A transformation of park (dq-abc) makes it possible to calculate the currents stator of references. These currents are compared with real currents to order of each inverter.

To adjust the speed, use the output of the fuzzy controller T^*_{em} , as well as the adjustment of gains to ensure stability of control. In our application, we used a PI-Fuzzy. The first concerns speed, where we assigned five membership functions to each entry of the model. Figure (2) illustrates the speed control structure of a doubly fed induction machine by applying the method of fuzzy adaptive control based on theory from Lyapunov.

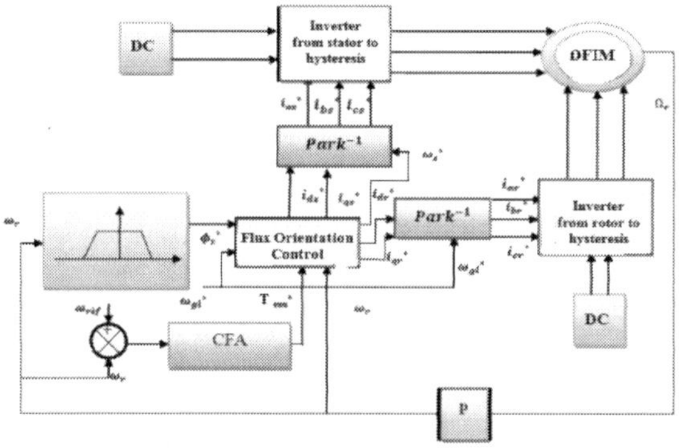

Fig 2. Structure of adaptive fuzzy control based on the Lyapunov theory

6. Simulation results

It appears, following the various results obtained, that the performance of speed control by the application of adaptive fuzzy control with the stability of control are very satisfactory. The coefficients γ_1 , γ_2 have respectively 0.005 and 70 respectively. So the initial values gains k_e , k_{dce} are respectively 7 and 5.8, To demonstrate the adaptability of the proposed control scheme to parametric variations, we introduce parametric variations on the resistance rotor, as well as the moment of inertia at the instant t = 1s . Resistance and moment of inertia are increased by 150%, and then the resistance and moment of inertia are decreased of 50%. The responses obtained are shown in Figure (4), (5), (6) . It is clear that this parametric variation did not affect the speed control performance, which proves the efficiency of the control algorithm used.

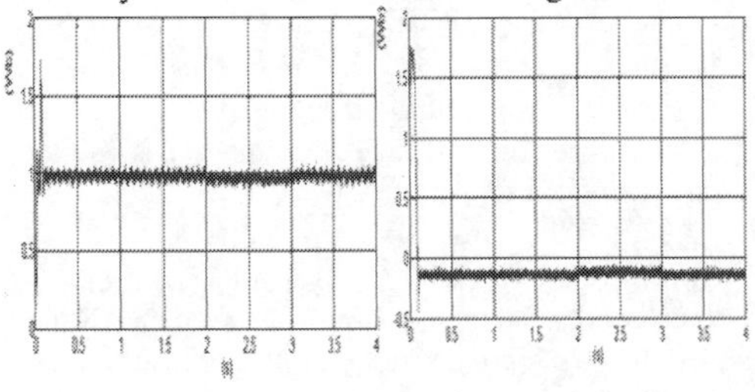

Fig (3.a). Components of direct stator flux φ_{ds} and in quadrature

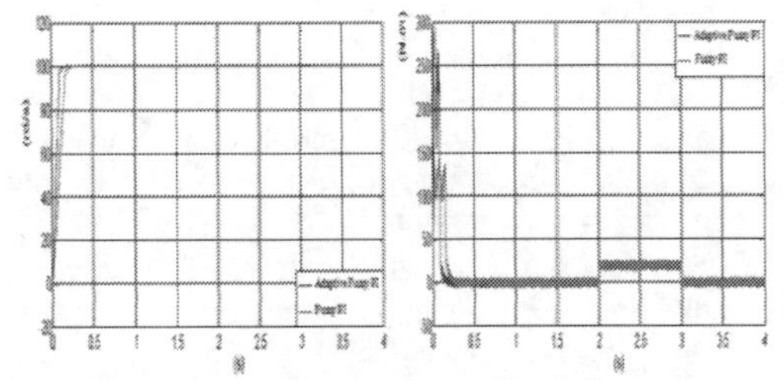

Fig (3.b). Dynamic response of speed and electromagnetic torque.

Figures (3.a) and (3.b), (3.c), show the performance of the speed control of the double-feed asynchronous machine for a set point of 100 rad / s with a variation of the load. The speed follows the reference speed despite the presence of charge. The response time is in the order of 0.1s. The overrun is zero

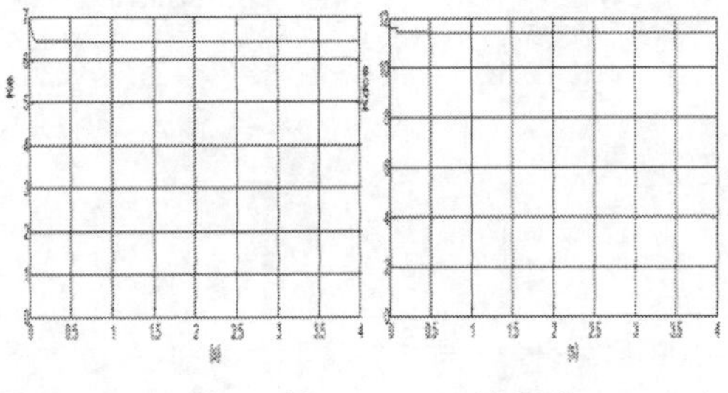

Fig (3.c). Dynamic machine response with PI-fuzzy control during the introduction of a load torque of (T_l= 20 N.m) at time t = 2 (s).

6.1 Tests of Robustness

The test consists of a variation of the rotor resistance in a range of 50% and 150%. The speed and the electromagnetic torque, the stator fluxes direct and quadrature as well as the evolution of the fuzzy gains

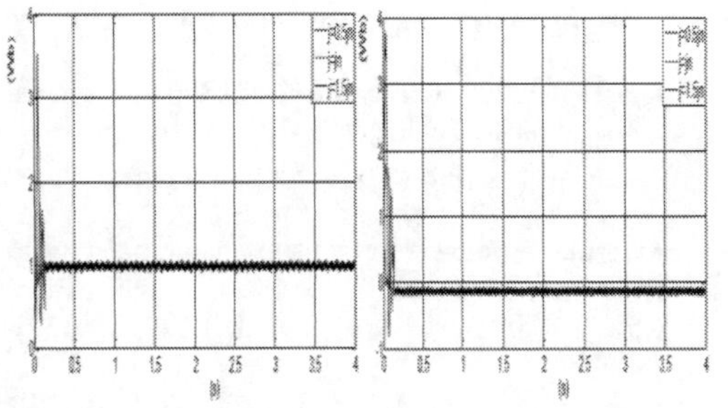

Fig (4). Components of direct stator flux φ_{ds} and in quadrature φ_{qs}

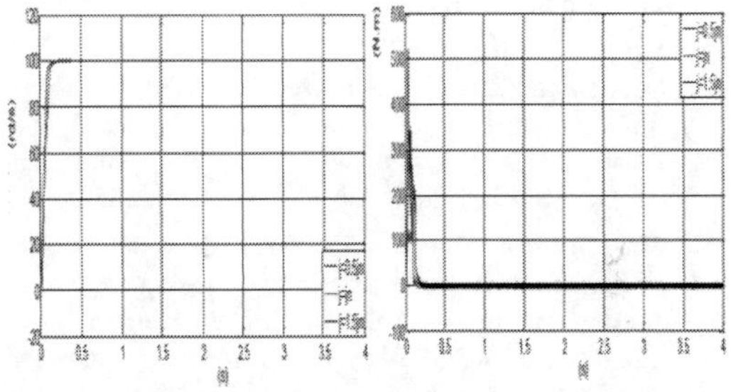

Fig (5). Dynamic response of speed and electromagnetic torque

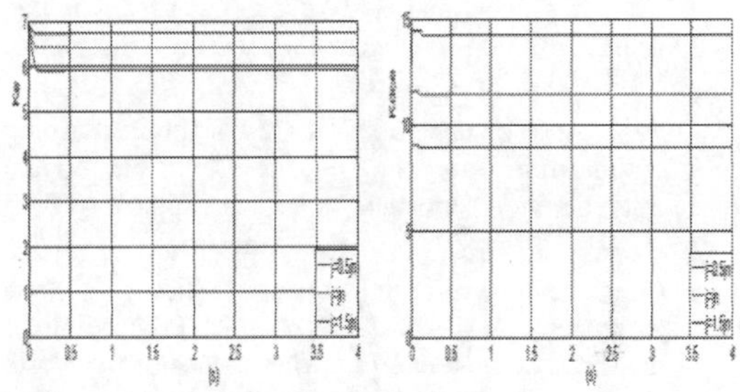

Figure (6). Evolution of the fuzzy gains (k_e , k_{dce})

7. Conclusion

The obtained results as well as the various tests have shown that this technique good performance in the presence of load disturbances and reversing the direction of travel, the speed follows its reference, the stator flux follows its reference value along the axis d with a zero component along the axis q. An increase in the electromagnetic torque in the presence of the resistive torque which leads to a good acceleration of the motor.

PI-fuzzy control with gain adaptation ke and kdce by the theory of Lyapunov is also tested for parametric variations of the system. The results have enabled us to judge that this new ordering strategy strong robustness in the presence of variations.

References

1. Saidi A, Naceri F. (2016), ' Comparative study between Sliding mode controller and Fuzzy Sliding

mode controller in a speed control for doubly fed induction motor', Proceedings of IEEE International Conference on Control Engineering & Information Technology (CEIT) Tunisia , pp. 1-6.

2. Payam F, Kashiha H. (2010), 'Robust Speed Sensorless Control of Doubly-fed Induction Machine Based on Input-output Feedback Linearization Control Using a Sliding-mode Observer ', World Appl. Sci. J, 10(11): pp1392-1400
3. Chafaa Kh , Salah Khireddine M , Slimane N , Ghanai M (2014), 'Fuzzy estimation based on type-2 fuzzy logic for adaptive Control', Proceedings of IEEE International Conference on Computer Applications & Research (WSCAR), Sousse, Tunisia.
4. Sundarapandian V, Taher Azar A. (2016), ' Takagi-Sugeno fuzzy logic controller for Liu-Chen four-scroll chaotic system', Int. J. of Intelligent Engineering Informatics, Vol.4, No.2, pp.135 – 150 .
5. H.F. Ho, K.W.E. Cheng. (2009) , 'Position Control of Induction Motor using Indirect Adaptive Fuzzy Sliding Mode Control', IEEE 3rd International Conference on Power Electronics Systems and Applications, pp. 1 - 5 .
6. Souli R , bali J,Minyar Sassi Hidri. (2015) , ' A cloud-based optimal fuzzy clustering of distributed data', Int. J. of Intelligent Engineering Informatics, Vol.3, No.1, pp.42 – 56.
7. Rabie A. Ramadan. (2017) , ' Fuzzy brain storming optimisation algorithm', Int. J. of Intelligent Engineering Informatics, Vol.5, No.1, pp.67 – 79
8. Calcev G ,Greco C , Bona B (1994), ' Fuzzy Controller Synthesis Using Sliding Mode Approach, 'NAFIPS/IFIS/NASA'94 Proceeding of the First International Joint Conference of the North American Fuzzy Information Processing Society Biannual Conference. The Industrial Fuzzy Control and Intelligent Systems Conference, and the NASA Joint Technology , IEEE,pp.255-258 ,18-21 Dec .
9. Minth. TA CAO. (1997) , 'Commande numérique de machines asynchrones par logique floue', thèse de Philosophie doctorat faculté des études supérieure de l'université Laval .
10. Annis Y, Karnavas L, Demetrios P. Papadopoulos .(2004), ' Excitation Control of a Synchronous Machine Using Polynomial Neural Networks ', Journal of Electrical Engineering, Vol. 55, No. 7-8, 169-179.
11. Chen B, Yingbi Zhang, J.Yen, Wei Zhao . (1999), 'Fuzzy Adaptive Connection Admission Control for Real-time Applications In ATM-Based Heterogeneous Networks', Journal of Intelligent and Fuzzy Systems 7 ,125-135.
12. Chih L -Min and Chun-Fei H. (2004) , 'Adaptive Fuzzy Sliding-Mode Control for Induction Servomotor Systems', IEEE Transactions On Energy Conversion, Vol. 19, No. 2, pp.362-368.
13. Chekkouri M, Catalài J Lopez, Rubira E, Martinez L (2003), ' Fuzzy Adaptive Control of an Induction Motor Drive' , Automatika 44, 113-122.
14. Rabie A. Ramadan, Basant R. El Samadony; Nevin M. Darwish . (2015), ' Fuzzy-based centroid localisation in WSNs', Int. J. of Intelligent Engineering Informatics, 2015 Vol.3, No.4, pp.293 – 312.
15. Fischle K and Schroder D 'An Improved Stable Adaptive Fuzzy Control Method'(1999), IEEE Trans on Fuzzy Systems,Vol.7,No.1,27
16. Yusupbekov NR , Marakhimov AR , Igamberdiev HZ , Umarov ShX . (2016), ' An Adaptive Fuzzy-Logic Traffic Control System in Conditions of Saturated Transport Stream', The Scientific World Journal.

On the Analysis Performance of Updating Weight for Estimation Target of Drone System

Kwan Hyeong Lee[1] and *Jae* Hoon Lee[2]

[1]Devisino of Human IT Convergence, 1007 University of Daejin, South Korea
[2]Devision of Human-Architectural Engineering, 1007 University of Daejin, South Korea

Abstract. In this paper, we propose the method which desired signal is estimated by updating the weight of the MVDR algorithm. The MUSIC algorithm is generally a lot of used in the direction of arrival estimation method. The MUSIC algorithm has a good resolution because of using subspace techniques consisting of a signal subspace and a noise subspace. The processor capability of drone system is required low power consumption and low computation complexity because it uses a microprocessor. If the drone system has a lot of computation complexity, the desired signal cannot be estimated. This paper study a method estimating the desired signal with a simple calculation. The proposed method is updated weight by the covariance matrix of MVDR algorithm. Through simulation, we analyse performance by comparing MVDR, MUSIC and the proposed method. In the simulation results, the proposed method is the same as the MUSIC algorithm in direction of arrival estimation. Since the proposed method has no subspace, it reduces computational complexity than MUSIC algorithm. The desired signal estimation of the proposed method is superior to the MVDR algorithm.

1 Introduction

Recently, the estimation of direction of arrival(DoA) for target has been many studied according to the development of wireless communication technique. Direction of estimation method have been widely used in many applications such as radar, sonar, biomedical, and communication systems. DoA estimation methods are Bartlett, Capon, Linear predict, MUSIC, and ESPRIT [1-3]. Also, the estimation method of target is divided into two methods. First, there is non-parameter method such as Bartlett and Capon, finally, there is parameter method such as MUSIC and ESPRIT. MUSIC estimation method has a super resolution because it uses the subspace method. There has a lot of computation complexity because the MUSIC method uses eigenvalue decomposition. In order to improve the DoA estimation, there are improved signal to noise ratio, higher transmission power, and adaptive array signal processing [4-6].

In this paper, we propose the method which is a low computation and an accurate DoA estimation. The proposed method is estimating the desired signal by updating the weight of the MVDR (Minimum Variance Distortionless Response) algorithm. Generally, the MVDR algorithm is called Capon method. The Capon method has poor resolution to estimate the desired signal due to low computation complexity and inaccurate weight. We propose a method how to improve the weight of MVDR algorithm. The proposed method for improving the weight use an adaptive array antenna and a beamforming technique. The adaptive array method finds the covariance matrix using Lagrange multiplier and applies the adaptive array algorithm to improve the resolution. And, we divided it into two step to find optimum weight. The source signal covariance matrix is obtained in two steps of the proposed method. As a result, we can obtain the covariance matrix of the received signals. The MUSIC method is mostly used to estimate for target in spatial. But, it is not effective for drone system to use the MUSIC method because of the processing capability of microprocessor. The processing capability of drone system using microprocessor degrades form computation complexity and power consumption, so that it cannot accurately estimate the desired signal.

The organization of this paper is as follow. In section 2, the signal mode analysis considered is described. The output power of MVDR algorithm and the proposed weight of covariance matrix are presented in Section 3. A performance analysis of the proposed method is provided in Section 4. Conclusions are drawn in Section 5.

2 Signal model analysis

Figure 1 shows an adaptive array system. We consider that the receiver is uniform linear array composed of an M-arrays antenna with adjacent array element spacing d deployed at the figure 1 and N narrowband signals. The N-dimensional column vector $a(\theta_K)$, the antenna array response vector is as follow[7-8]

$$a(\theta_k) = [1, a(\theta_1), \cdots, a(\theta_K)]^T \quad (1)$$

Where $a(\theta_k) = \exp(-\frac{j2\pi d\cos\theta_k}{\lambda})$. Bothd λ and k are the the weavelenght and incident signal on array antenna, respectively. Then, output signal of the array antenna is as follow

$$X(t)=A\,s(t) + N(t) \quad (2)$$

$X(t)$: the incident signal of the array antenna.
$s(t)$: the source signal of the array antenna.
$N(t)$: the noise signal of the array antenna.
$A = [a(\theta_1), a(\theta_2), \cdots, a(\theta_K)]$: steering matrix

Where $a(\theta)$ is steering vector corresponding to the direction of arrival of kth signal. The covariance matrix of the received signal vector on the array antenna can be written as follow

$$R = E[X(t)\,X^H(t)] = \frac{1}{L}\sum_{k=1}^{N} X(t)\,X^H(t) \quad (3)$$

Where L is a number of snapshot.

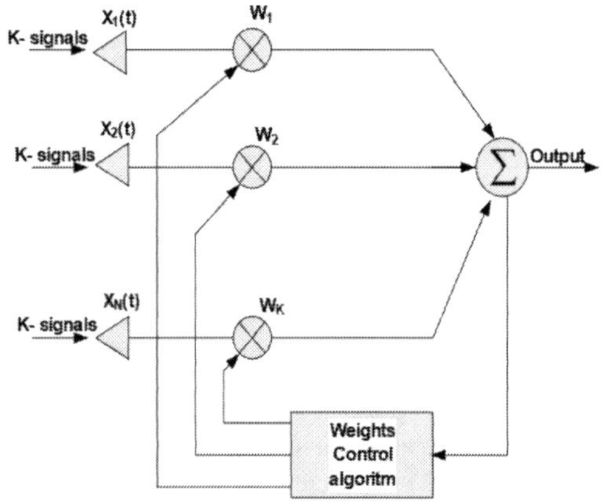

Fig 1. Block diagram of the adaptive array system.

3 Covariance matrix of estimation signal

3.1. Output spectrum of MVDR Algorithm

In this chapter, we discuss the MVDR algorithm for estimating the direction of arrival. Output signal of incident wave at receiver can be written as follow[9-11]

$$X(t)=x(t) + N(t) \quad (4)$$

where, $x(t) = s(t)a(\theta)$. $a(\theta)$ and $s(t)$ is called the array steering vector and the source signals on the array antenna, respectively. $N(t)$ is a zero mean complex Gaussian random signal($\sigma^2 I$). σ and I are a variance and unit matrix, respectively. Output signal is represented by multiplier of weight(W) and the receiver signal on array element. When the weight vector is 1 x N, output signal to minimize the variance of $Y(t)$ in the noise can be written as follow

$$Y_n = W^H X(t) \quad (5)$$

In order to minimize output noise $E[|Y_n|^2]$, the constraint of no distortion can be written as follow

$$Y_n = W(AR_sA^H + \sigma^2 I)\,W^H \quad (6)$$

Where $R_s = E[s(t)s^H(t)]$ is a source signal correlation matrix. The constraint condition is as follow

$$W^H a(\theta) = 1 \quad (7)$$

The output noise power is can be as follow

$$E[|Y_n|^2] = W^H\,E[X(t)X^H(t)]W \quad (8)$$

Where $()^H$ is Hermit matrix. We would like to minimize the output noise power in subject to the constraint at the equation (7). The output power is as follow

$$E[Y(t)] = E[X(t)] + E[N(t)] = E[s(t)] \quad (9)$$

Let us assume that

$$E[N(t)N^*(t)] = \begin{bmatrix} \sigma_1 & 0 & \cdots & 0 \\ 0 & \sigma_N & \cdots & 0 \\ \vdots & \vdots & \ddots & \vdots \\ 0 & \cdots & \cdots & \sigma_N \end{bmatrix} \quad (10)$$

$$R = E[X(t)\,X^*(t)] = \sum_{k=1}^{N} s_k a_k a_k^* + \begin{bmatrix} \sigma_1 & 0 & \cdots & 0 \\ 0 & \sigma_N & \cdots & 0 \\ \vdots & \vdots & \ddots & \vdots \\ 0 & \cdots & \cdots & \sigma_N \end{bmatrix} \quad (11)$$

$$= [a_1, \cdots, a_N] \begin{bmatrix} s_1 & 0 & 0 & \cdots & \cdots & 0 \\ 0 & \ddots & \cdots & 0 & 0 & 0 \\ 0 & 0 & s_N & \cdots & \cdots & 0 \\ 0 & \vdots & \vdots & \sigma_1 & \vdots & 0 \\ 0 & 0 & \cdots & \cdots & \ddots & 0 \\ 0 & 0 & \cdots & 0 & \cdots & \sigma_N \end{bmatrix} \begin{bmatrix} a_1^* \\ a_2^* \\ \vdots \\ a_N^* \end{bmatrix} \quad (12)$$

We find it the solution minimizing the variance of the output signal by using a Lagrange multiplier. The solution can be as follow

$$S \triangleq W^H(t) R\,W + \lambda(t)\,[W^H a(\theta) - 1] + \lambda^*(t)\,[a(\theta)W - 1] \quad (13)$$

Taking the gradient with respect to W^H in equation(14) as follow

$$W_{op}^H = -\lambda(t)\,a^H(\Theta)\,R^{-1} \quad (14)$$

Substituting equation(13) into equation(7) to find $\lambda(t)$, which give by

$$\lambda(t) = -[a^H(\theta)R^{-1}(t)a(\theta)]^{-1} \quad (15)$$

Thus, optimum weight can be written as follow

$$W_{op}^H(t) = B(\theta)a^H(\theta)R^{-1} \quad (16)$$

Where

$$B(\theta) = [a^H(\theta)R^{-1}\,a(\theta)]^{-1} \quad (17)$$

The equeation(16) is called the MVDR algorithm. Array output power is as follow

$$P_{MVDR} = \frac{1}{a^H(\theta)\ R^{-1}\ a(\theta)} \tag{18}$$

3.2 Proposed covariance matrix in mutual coupling

We consider that all signals of the receiving on array antennas are coherent. Source signal of each array antenna are amplitude and phase delayed due to multipath. First array antenna is reference signal. In the case of s_k $(k = 1,2,\cdots,K)$narrowband sources, Replicas of the first array antenna source signal can be as follow

$$s_k(t) = h\, s_1(t) \tag{19}$$

Where h represents the complex attenuation of the kth signal with respect to the first signal. The signal correlation matrix is as follow

$$R_s = H\, H^H \tag{20}$$

Where $H= [h_1, h_2, \cdots, h_K]^T$. We have to remove the effects of mutual coupling before estimating desired signals because it is impossible to estimate desired direction of arrival signals. We have the following steps:

Step 1. eigenvalue decomposition of R_s can be written as follow

$$R_s = u_s \lambda_s\, u_s^H + u_c\, \lambda_c\, u_c^H \tag{21}$$

Where u is $M \times 1$ eigenvector corresponding to the largest eigenvalue, and u is eigenvector corresponding to the smallest eigenvalue by $M-1$. According to subspace method, we have as follow

$$A \cdot H = b\, C^{-1}\, u_s \ \perp span\{\, C^{-1}\, u_c\} \tag{22}$$

Where b and C are constant and mutual coupling matrix, respectivly. Mutual coupling matrix is as follow

$$C = \begin{bmatrix} 1 & c_1 & \cdots & c_m & \cdots & 0 \\ c_1 & 1 & c_1 & \cdots & \ddots & 0 \\ \vdots & c_1 & 1 & \ddots & \cdots & c_m \\ c_m & \cdots & \ddots & \ddots & c_1 & \vdots \\ 0 & \ddots & \cdots & \ddots & 1 & c_1 \\ 0 & 0 & c_m & \cdots & c_1 & 1 \end{bmatrix} \tag{23}$$

Step 2. Reconstruct the covariance matrix of the array output is as follow

$$R_s = b\, C^{-1}\, u_s\, u_s^H\, C^H\, b^H \tag{24}$$

Spatial smoothing is represented from subarray method which the nature array is divided uniform overlapping subarray.

4 Simulation & performance analysis

In this chapter, we analyse the performance to compare classical direction of arrival method with the proposed method. The classical method used which is MVDR and MUSCI. With the simulation condition, Snapshot, SNR, and targets were 100times, 10dB, and objects 2, respectively. Figure 2 shows the angle estimated by the MVDR algorithm using 9 array elements, and the desired signal estimate at (-20°, 20°). In figure 2, the desired signal of two targets is estimated accurately at (-20°, 20°). Figure 3 shows the angle estimated by the MVDR algorithm with 6 array elements, and the desired signal estimate at (-5°, 5°) as two targets. In figure 3, the desired signals could not be estimated accurately at (-5°, 5°). The desired signal estimation in figure 3 represent one signal at (0°). Figure 4 shows the angle estimated by the MUSIC algorithm with 6 array elements, and the desired signal estimate at (-5°, 5°). In figure 4, the desired signals of two targets could were estimated accurately at (-5°, 5°). Figure 5 shows the angle estimated by the proposed method with 6 array elements, and the desired signal estimate at (-5°, 5°). In figure 5, the desired signals of two targets were estimated accurately at (-5°, 5°).

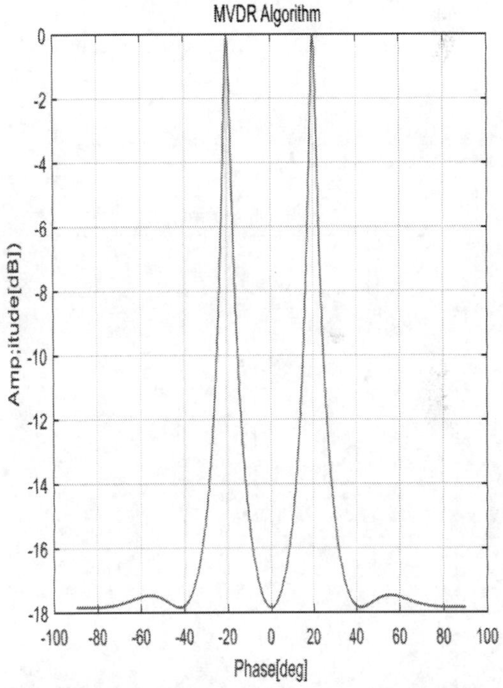

Fig 2. Angle of estimation (-20°, 20°) of the MVDR algorithm SNR=10dB, array element=9

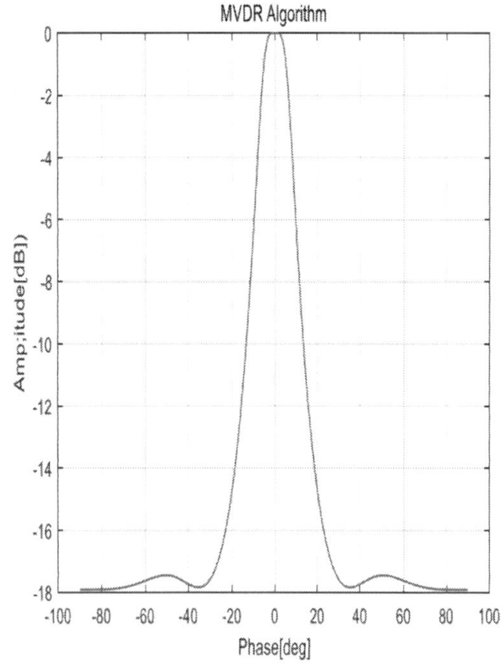

Fig 3. Angle of estimation (-5°, 5°) of the MVDR algorithm, SNR=10dB, array element=6

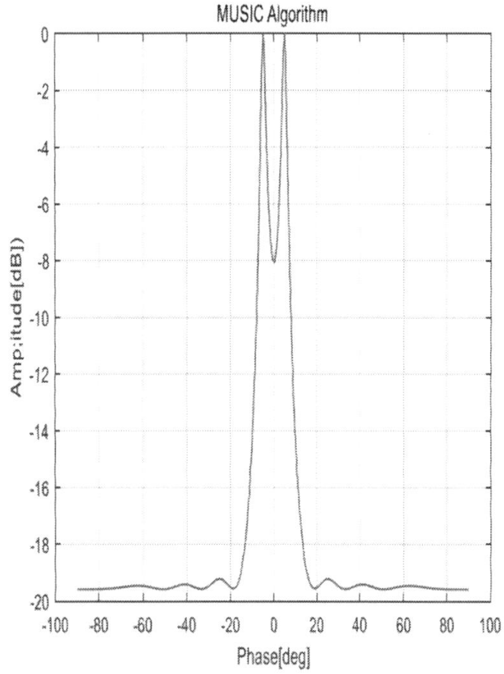

Fig 4. Angle of estimation (-5°, 5°) of the MUSIC algorithm, SNR=10dB, array element=6

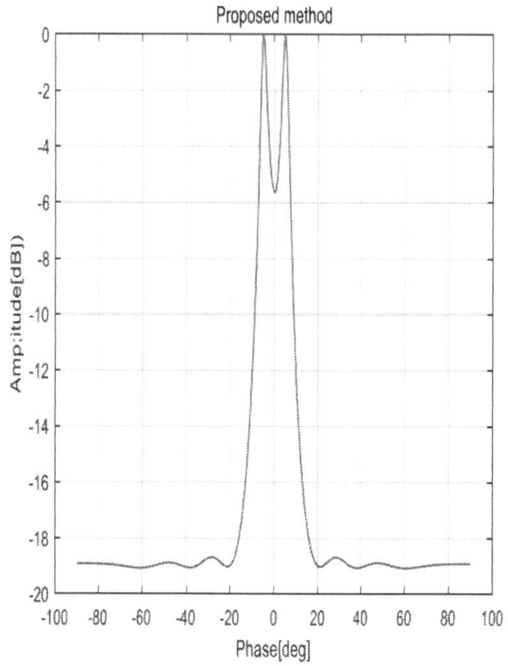

Fig 5. Angle of estimation (-5°, 5°) of the proposed algorithm, SNR=10dB, array element=6

5 Conclusion

In this paper, we studied the proposed method to accurately estimation the desired signal with the modified MVDR algorithm. The proposed method is to estimate the direction of arrival by updated weight. First, we studied the MVDR algorithm to obtain the weight. Second, the obtained weight is updated by covariance matrix. Final, the covariance matrix is obtained by the mutual coupling matrix of Toeplitz matrix. In simulation, the classical MVDR algorithm have a poor resolution. Comparing figure 4 and figure 5, the resolution of the proposed method is the same as the MUSIC algorithm. The proposed method has much better resolution than the convention MVDR algorithm. Since the proposed has no subspace, the calculation complexity decreases than the MUSIC algorithm. The proposed method is suitable for the drone system using microprocessors to detection others targets.

Acknowledgment

This work was supported by the National Research Foundation of Korea grant funded by the Korea government (NRF-2017R1D1A1B03032556).

References

1. Harry L. Van Trees, *Optimum array processing* (John Wiley&Sons,2002)
2. S. Unnikrishna Pillai, *Array signal processing* (Springer-Verlag , 1989)

3. Frank B. Gross, *Smart antenna for wireless* communications *with matlab* (McGraw-Hill,2005)
4. B. Allen ,M.Ghavami, *Adaptive array systems* (John Wiley&Sons,2005)
5. Don H.Johanson, Dan E. Dudgeon, *Array signal* processing (Prentice-Hall, 1963)
6. Mark C. Sullivan *Practical array processing* (McGraw-Hill 2009)
7. Youssef. Khmou, Said. Safi, Miloud. Frike, Comparative Study between Several Direction of Arrival Estimation Methods, Journal of Telecommunication and Information Technology, **1**,41 (2014)
8. Khairy A. El barbary, Tawil S. Mohamed, Mohamed.S.Melad, High Resolution Direction of Arrival Esitmation(Coherent Signal Source DOA Estiamtioin), IJERA, **3**, 132 (2013)
9. Petre Stoica, Prabhu Babu, Jian Li, SPICE: ASparse Convariance-Based Esitmation Method for Array Processing, IEEE Trans. SP, **59**, 629 (2011)
10. Liran Tzafri, Anthony j, Weiss, High-Resolution Direct Position Determination using MVDR, IEEE Trans. WC, **15**, 6449 (2016)
11. Ying Liu, Chi Xie, Yourun Zhang, Direction of arrivals estimation for correlated broadband radio signals by MVDR algorithm using wavelet, IEEE JM, 190 (2017)

MATEC Web of Conferences **208**, 03010 (2018) https://doi.org/10.1051/matecconf/201820803010
ICMIE 2018

Research on Ultrasonic Machining Gap Control Based on Fuzzy Self-tuning PID Parameters Technology

Chen Chun-hui[1], *Chen* Wei-you[2]

[1]Naval Petty Officer Academy, Mechanical Department, lecturer, China
[2]Naval Petty Officer Academy, Foundation, associate professor, China

Abstract. The working principle and measurement of ultrasonic machining gap is analyzed in this paper. A stable machining gap between the tool and workpiece must be maintained for improving processing accuracy and efficiency. Therefore, a fuzzy self-tuning PID parameters controller for ultrasonic machining tool feed system is designed, by means of using Matlab /Simulink simulation experiment, in this paper.

1 Formulation of issues

Onf the major concerns in any kinds of processing method is the producing proficiency. Ultrasonic machining is a method which process material by suspension abrasive material under the condition of a high-frequency vibration of the tool, thus, it is one of the most effective methods in processing hard and brittle materials such as glass and ceramics. Tool vibration amplitude and frequency, machining gap (that is, workpiece feed pressure produced by the tool) between the tool and the workpiece, abrasive suspensions, materials of the tool are the main factors that affect the processing speed. With other conditions remain unchanged, when the machining gap is too big, the impact of machining abrasive on the workpiece will decrease, resulting in decreased processing speed and reduced efficiency; when machining gap is too small, machining abrasive will be squeezed out, reducing the amount of abrasives, also resulting in decreased processing speed and reduced efficiency. To ensure accuracy and efficiency, we must maintain a stable machining gap (usually slightly larger than the average diameter of abrasive) between the tool and the workpiece. Therefore, under certain conditions, the controlling ultrasonic machining gap is a critical issue to improve ultrasonic machining productivity. In this paper, we mainly discuss the ultrasonic machining gap control based on fuzzy self-tuning PID parameter technology.

2 Basic principle of ultrasonic machining

Ultrasonic machining is a forming processing method in which the end face of an ultrasonic machine tool (Fig:1) is made ultrasonic vibration, by use of abrasive suspension to process brittle materials. Ultrasonic generator produces more than 16000 Hz high frequency AC power supplied to the ultrasonic transducer to generate ultrasonic

Vibrations, and the amplitude is amplified up to about 0.05 ~ 0.1mm by means of the horn, so that the lower end of the tool horn generates strong vibration. The suspension, containing water and an abrasive material, driven by the

Tool, also produces a strong vibration, impacting the surface of the workpiece. When machining, the tool puts a tiny pressure on the workpiece and the abrasive surface, continuously impacted by abrasive material with great speed and accelerated velocity, is crushed into small particles, falling down from the surface. The tool continuous feeds and the processing goes on until finally the desired size is reached and the shape of the tool is "Copied" on the workpiece.

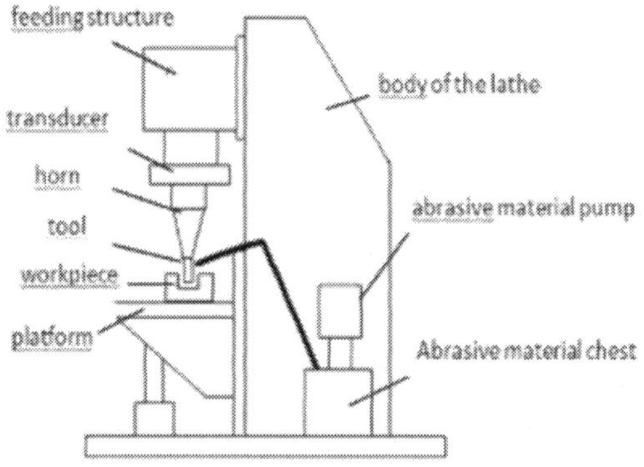

Fig. 1. Structure of Ultrasonic Machine Tool

3 Measure and control principle of machining gap

Ultrasonic resonance principle can be used to measure the gap between the tool and the work piece. Figure 2 is an ultrasonic machining gap measure and control system illustrative diagram.When machining, the ultrasonic is transformed into mechanical oscillation by machining transducer, then amplified by the horn and passed to tools.In order to improve the processing precision and efficiency,we expect the gap between tool and workpiece to be maintained at the set valuee = λ / 2(λ is ultrasonic propagation wavelength measured in the abrasive suspension). When measured, pulse excitation transducer converts signal into measured ultrasound and launched and spread to abrasive suspension by the horn.Part of the ultrasonic waves are reflected at the interface, others pass through the suspension and reach the surface of the work piece, resulting in reflection and transmission. When the gap between the workpiece and the tool is λ / 2, or nλ / 2 (n is an integer), the ultrasonic waves and its reflected waves transmitback and forthin the suspension and interfere with each other precisely in the same phase.According to the principle of resonance, as the ultrasound energy transmitted from the horn to abrasive suspension is reduced, ultrasonic amplitude is reduced accordingly. Based on the circuit feedback signal, CNC system then decideswhether the gap between the workpiece and the tool is equal to λ / 2 or nλ / 2.If it is equal to λ / 2 or nλ / 2, the gap information will be sent back to CNC system;if it is not equal to λ / 2 or nλ / 2, the feed structure will keep feeding until the machining gap is equal to λ / 2 or nλ / 2.

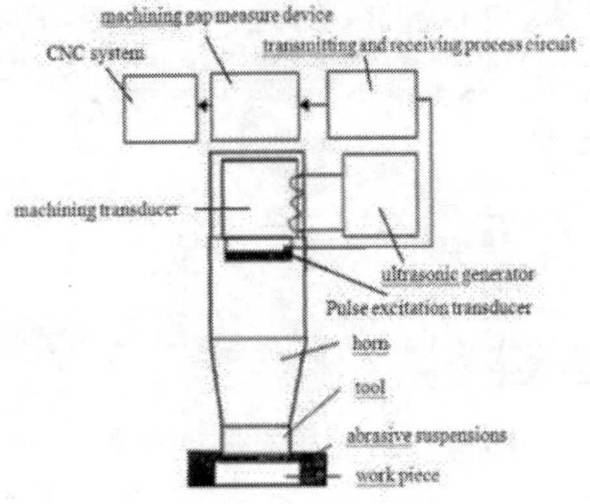

Fig. 2. Ultrasonic Machining Gap Measure and Control System

4 Design of fuzzy self-tuning PID parameters controller

4.1 Design

Conventional PID control is simple in principle and easy to use, but its control methods are mostly to be fixed a set of adjustable parameters under certain performance requirements. Such controls cannot often give attention to both static and dynamic performance, therefore the control of the system is not optimal. As to the fuzzy control, in which the mathematical model of the controlled objectis not necessarily to be known, it is easy to implement the control on uncertain or nonlinear systems and has a strong robustness to the parameters of the controlled objects, also it has strong suppression to external interference, and other similar characters. We, in this paper, try to combine the advantages of PID fuzzy control and design the fuzzy self-tuning PID parameters controller, with the function of identifying fuzzy relations between the three parameters, scale factor, integration constant and differential constant, and the error as well as error rate of change. It also keeps detecting error and error rate of change, and then revises the three parameters according to the fuzzy control principle, to meet the requirements of errors and error rate of change to the three parameters, thus the control system processes good dynamic and static performance.

4.2 Structure

The fuzzy self-tuning PID parameters controller consists of PID controller in which the parameters are adjustable and fuzzy controller. The fuzzy controller takes the deviation e and deviation rate of change ec of the ultrasonic machining gap as input value, the three parameters K_p , K_i , K_d of the conventional controller as output value, and then conducts real-time adjustments to K_p , K_i , K_d with fuzzy inference methods so as to meet the requirements of deviation e and deviation rate of change to PID parameters in different time.The structure of the composition of the controller is shown in Fig.3.

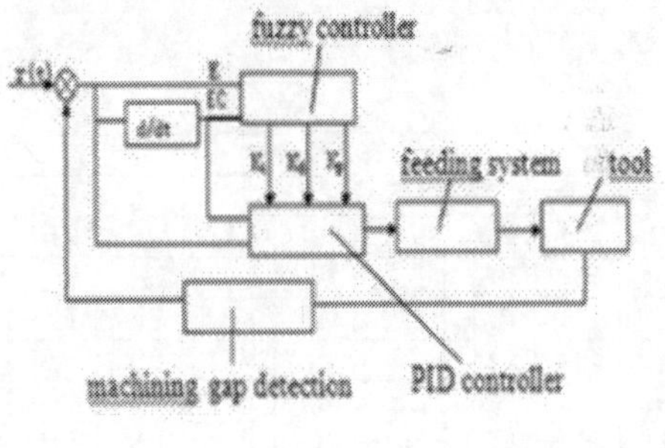

Fig. 3. Structure of Fuzzy Self-Turning PID Parameters Controller

4.3 Design of fuzzy control

The design of fuzzy control system employs two inputs (E, EC), triple-outputs (ΔK_p , ΔK_i , ΔK_d). The fuzzy domain of error e and error change rate ec is:E, EC={-6, -5, -4, -3, -2, -1,0,1,2,3,4,5,6}.The language variable value is taken as E, EC={NB, NM, NS, ZE, PS , PM, PB}. The fuzzy domain of ΔK_p ,ΔK_i , ΔK_d of is: ΔK_p , ΔK_i, ΔK_d ={-1, -0.8, -0.6, -0.4, -0.2,0,0.2,0.4,0.6,0.8,1 }, and the language variable value is taken as ΔK_p , ΔK_i, ΔK_d={NB,NM, NS, ZE, PS, PM, PB}. The membership functions of input and output variables are taken evenly distributed triangular membership function. The membership function distribution curve of E, EC is shown in Figure 4 andthe membership function profiles of ΔK_p, ΔK_i, ΔK_d is shown in Figure 5.The control rules of parametersΔK_p, ΔK_i, ΔK_d are shown in Table 1, Table 2 and Table 3.The control rules of fuzzy PID parameter self-tuning controller is shown in Figure 6.

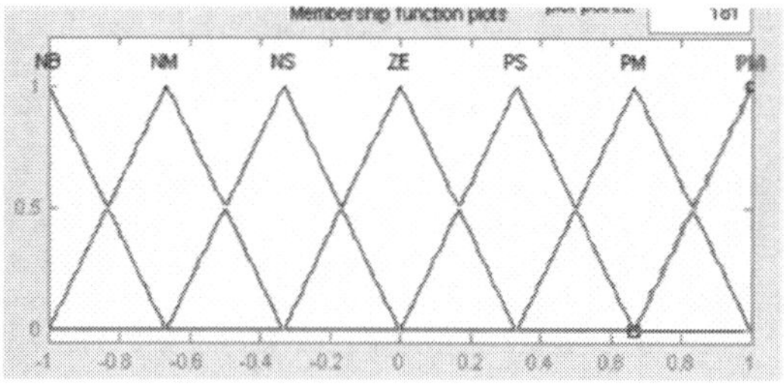

Fig. 4. Membership Function Distribution Curve Of E, EC

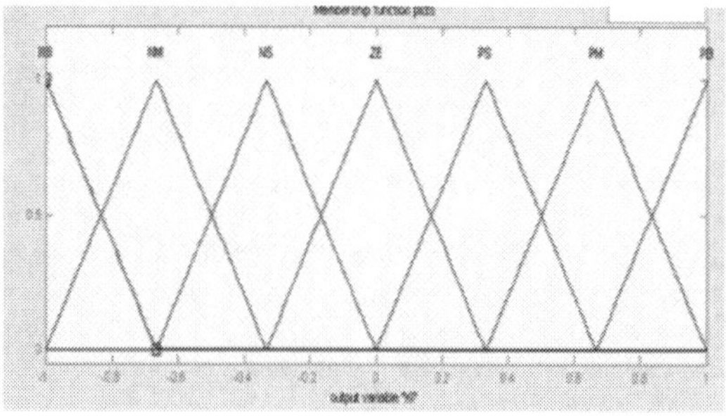

Fig. 5. Membership Function Profiles Of ΔKp,ΔKi,ΔKd

Table 1. Control Rules of Parameters ΔKp

EC \ E	NB	NM	NS	ZE	PS	PM	PB
NB	PB	PB	PM	PS	NM	ZE	ZE
NM	PB	PB	PS	PS	NS	ZE	PS
NS	PB	PM	ZE	ZE	ZE	PS	PM
ZE	PB	PM	ZE	ZE	ZE	PS	PB
PS	PB	PS	ZE	ZE	ZE	PS	PB
PM	PS	PS	NS	NS	PS	PM	PB
PB	ZE	ZE	NM	NS	PM	PM	PB

Table 2. Control Rules of Parameters ΔKi

EC \ E	NB	NM	NS	ZE	PS	PM	PB
NB	PB	PB	PB	NM	NM	ZE	ZE
NM	PB	PB	PM	NS	NS	PS	PS
NS	PB	PM	PS	NS	ZE	PS	PM
ZE	PB	PM	PS	ZE	PS	PM	PB
PS	PB	PS	ZE	NS	PS	PB	PB
PM	PS	PS	NS	NS	PM	PB	PB
PB	ZE	ZE	NM	NM	PB	PB	PB

Table 3. Control Rules Of Parameters ΔKd

EC \ E	NB	NM	NS	ZE	PS	PM	PB
NB	PB	PM	PB	PB	PB	ZE	NB
NM	PB	PS	PM	PM	PM	NS	NB
NS	PM	PS	PS	PS	PS	NB	NB
ZE	ZE	ZE	ZE	ZE	ZE	ZE	ZE
PS	NB	NM	PS	PS	PS	PS	PS
PM	NB	NS	PM	PM	PM	PS	PM
PB	NB	ZE	PB	PB	PB	PM	PB

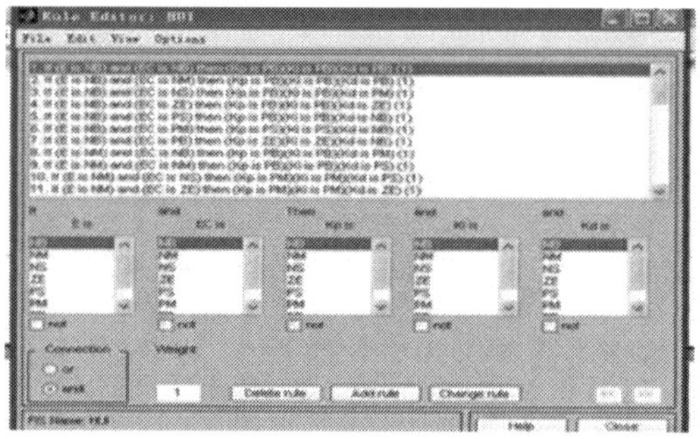

Fig. 6. Control Rules of Fuzzy PID Controller

5 Simulation experiment based on simulink simulation

In order to verify the control effect of fuzzy self-tuning PID parameters controller, and for comparison, we apply the software Mtlab/Simulink to conduct the simulation experiment and analysis on conventional PID control and fuzzy self-tuning PID control. The common second order delay in actual control tasks is selected as the control object. Suppose the transfer function is:

$$G(s) = \frac{20\ e^{-2s}}{S^2+3S+5}$$

5.1 Conventional PID control system simulation experiment

Conventional PID control system simulation model shown in Figure 7.

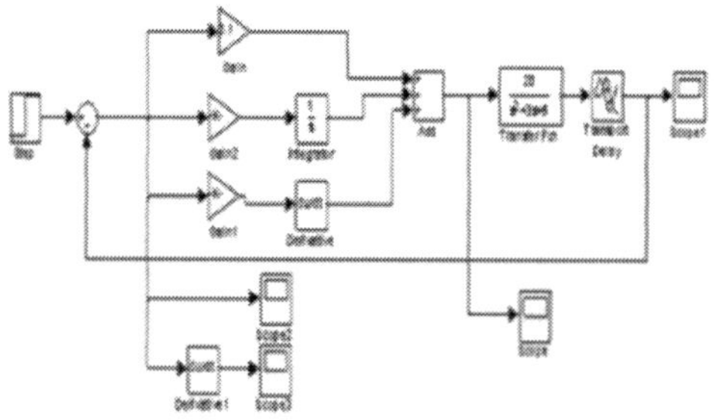

Fig. 7. Conventional PID Control System Simulation Model

In the role of unit step signal, we take sampling time T=0.1 seconds, the simulation time t = 60 second, PID controller parameters K_p=0.1, K_i=0.12, K_d=0.01, then we get the output waveform of the system shown in Figure 8, system deviation e and deviation change rate e care shown in Figure 9 and Figure 10, respectively, andthe curve of the PID controller output u changes in pace with time t is shown in Figure 11.

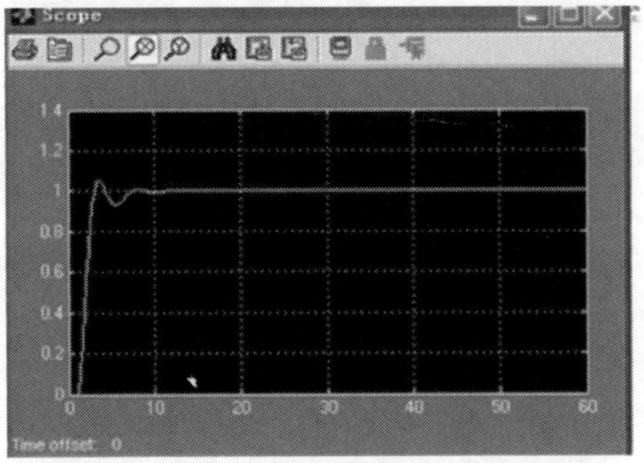

Fig. 8. Simulation Effect of Conventional Control System

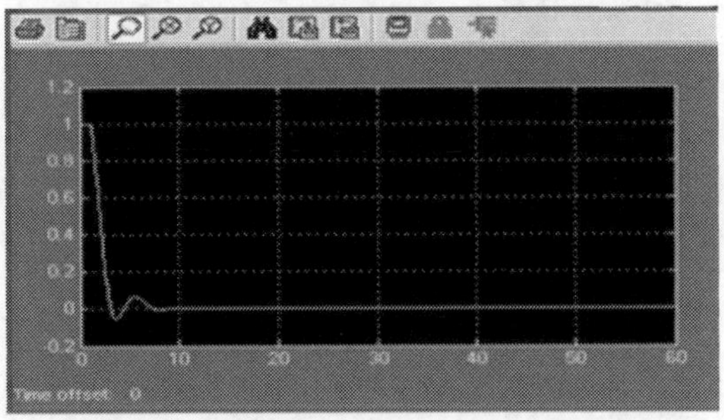

Fig. 9. Curve of Deviation E In Pace With Time T

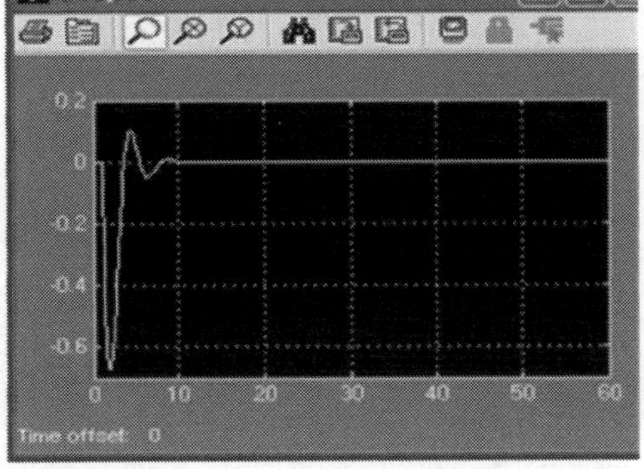

Fig. 10. Curve of Deviation Change Rate E in Pace with Time T

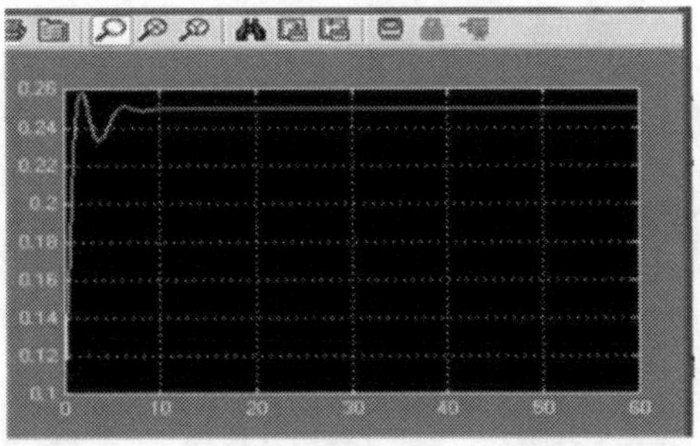

Fig. 11. Curve of PID Controller Output U in Pace with Time T

5.2 Fuzzy PID self-turning parameters controller simulation experiment

Based on Figure 8, Figure 9, Figure 10, we decide the basic domain to be e∈(-0.2,1), ec∈(-0.8,0.2), $\triangle K_p$∈(0.22,0.26),$\triangle K_i$∈(0.22,0.26), $\triangle K_d$∈(0.22,0.26). In the role of the unit step signal through repeated experiments, we take quantitative factors to be K_e=0.8, K_{ec}=0.35, scale factor G_{kp}, G_{ki}, G_{kd} were 0.01,0.001,0.01. Then fuzzy self-tuning PID parameters controller simulation model is shown in Figure 12. The simulation result of the system in the role of unit step signal is shown in Figure 13.

To compare the simulation results in Fingure 8 and Figure 13, we can see that the fuzzy PID parameter self-tuning controller discussed in this article can be well used to control machining gap, to improve processing quality and efficiency, at same time free of such shortcomings as large overshoot amount and long time for adjustment that exist in conventional PID controlling.

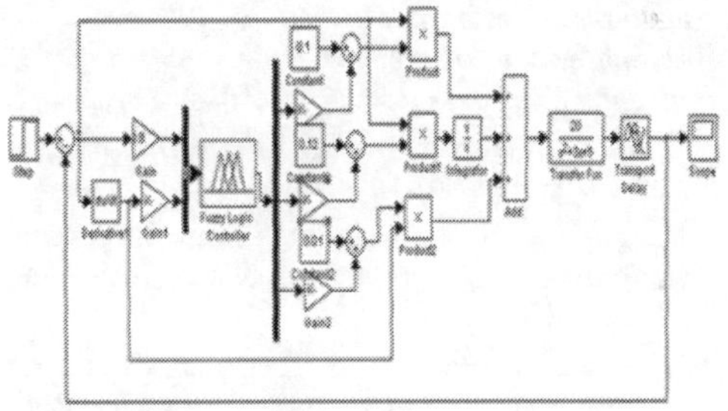

Fig. 12. Simulation Model of Fuzzy PID Parameter Seif-Turning Control System

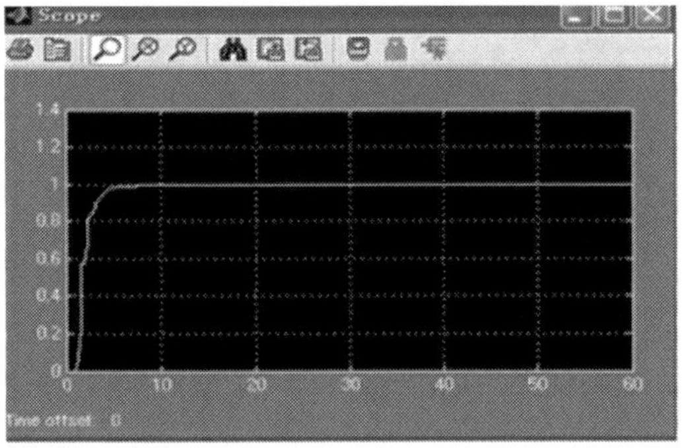

Fig. 13. Simulation Result in the Role of Unit Step Signal

6 Conclusion

Machining gap is a critical parameter in ultrasonic CNC machining. This article puts forward that fuzzy self-tuning PID parameters controller is used to control the feed system of ultrasonic CNC machine tool to ensure a stable machining gap between the tool and the workpiece. Matlab/Simulink simulation software is applied to conduct the experiment on the designed system, and proved to have good results for it achieves an on-line control of the process.

References

1. Zhong Jin'an. Research on Ultrasonic Machining gap Online Measurement and control[J]. Chinese Journal of Instrument. 2002(6).vol.23.
2. Song Zhen. Simulation Research on the Establishment of ultrasonic vibration cutting system and horn performance. Master Proceedings of Harbin Institute of Technology[C]. 2006.6:1-12.
3. Luo Yue, Zhang Mingjun, Kong Jinxing. Design on Ultrasonic Vibration Turning Device. Teaching and Technology[J]. 2005(J).
4. Yang Yanglin, Yu Xingzhi, Zhang Chengguang. Application of Ultrasonic Machining Technology. Modern Machinery [J].2009.
5. Hu Chuanxin, Huang Yan, The Research of Fuzzy Control Method Based on Self-Adjusting Rules. Master Thesis of Shantou Uninversity[C]. 2003.
6. Feng Dongjv. A Simple Measurement of Tool Amplitude for Ultrasonic CNC machining. Electrical Processing and Model[J]. 2004(5).
7. Zhang Li, Li Zhong, Liu Yanquan. Expert Self-tuning Fuzzy Control System. Chinese Journal of Scientific Instrument [J]. .2004(3):34-37.
8. Li Zhuxin. A Double Model Controller with Self-adjusting Base Sets Using Fussy Switching Based on Fussy Rules. Journal of Yunnan University[J]. 2003(4):313-316.
9. Hu Chuanxin. Special Processing Handbook[M]. Beijing University of Technology Press.2011.
10. Feng Dongqing, etc. Design of PID-Controller with Fuzzy Self-adjusting Parameter and Its Robustness Analysis. Proceedings of Progress of Computer Technology and Applicationin the 18th CACIS[C]. 2007.
11. FangGenSun， LongJingFan. N-widths on the classes of multivariate bandlimited unctions.Progressin Natural Science.2004,(10):863-866.
12. D.M .Alter， etc. Control of Linear Motors for Machine Tool Feed DrivesDesign and Implementation of H∞ Optimal Feedback Control.Journal of Dynamic Systems.Measurement and control, 1996, (1):18.
13. T.H.Lee，etc.Intelligent Control of Precision Linear Actuators.E ngineering Applications of Artifical Inteligence.2000, (13):671-684
14. David M.Alter，etc.Control of Linear Motors for Machine Tool Feed Drives Experimental Investigation of Optimal Feed-forward Tracking Control.Journal of Dynamic System，Measurement and Control.1998, (3):221-225
15. Velthuis ， etc.Experimental Verification of the Stability Analysis of Learning Feed-Forward Control.Proceedings of the IEEE Conference on Decision and control.1998, （2）: 1225-1229
16. Bassi，etc.Force Disturbance Compensation for An A.C.Brushless Linear Motor.IEEE International Symposium on Industrial Electronics.1999, （3）: 1350-1354
17. Su Li ， etc.Adaptive Robust Precision Motion Control of Linear Motors with Negligible Electrical Dynamics Theory and Experiments.Proceeding of the A merican Control Conference.IEEE ， Piscataway,NJ,USA,OOCB36334.2000,4: 2583-2587
18. Kim， etc . Robust Internal Loop Compensator Design for Motion Control of Precision Linear Motor. IEEE International Symposium on Industrial Electronics.1999,3: 1045-1050.
19. Layne J R，Passino K M. Fuzzy Model Reference Learning Control for a Cargo Ship Steering .IEEE Control Systems.1993:23-34
20. GCGoodwin， S F Graebe,M E Salgado.Control System Design.Prentice Hall,2001.
21. LennartLjung.System Identification Toolbox User's Guide.The MathWorks Inc,2001.
22. Jean-JacquesE.Slotine.Applied Nonlinear Control.China Machine Press,2004.
23. WangLei,XiaoYun-shi,WuQi-di,ZhouGuo-Xing.Neural nerwork based p arameters identification and adaptive speed control of AC drive system.P roceedings of the IEEE International conference on industrialte chnology,1996.

24. D.Renton,etc. High Speed Servo Control Multi-axis Machine Tools. International Journal of Machine Tools&Manufacture.2000, (40): 539-559.
25. M.Gringel, etc.Machine Tool Design Requirements for High-Speed M achining.Annals of the CIRP.1996, (1):1-5.
26. Kim , etc.Robust Internal Loop Compensator Design for Motion Control of Precision Linear Motor.IEEE International Symposium on Industrial Electronics.1999,3: 1045 -1050.
27. Lin C T and Lu Y C .Aneural fuzzy system with fuzzy supervised learning.IE EE Trans.Syst., Man and Cybem.1996, (5):744-763.
28. Rajani K. Mudi and Nikhil R.Pal. A Robust Self-Tuning Scheme for PI-and PD-Type Fuzzy Controllers.IEEE Trans.Fuzzy Syst.1999, (7):3-16.

Educational 110 kV step-down substation in the scale of 1/10

Alexander Egorov, *Anastasia* Larionova and *Pavel* Kovalenko

Automated Electrical Systems Department, Ural Federal University named after the first President of Russia B.N. Yeltsin, 19, Mira St., Yekaterinburg, Russia

Abstract. The educational 110 kV step-down substation in the scale of 1/10 is intended for material supplementation of educational processes in higher education institutions training power industry specialists. Lack of continuous and unlimited access of students to power facilities as well as absence of an opportunity to carry out any practical activities there results in severe deficiency in practical knowledge and skills among students. To solve this problem operating material models of substations and separate power equipment were created, both as the visual aid and as laboratory unit scaled in size. The actually operating substation in the scale of 1/10 allows performing and practicing a wide range of educational activities such as designing, constructing, installing, routine switching, etc.

1 Introduction

"Electrical power systems equipment designer" project has been being in operation within Automated Electrical Systems dept. (Ural Power Engineering Institute, Ural Federal University, Yekaterinburg) since 2013. The primary aim of the project is to improve the material-and-technical and informational infrastructure of the educational processes with the application of scale modeling methods and technologies. Creating the prefabricated scale models of primary and secondary power systems equipment is intended to address various educational challenges associated with the quality of students and technical specialists training in higher education institutions for power engineering industry. More to that, the project framework involves teaching students to design, model, engineer and analyze, opening the opportunities to design and create new proprietary equipment [1].

The designer is implemented based on the scale modeling methods and technologies within the industry-recognized 3D-modeling CAD/CAM systems, e.g. AutoCAD. To date methods and technologies of laser 3D-scanning, reverse engineering and 3D-printing are widely introduced into the designer production process. This kind of approach to the problem allows to prototype the scale models of substations as well as separate primary and secondary power system equipment using different materials with ensuring high external consistency and geometrical equivalence. It opens the perspective of utilizing the models as the actually operating laboratory setup of small dimension and low cost. Another advantage of scale modeling is maintaining the clearness of the visualization of an object under study or under control, providing the "participation effect" or impression of being at power facility.

2 Goals And Major Challenges

The main goal of the project is the improvement of students' education quality by means of expanding the material-and-technical and informational infrastructure of the educational processes with new laboratory classes and practical training being developed to take into account typical as well as non-typical power engineering cases. In other words, the project implies preparing and utilizing the power equipment models along with power facilities based on the models produced using original materials as well as digital and printed prototypes of power plants, substations and other power facilities for educational purposes within multiple related courses (electrical machinery, electrical apparatus, power system equipment, automated control systems etc.). Nowadays the study guides publishing and high-volume manufacturing launch of an educational 110 kV step-down substation are underway in order to teach the students about the equipment external configuration and the corresponding challenges, its hardware components, principle of operation, dedication and functionality, transporting, mounting, assembly, operation, maintenance, and control.

Designing, assembly and printing of the scale models plays an important role at current stage since seeing a demonstrative model of external and internal arrangement of power equipment ensures the tangible conception of the processes physics and dynamics, together with letting the student into the procedures of a power facility construction and O&M.

"Electrical power systems equipment designer" is described in detail on the example of a 110 kV unit-type substation transformer bay outdoor switchgear.

 https://doi.org/10.1051/matecconf/201820804001

3 Technological Stages of Scale Models Creating

Creating the scale model of any equipment piece comprises several technological stages by applying the reverse engineering approach.

The first stage is the investigation including the following steps: selecting the best suitable equipment piece of the various models supplied by different manufacturers; collecting the complete set of initial data (photographs, dimensional drawings, equipment registers, technical datasheets etc.). To date the first stage has been successfully accomplished for 110 kV equipment. "Mednaya" 110/20 kV substation (EESK, www.eesk.ru) with 40 MVA TDN transformers (SVEL, www.svel.ru) was selected to be the prototype of the educational 110 kV substation since it is supplied with modern equipment. The investigation was performed using three-dimensional laser scanning methods on the substation site in order to proceed with equipment reverse engineering [2]. The technical documentation and dimensional drawings were requested supplemented by the photographs made for the purpose of subsequent digital modeling of the educational 110 kV substation and then manufacturing its physical model in the scale of 1/10. The fragments of the photographs and the drawings are shown in figures 1–4.

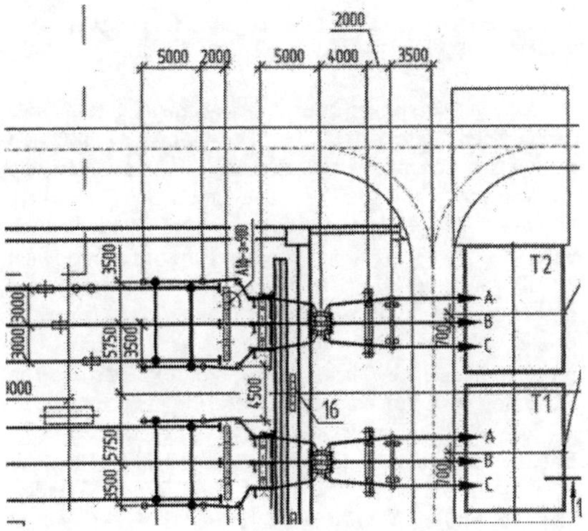

Fig 1. 110/20 kV unit-type substation outdoor switchgear, top view

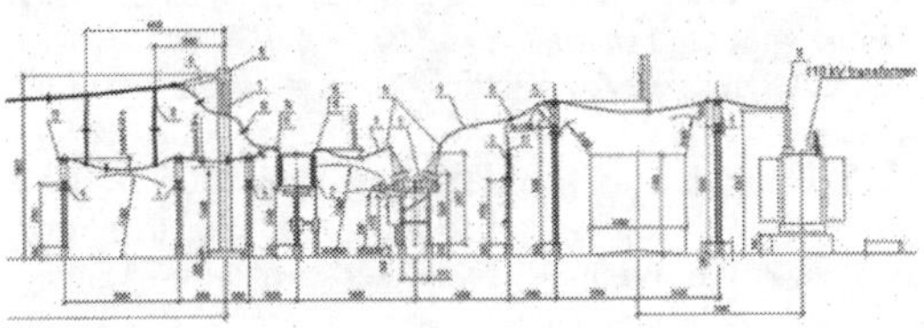

Fig 2. 110/20 kV unit-type substation bay, side view

Fig 3. 110/20 kV unit-type substation outdoor switchgear

Fig 4. 110/20 kV unit-type substation 40 MVA transformer

The second stage features the collected initial data processing and modeling the power facility exterior and – partially – interior arrangement of the outdoor switchgear components. There were experiments conducted in order to select the optimum scale of the 110 kV outdoor switchgear transformer bay, the results are shown in figure 5 and in table 1.

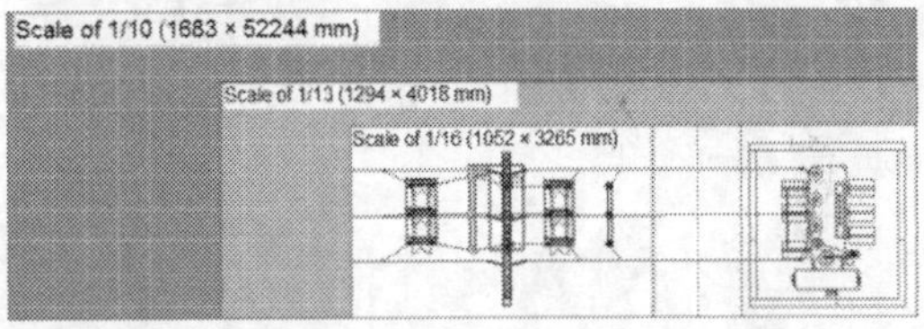

Fig 5. 110 kV outdoor switchgear transformer bay land allocation

Table 1. Area Ratio of the 110 kV Educational Substation

Scale	Width	Length	Area	Area ratio
1/10	1683 mm	5224 mm	$8{,}79 \cdot 10^6$ mm^2	256 %
1/13	1294 mm	4018 mm	$5{,}19 \cdot 10^6$ mm^2	151 %
1/16	1052 mm	3265 mm	$3{,}43 \cdot 10^6$ mm^2	100 %

The figure and the table clearly show that the transformer bay area in the scale of 1/10 is 2.5 times bigger than the corresponding value in the case of 1/16 scale. However, the scale of 1/10 fits the requirements of introducing the clockwork drives and fitting assortment into separate equipment pieces. Hence the scale of 1/10 was adopted as the generic for the 110 kV outdoor switchgear transformer bay with the size of 1.68×5.22 m. Modeling process includes three-dimensional drawing in the scale of 1/1 of separate equipment pieces (see figures 6–12) with full specification and replication of all components, connections, elements and units. This approach preserves the future opportunity to switch to any typical as well as non-typical scale if necessary and if the circumstances require such a transition.

To date the scale range is well-developed and covers the bigger – 1/2, 1/4, 1/8, 1/10, 1/13, 1/16 – and the smaller ones – 1/20, 1/24, 1/35, 1/48, 1/72, 1/100 и 1/144. The research results defined the decision to continue the project using three scales – 1/10, 1/13, 1/16 – since it, beside the reasons given above, ensures the opportunity to install microelectronic components into the equipment pieces.

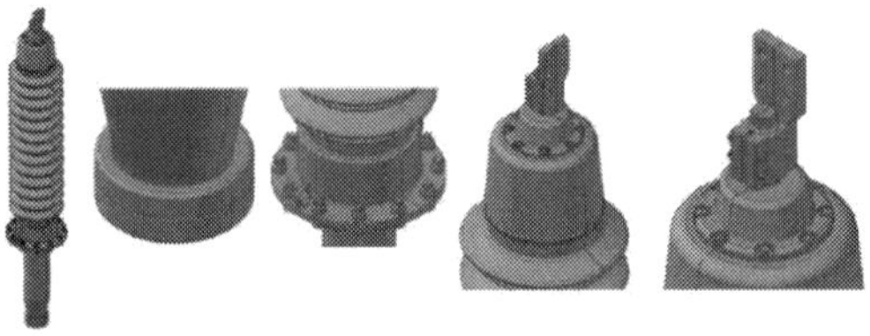

Fig 6. Mosizolyator 110 kV bushing. Detailed: flux shield, bearing flange, bushing cap, clamping arrangement and terminal

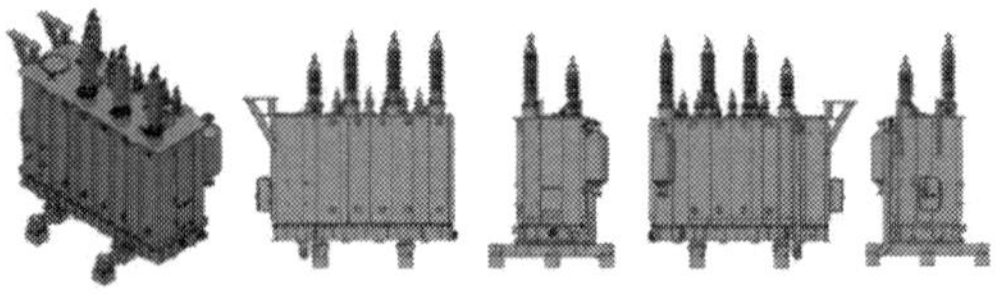

Fig 7. SVEL TDN 110/20 kV 40 MVA transformer (oil conservator removed)

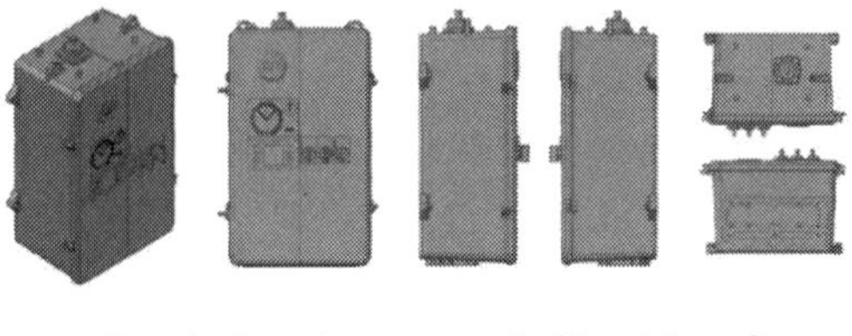

Fig 8. MR on-load tap changer control cabinet (views: front, left, right, top, bottom)

Fig 9. "D" model EuroCooler heatsink (views: left, front, top, bottom)

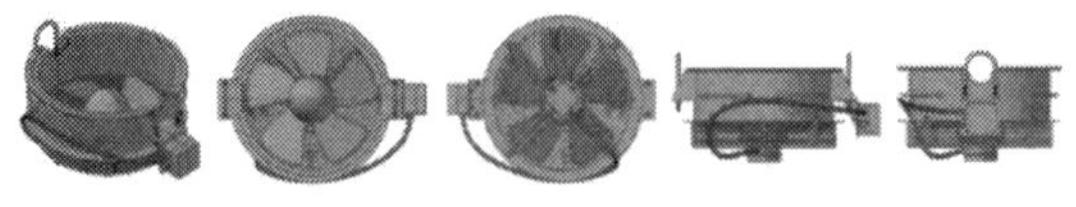

Fig 10. EuroCooler fan (views: top, bottom, left, front)

Fig 11. AIZ 110 kV busbar post insulator (2 different versions) with Arutti rigid busbar mounted

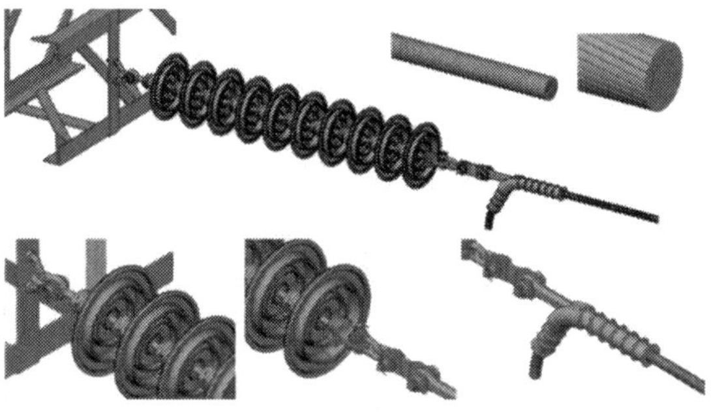

Fig 12. GIG 110 kV outdoor switchgear aluminium-stell 240 mm^2 bus system insulator string

After all equipment pieces three-dimensional models have been complete, the substation construction elements – concrete products, metal structures carrying the power equipment – are modelled as well. "BaltTeplostroy" concrete products were adopted as the basis lineup, SVEL group provided the information on metal structures [3]. The third stage includes three-dimensional printing of the substation parts using 3D-printers. 3D-printing technology are utilized for prototyping since they are quite expensive. The technology is involved in manufacturing the prototype of a single 110 kV substation outdoor switchgear bay. Batch production of the substation elements requires other high-tech solutions, namely non-ferrous foundry. Two machines (3D-printers owned by Automated Electrical Systems dept.) are used for the purpose of manufacturing the substation prototype: StarLight 3D-printer (made in Russia) employed for printing small details with high accuracy, and CubeX 3D-printer (made in USA) utilized for printing large elements, flashing avtions

The fourth stage implies components surface finish processing – mechanical peeling-off and deburring, removing the printing frame necessary for printing, cleaning, light-striking and drying in the UV-camera. After that the primary base coating and the painting processes are performed, the substation bay is prepared for assembly and then assembled.

4 Conclusion

To date the technical drawings have been completed and 3D-printing has been partly complete for the following equipment: SVEL TDN 110/20 kV 40 MVA transformer, Mosizolyator 110 kV, 66 kV and 20 kV bushings, UETM VEB-110 kV circuit breaker, UETM TRG-110 current transformer, AIZ SO-110 kV busbar post insulator, Global Insulator Group overhead line hardware and insulation, Arutti rigid busbar. The following equipment is at the stage of three-dimensional drawings UETM RPD-110 kV switch, UETM ZNG-110 kV voltage transformer, SVEL high-frequency choke, and DECRA CMA coupling capacitor. The 110 kV substation bay was developed with maximum elaboration and detailing. Its prototype model is in production using heliocomposite and plastic materials in the scale of 1/10.

5 Future Opportunities

The search is underway for high-tech solutions addressing the requirements of batch production. At the same time, the substation bay systematic assembly tutorial is in progress. The non-ferrous metals types and alloys research is conducted. The overall 110 kV outdoor switchgear bay project scope includes:

1) 110 kV substation three-dimensional technical drawings;
2) 110 kV substation two-dimensional technical drawings;
3) 110 kV substation design educational guidance;
4) 110 kV substation design information graphics posters;
5) 110 kV substation static prototype model in the scale of 1/10;
6) 110 kV substation static prototype assembly tutorial;
7) 110 kV substation operational model in the scale of 1/10.

In the future 110 kV substation outdoor switchgear bay in the scale of 1/10 will allow to aid numerous educational activities – laboratory classes and practical training: design, construction, installation, routine switching etc. The technical drawings and static prototype are expected to be complete by the end of 2018.

Acknowledgement

The work was supported by Act 211 Government of the Russian Federation, contract № 02.A03.21.0006 and the Ministry of Education and Science of Russian Federation (in the framework of state assignment, № 13.1928.2014/K (project № 1928)).

References

1. O. Vozisova., A. Egorov, A. Trembach. Designer of electric power systems. Siemens 3API DT 145kV Circuit-Breaker // Applied Mechanics and Materials. Vol. 698 (2015). pp. 699-703. Submitted © 07.10.2014 (2015) Trans Tech Publications, Switzerland
2. Fröhlich C. and Markus M. Terrestrial laser scanning–new perspectives in 3D surveying. International archives of photogrammetry, remote sensing and spatial information sciences 36.Part 8 (2004): W2.
3. A. Tavlintsev, S. Eroshenko, P. Kuzin, A. Kazantsev, A. Trembach, Vozisova O., E. Koksharova, N. Zhavrid, V. Kunshin, P. Bannih, Baltin D. 220/110 kV 250 MVA Autotransformer kit. Advanced Materials Research, 1008-1009. pp. 1162-1165

Defining the power system load frequency static response slope based on transient synchrophasor data

Alexander Berdin and *Pavel* Kovalenko

Automated Electrical Systems Department, Ural Federal University named after the first President of Russia B.N. Yeltsin, 19 Mira St., Yekaterinburg, Russia

Abstract. The method is proposed for defining the power system load frequency static response based on synchrophasor measurements during transients accompanied by frequency deviations. The method was successfully validated involving four events resulting in frequency deviations in the Northern part of Tyumen region of up to 0.06 Hz magnitude and recorded by means of the System operator WAMS.

1 Introduction

Frequency and active power changes in bulk power systems are characterized by a power system or a power region active power frequency static response. Resulting power system response depends upon the corresponding response of the generation and the loads [1].

Different from the nodes load voltage static response for a power system or a power region, load frequency static response might be derived only as an aggregate of the nodes responses. It is due to the fact that load nodes voltage is a locally controlled parameter while the frequency might be assumed similar or close to similar for all power region nodes at any time.

Power region aggregate load is defined as the sum of the region generation and net external power flow:

$$P_{\text{load}}(t) = P_{\text{gen}}(t) + P_{\text{flow}}(t) \quad (1)$$

However, not all generation sources provide frequency and power measurements, that is generally applicable to low-capacity generation, mostly distributed (distributed energy resources, DER).

Hence, due to the impossibility to measure overall generation and evaluate the actual influence of the nodes voltage on their consumption, the aggregate power region static response is to be considered, being characterized by the load value.

What should be noted here is the importance of the frequency static response slope. Frequency static response is an inherent parameter of a power system computational model along with other parameters. Utilizing the power system computational model enables the calculation of power system operation conditions, which serve as the basis for solving the power system dispatch control tasks, such as defining the admissible power flow over the power system sections, required minimum generation capacity margins, effectiveness of automatic control actions etc. [3, 4, 5] In case the actual frequency static response is unknown, the computational model parameters are defined based on rated values with the most undesirable values being considered in order to ensure the results account for the heaviest power system operation conditions. If the frequency static response is better than the assumed, this leads to excessive power system operation restrictions, which, in turn, reduces the overall power system efficiency.

In 1970s-1980s frequency and voltage static response as well as frequency static response slope for the load nodes with different consumption mix were defined experimentally involving actual power systems in operation. To date conducting these experiments is difficult and expensive, moreover, their consequences are unpredictable [1, 6]. Hence, the existing frequency static responses have become obsolete due to significant changes in consumption structure and steadily widening variety of load components (storage systems, electric vehicles, etc.).

The introduction of the new generation measurement devices, providing the possibility to obtain the time-synchronized high-accuracy measurements of power and frequency corresponding to the external power flows and high-capacity generators, allows to define the external power flows and generators frequency static response based on the transient recordings during regular operation (i.e. conduct passive experiments) under random disturbances conditions in power system. Current work proposes the investigation of the possibility to define a power region or a load node frequency static response slope based on the transient recording corresponding to the predefined short periods of time:

$$P_{\text{load}}(t,f) = P_{\text{gen}}(t,f) + P_{\text{flow}}(t,f). \quad (2)$$

2 Defining load frequency static response slope

The methods and algorithms have been developed in order to define and monitor the response based on the actual recordings of the power system performance parameters in the nodes equipped with PMUs under deviating frequency and voltage conditions in a load node driven by disturbances of the following types:

- emergency generation or load trip;
- external grid parameters change (transmission line or generator trip);
- load terminals or external grid fault;
- several seconds power outage (automatic supply restoration delay);
- out-of-step operation;
- low-frequency oscillations.

Hence the nodal load static and dynamic response can be defined under changing voltage and/or frequency conditions $\{P_{load}(U),\ Q_{load}(U),\ U(t),\ f(t),\ P_{load}(t,\ U),\ Q_{load}(t,\ U)\}$ along with the performance parameters mutual influence $\{P_{load}(U,\ dU/dt),\ Q_{load}(U,\ dU/dt),\ P_{load}(f, df/dt),\ Q_{load}(f,\ df/dt)\}$.

Defining the interconnection between various monitored parameters is based on the modified Gauss technique [7].

The aggregate power system or single node frequency static response is evaluated by identifying the expression parameters on a simulation time period t_m based on the active power and frequency measurments:

$$P_{load}(t_i, f(t_i)) = a_0 + a_1 t_i + \\ + a_2 t_i^2 + a_3 f(t_i) + a_4 f^2(t_i), \tag{3}$$

where $t_i \in T_m = \{t_1, t_2, ..., t_m\}$;

$$f(t_i) \in F_m = \{f(t_1), f(t_2), ..., f(t_m)\};$$

$$P_{load}(t_i, f(t_i)) \in P_m =$$

$$= \{P(t_1, f(t_1)), P(t_2, f(t_2)), ..., P(t_m, f(t_m))\}.$$

By differentiating the expression with respect to frequency, one can obtain $\partial P_{load} / \partial f = a_3 + 2a_4 f(t_i)$, and hence

$$k_{si} = (\partial P_{load} / \partial f)(f_b / P_b) = \\ = (a_3 + 2a_4 f(t_i)) f_b / P_b, \tag{4}$$

where P_b, f_b – active power and frequency basis values.

Load frequency static response slope k_s for a single node can be derived by taking into account the voltage values at the node corresponding to the simulation time period:

$$P_{load}(t_i, f(t_i), U(t_i)) = a_0 + a_1 t_i + a_2 t_i^2 + \\ + a_3 f(t_i) + a_4 f^2(t_i) + a_5 U(t_i) + a_6 U^2(t_i) \tag{5}$$

Then the load derivative with respect to frequency is:

$$\frac{dP_{load}}{df} = a_3 + 2a_4 f(t_i) \tag{6}$$

Under small frequency deviation it is feasible to reduce the expression to linear components and assume the load change over time to be insignificant, therefore (3) can be written over as follows:

$$P_{load} = a_0 + a_3 f(t_i), \tag{7}$$

and

$$\frac{dP_{load}}{df} = a_3. \tag{8}$$

Then the frequency static response slope k_s, according to [1] and (4), is expressed as

$$k_s = a_3 \frac{f_b}{P_b}. \tag{9}$$

3 Case study

Current research presents the example of utilizing the proposed technique for defining the frequency static response slope of one of the Ural interconnection power regions.

3.1 Cases background

The majority of the region consumption is constituted by oil and gas industrial facilities, including refining, transport, waste handling, along with residential load.

The considered region is deficient in terms of active power. Its demand is partially met by several power plants and numerous CCGTs deployed at several oil and gas companies' production sites.

According to legal regulations [8], the governors controllers deadband must not exceed 0.3 % (which is ±0.15 % or, in case rated frequency of 50 Hz is assumed, ±0.075 Hz) [9]. Frequency deviations do not exceed 0.6 Hz in all considered cases. Moreover, since the region power plants do not participate in primary frequency control, their response to those deviations can be neglected. Therefore, the obtained frequency static response slopes reflect the actual load response to frequency deviations.

In case the investigated power system includes power plants involved into primary frequency control, their frequency response can be defined according to (2) if there are PMUs there. The monitored parameters (active power and frequency) were recorded for the complete section separating the considered region from the Ural interconnection. The base frequency and load power were assumed to be equal to, correspondingly, 50 Hz and the region consumption prior to the disturbance, according to the System operator SCADA data, with local generation taken into account.

The key issue of utilizing the frequency static response slope definition technique for the power system region at hand is the considered time period selection. Different periods of the recordings shown in Figures 1–5 were selected for the detailed investigation. The slopes defined for the whole transient are given in **bold** in tables.

3.2 02/24/2017 event

The first considered transient resulted from the disturbance, which happened February 24th, 2017. The active power and frequency plot is presented in Figure 1. Five indicative intervals were selected for the detailed analysis the pre-disturbance steady-state operation, the period of frequency drop to 49.93 Hz and its subsequent restoration to 49.99 Hz. The calculated frequency static response slope are summarized in Table 1.

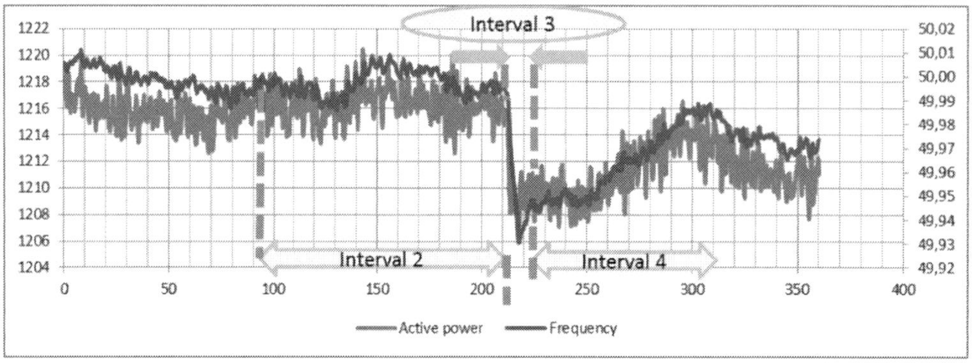

Fig 1. 02/24/2017 event, considered intervals

Table 1. 02/24/2017 event P(f) coefficients, P_b=2182 MW

Object	Int 1 100–300 s	Int 2 100–210 s	Int 3 210–220 s	Int 4 220–300 s	Int 5 212–214 s
Section aggregate	165.0	97.6	107.8	136.1	345.5
Line 1	88.2	36.4	66.2	96.2	186.3
Line 2	28.1	28.8	11.9	2.0	94.4
Line 3	12.6	5.9	7.7	3.6	19.6
Line 4	35.7	38.0	27.2	37.7	113.8
Section aggregate dP/df	165.0	97.6	107.8	136.1	345.5
Section aggregate response slope	**3.78**	2.24	2.47	3.12	7.92

3.3 06/27/2017 event

The next case comprises two sequential transient, which took place June 27th, 2017. The graphs showing the active power and frequency are shown in Figures 2 and 3. There are three distinctive intervals selected for the first transient – the pre-disturbance operation, the oscillatory transient and the complete process. The second transient is subdivided into two intervals: the oscillatory transient ant the complete process including the pre-disturbance state. The frequency static response slope calculation results are presented in Table 2.

Table 2. 06/27/2017 event P(f) coefficients, P_b=1856 MW

Object	Int 1 175–190 s	Int 2 173–179 s	Int 3 179–191 s	Int 4 1614–1622 s	Int 5 1615.6–1622 s
Section aggregate	138.0	68.6	156.3	189.8	186.5
Line 1	65.9	28.8	67.2	79.6	77.2
Line 2	35.9	12.7	43.2	41.6	39.6
Line 3	3.9	-0.1	4.9	9.2	10.2
Line 4	48.0	30.2	49.1	59.6	59.5
Section aggregate dP/df	138.0	68.6	156.3	189.8	186.5
Section aggregate response slope	3.72	1.85	**4.21**	5.11	**5.02**

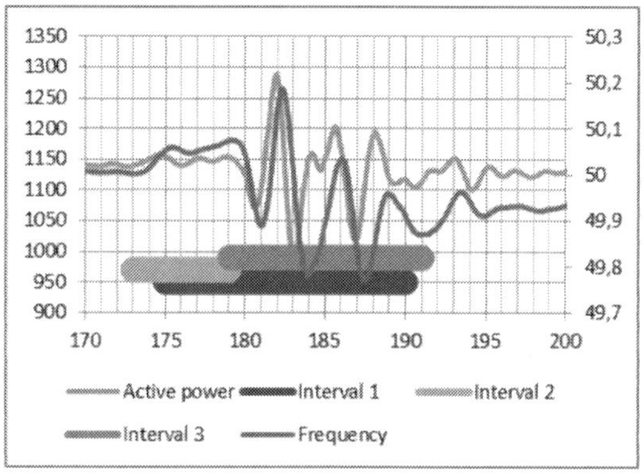

Fig 2. 06/27/2017 event (part 1), considered intervals

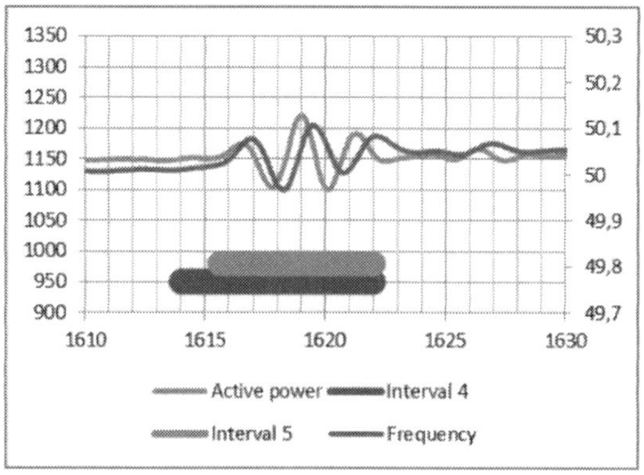

Fig 3. 06/27/2017 event (part 2), considered intervals

3.4 07/25/2017 event

The other considered transient is relatively long-standing, it took place July 25th, 2017. The recordings of active power and frequency are plotted in Figure 4. Three indicative intervals are selected for the analysis: the steady increase of frequency and its rapid drop by 0.08 Hz. The frequency static response slope calculation results are shown in Table 3.

Table 3. 07/25/2017 event P(f) coefficients, P_b=1764 MW

Object	Int 0 0–1200 s	Int 1 820–1160 s	Int 2 700–1160 s	Int 3 0–700 s
Section aggregate	142.7	184.3	184.3	55.0
Line 1	-83.3	-97.6	-97.1	-49.2
Line 2	-0.2	-19.2	-20.4	30.1
Line 3	-3.2	-5.0	-4.8	-0.9
Line 4	56.5	62.5	62.0	42.0
Section aggregate dP/df	142.7	184.3	184.3	5.0
Section aggregate response slope	4.05	**5.22**	**5.22**	1.56

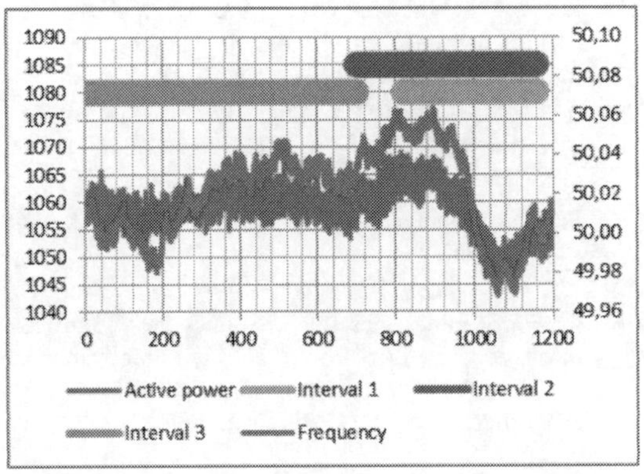

Fig 4. 07/25/2017 event, considered intervals

3.5 07/30/2017 event

The transient involving frequency deviation happened July 30th, 2017 is considered. The frequency and active power plots are shown in Figure 5. Four indicative intervals are selected for the analysis the pre-disturbance conditions, the period of frequency decrease to 49.94 Hz, the subsequent frequency restoration to its rated level and the combination of the last two intervals. The calculation results are shown in Table 4.

Table 4. 07/30/2017 event P(f) coefficients, P_b=1847 MW

Object	Int 1 0–1200 s	Int 2 0–400 s	Int 3 400–1200 s	Int 4 400–800 s	Int 5 800–1200 s
Section aggregate	154.8	80.7	137.6	161.1	137.6
Line 1	87.6	53.2	80.1	103.9	76.3
Line 2	8.9	5.0	6.0	-1.2	22.2
Line 3	3.7	2.2	4.0	2.3	4.9
Line 4	54.7	22.1	54.7	56.2	52.1
Section aggregate dP/df	154.8	80.7	137.6	161.1	137.6
Section aggregate response slope	**4.19**	2.18	3.72	4.36	3.72

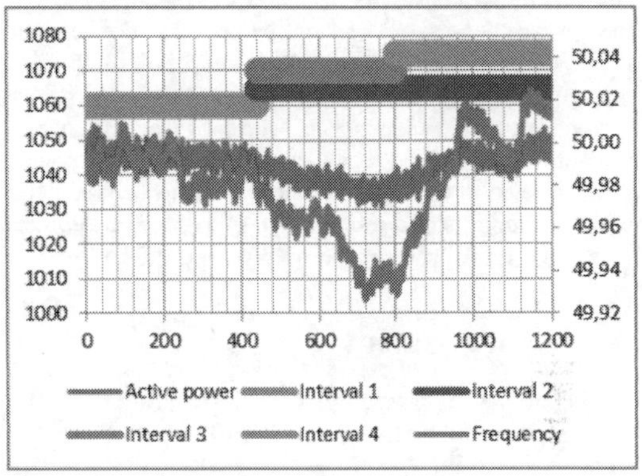

Fig 5. 07/30/2017 event, considered intervals

3.6 Case study summary

The case study results analysis is summarized in Table 5. Frequency static response slopes were selected from Tables 1–4 for the time periods corresponding to the whole transient in accordance with [6]. Overall consumed power is obtained from the System operator SCADA system for the moment prior to the considered disturbance.

Table 5. Frequency static response slopes, summarizing table.

Event	Load, MW	Response slope
02/24/2017 event	2182	3.78
06/27/2017 event (part 1)	1856	4.21
06/27/2017 event (part 2)	1856	5.02
07/25/2017 event	1764	5.22
07/30/2017 event	1847	4.19
Average		4.48

It can be seen from the table that the resulting frequency static response slopes defined for different time periods differ significantly. The value obtained for the only considered winter period (3.78) stands out from the overall range. In case it is not taken into account, the average slope equals to 4.66. at that, the value adopted in actual practical tasks is the same over the year and equals 5.

4 Conclusion

The proposed technique for defining the frequency static response slope has been validated using actual WAMS data for one of the Ural interconnection power regions under frequency deviations of up to 0.06 Hz. The results obtained for different time periods closely match, and the average slope is close to the one utilized in practical tasks. At that it is worth noting that partial time periods do not provide valid results, which renders their consideration undesirable for the corresponding task. On the contrary, one should be inclined to take into account the complete transient time period, which accords with [6] summary.

The future development of current research implies in-depth investigation using computational models and physical electrodynamic simulator equipped with PMU, with frequency static response being predefined, which makes the study environment totally controlled.

Acknowledgement

The work was supported by Act 211 Government of the Russian Federation, contract № 02.A03.21.0006 and the Ministry of Education and Science of Russian Federation (in the framework of state assignment, № 13.1928.2014/K (project № 1928)).

References

1. Experimentalnye issledovaniya rezhimov energosistem / L.M. Gorbunova, M.G. Portnoy, R.S. Rabinovich et al. under S.A. Sovalov edition. Moscow, Energoatomizdat, 1985.
2. Pravila predotvrashcheniya razvitiya i likvidatsii narusheniy normalnogo rezhima elektricheskoy chasti energosistem. / "SO UPS" JSC standard. Approved by "SO UPS" JSC order #114 dated 09/24/2008. Moscow, 2008.
3. Avtomaticheskoe protivoavariynoe upravlenie rezhimami energosistem. Protivoavariynaya avtomatika energosistem. Usloviya organizatsii processa. Usloviya sozdaniya obyekta. Normy i trebovaniya. / "SO UPS" JSC standard. Approved by "SO UPS" JSC order #102 dated 04/19/2011. Moscow, 2011.
4. Kochneva, E., Pazderin, A., Sukalo, A. Testing Equation Method Modification for Demanding Energy Measurements Verification. MATEC Web of Conferences, 2016.
5. Mukhlynin, N.D., Pazderin, A.V. The model of distribution grids state estimation and optimization based on the energy balances equations. 2nd International Conference on Industrial Engineering, Applications and Manufacturing, Proceedings, 2016.
6. Metod opredeleniya krutizny staticheskih kharakteristik uzlov nagruzki po napryazheniyu putem approksimatsii perehodnyh protsessov, voznikayushchih v ekspluatatsionnyh rezhimah / D.A. Myshlyanikov, A.G. Fishov // Elektroenergetika glazami molodezhi: 5 international conference. Tomsk, 2014.
7. A.S. Berdin, P.A. Kryuchkov. Formirovanie parametrov modeli EES dlya upravleniya elektricheskimi rezhimami. Yekaterinburg, 2000.
8. Pravila tehnicheskoy ekspluatatsii elektricheskih stantsiy i setey Rossiyskoy Federatsii. Energoservis, Moscow, 2003.
9. Regulirovaniye chastity i peretokov aktivnoy moshchnosti v EES Rossii. Normy i trebovaniya. / "SO UPS" JSC standard. Approved by "SO UPS" JSC order dated 12/05/2012. Moscow, 2012.

Modeling and simulation for low-frequency vibration energy harvesting based on piezoelectric unimorph cantilever beam

Binjie Song[1], *Jianhai* Yue[1] and *Zhunqing* Hu[1]

[1]School of Mechanical, Electronic and Control Engineering, Beijing Jiaotong University, No.3 Shangyuancun Haidian District, 100044, Beijing, China

Abstract. In order to solve the problem of sustainable energy supply for low-power electronic products used in low-frequency vibration environment, the mathematic model was established based on the theory of piezoelectricity and Euler-Bernoulli beam. Also, the effects of different parameters of PZT unimorph beams such as the length, width, and tip mass on generating capacity were studied by FEM. The results show that the energy harvester with PZT unimorph beam and tip mass is suitable for low-frequency vibration environment. Increasing the length or reducing the width of the beam can significantly lower the first-order modal frequency of energy harvester when other conditions remain the same. Within certain range, the first-order modal frequency of the beam also gradually reduced as the tip mass increasing. When the size of the PZT unimorph beam is 60x60x0.33mm, the tip mass is 8.92g and an exciting force of 0.01N is applied to it along z axis, an output of 8.1V can be obtained. Meanwhile, the PZT unimorph beam is under the first vibration mode and the resonant frequency is 16.296Hz.

1 Introduction

With the development of MEMS technology, wide application of small electronic components and Wireless Sensor Networks (WSNs), many disadvantages of the traditional way using chemical battery as the main energy supply have been revealed, one of the biggest defects is the short service life, which means that batteries are difficult to fundamentally solve the problem of providing sustainable energy supply [1]. In recent years, the technology of collecting vibration energy from ambient environment and transforming it into usable electricity has been a research hot-spot [2,3].Besides, the research on the power generator based on vibration is trending to three common orientations: piezoelectric power generator, electrostatic power generator and electromagnetic power generator. [5,6], compared with the other two methods, the piezoelectric power generator based on vibration has long life, simple in structure, easy to miniaturize, no electromagnetic interference and clean environment. It can provide and meet the electric energy demand for low power electronic products and WSNs working in low-frequency vibration ambient environment. Therefore, the research on piezoelectric power generator has drawn an increasing attention [7].

Previous studies have shown that the power generating capacity of energy harvester with PZT beams depends mainly on the structural parameters of the PZT beams and the vibration energy in ambient environment [8, 9]. Generally, the vibration energy is very limited and the frequency is relatively low in the environment. Under this circumstances, in order to make the piezoelectric energy harvester has good capacity to meet the power demand of small electronic components and WSNs working in low-frequency vibration environment (10-30Hz), we must adjust the structure and shape parameters of the piezoelectric vibrator to reduce the first-order modal frequencies. Most studies have shown that the PZT beam has a relatively larger power density in low-frequency vibration environment, and can obtain larger energy output [10]. Thus, through the establishment of mathematic model and FEM, the paper studied the influence on the first-order modal frequency and power generating capacity of the piezoelectric energy harvester, when the PZT beams has different length, width and tip mass with the aim of better meeting the requirements for energy supply using in low-frequency vibration ambient environment.

2 Mathematic model

The PZT unimorph beam is composed of metal substrate and piezoelectric layer, and these two layers are stuck together. Figure 1 shows the structural schematic diagram of the PZT unimorph beam. In the Cartesian coordinate system, we take the length direction as x axis and take the vertical direction as z axis.

When the PZT beam is driven by external excitation, the bending deformation along the z direction will appear. Usually, the ratio of length to thickness of the PZT beam is very large, so that the effects of shear deformation and rotary inertia are ignored. The bending deformation in the PZT beam will cause the change of stress and strain in piezoelectric body. Based on the direct piezoelectric effect, the piezoelectric body will be polarized on the upper and lower surfaces, and free charge is generated, thus the voltage output is formed between the two surfaces.

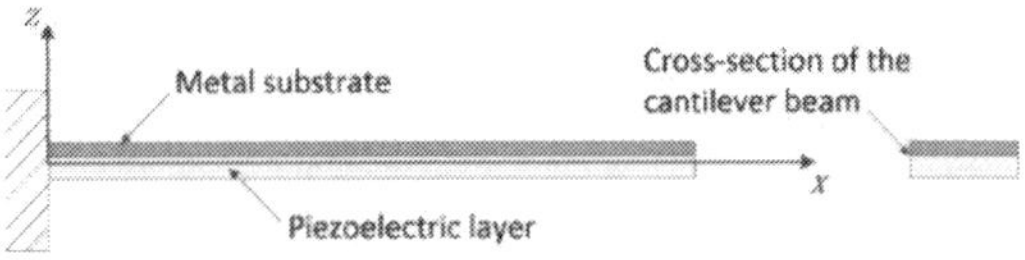

Fig 1. Structure schematic diagram of the PZT unimorph beam

We respectively use L , w and h to indicate the length, width and total thickness of the PZT unimorph beam. The subscript p and m stand for piezoelectric body and metal substrate. The ratio of h_m and h is defined as thickness ratio α , it is obvious that $\alpha = h_m / h$ and the thickness of the piezoelectric layer is $h_p = (1-\alpha)h$.

When a force is applied at the free end of the beam, the stress and strain generated in piezoelectric layer are obviously functions of α . According to the elastic theory of materials, the relationship between the stress and strain can be given as [11]:

$$T_p = E_p\left(S_1 - g_{31}D_3\right) \quad (1)$$

$$E_3 = -g_{31}T_p + \beta_{33}^T D_3 \quad (2)$$

where, $S_1 = -\rho z$ and T_p are the strain and stress generated in x direction, E_p and g_{31} are the young's modulus and piezoelectric voltage constant of the piezoelectric material, D_3 is the electric displacement in z direction, E_3 is the electric field strength in z direction, $\beta_{33}^T = 1/\varepsilon_{33}^t$ is the dielectric impermeability, and ε_{33}^t is the dielectric constant.

Because the metal substrate and piezoelectric body are different materials, the neutral layer of the PZT unimorph beam may not be the geometric center layer. Therefore, in this paper, the location of the neutral layer is determined by Euler-Bernoulli bean [4]. The PZT unimorph beam will bend along its neutral layer when an external force acts on the free end of the beam, in this situation, the distance between the top surface of the piezoelectric layer to the neutral layer can be written as:

$$Z_S = \frac{E_p \dfrac{h_p^2}{2} + E_p h_m\left(h_p + \dfrac{h_m}{2}\right)}{E_p + h_p + E_m h_m} = \frac{h}{2}\cdot\frac{(1-\alpha)^2 + \alpha\beta(2-\alpha)}{1-\alpha+\alpha\beta} \quad (3)$$

where, E_m is the young's modulus of metal substrate, $\beta = E_m / E_p$ is the elastic modulus ratio.

When an external force acts on the free end of the beam, the moment equation is given in [12] as:

$$M = n\int_0^w\int_{Z_S-h_p}^{Z_S} zT_p\,dydz + \int_0^w\int_{Z_S-h}^{Z_S-Z_p} zT_m\,dydz = (x-L)F \quad (4)$$

where n is the number of piezoelectric layer, $T_m = E_m S_1$ is the stress in the metal substrate along x direction.

From formula (1) and (4) can obtain equation:

$$\int_0^w\int_{Z_S-h_p}^{Z_S} zE_p\left(S_1 - g_{31}D_3\right)dydz + \int_0^w\int_{Z_S-h}^{Z_S-Z_p} zE_m S_1\,dydz = (x-L)F \quad (5)$$

By $S_1 = -\rho z$ and the formula (3) and (5), the curvature radius of the beam can be obtained as:

$$\rho = -\frac{6}{AE_p wh^3}\left[2(1-\alpha+\alpha\beta)(x-L)F + \alpha(1-\alpha)\beta wh^2 g_{31}E_p D_3\right] \quad (6)$$

and the coefficient A can be given as:

$$A = \alpha^4(1-\beta)^2 - 2\alpha\left(2\alpha^2 - 3\alpha + 2\right)(1-\beta) + 1 \quad (7)$$

The electric field strength can be obtained by combining (6) with (2), by integrating the electric field strength with z and then voltage can be expressed as:

$$V = \int\int_{Z_{S1}-h_p}^{Z_{S1}} E_3\,dz = \frac{1-\alpha}{8Awh}\left[6\alpha\beta g_{31}(L-x)F + \frac{Bwh^2\beta_{33}^T D_3}{1-\alpha+\alpha\beta}\right] \quad (8)$$

The coefficient B and k_{31}^2 can be written as:

$$B = A(1-\alpha+\alpha\beta)\left(1+k_{31}^2\right) - 3\alpha^2(1-\alpha)\beta^2 k_{31}^2 \quad (9)$$

$$k_{31}^2 = E_p g_{31}^2 / \beta_{33}^T \quad (10)$$

The formula (8) can be reduced to:

$$D_3 = \frac{(1-\alpha+\alpha\beta)}{(1-\alpha)\beta_{33}^T Bwh^2}\left[6\alpha(1-\alpha)\times\beta g_{31}(x-L)F + AhwV\right] \quad (11)$$

The charge generated on the piezoelectric electrode is obtained by integrating the electric displacement to the area. As the electrode is on the equal potential surface, so the charge generated on the piezoelectric surface is:

$$Q = \int_0^w\int_0^L D_3\,dydx = \frac{(1-\alpha+\alpha\beta)L^2}{\beta_{33}^T Bh^2}\left[-3\alpha\beta g_{31}F + \frac{AwhV}{(1-\alpha)L}\right] \quad (12)$$

The last formula is a general formula about the charge when external force and self-excited electric field of the PZT unimorph beam both exist. As for piezoelectric power generator, the impacts of voltage generated by its own to the power generator are ignored,

https://doi.org/10.1051/matecconf/201820804003

which means that $V=0$. Therefore the electric charge generated only by the external excitation can be expressed as:

$$Q_g=-\frac{3\alpha\beta(1-\alpha+\alpha\beta)g_{31}L^2}{\beta_{33}^T Bh^2}F \tag{13}$$

According to the relation between charge and voltage: $Q=CV$ and formula (12), the free capacitance of the piezoelectric vibrator is obtained as:

$$C_f=\frac{(1-\alpha+\alpha\beta)AwL}{(1-\alpha)\beta_{33}^T Bh} \tag{14}$$

The open circuit voltage can be given according to the formula (13) and (14).

$$V_g(F)=-\frac{3\alpha(1-\alpha)\beta g_{31}L}{Awh}F \tag{15}$$

At last, on the basis of formula: $U_g=Q_gV_g/2$, the U_g can be written as:

$$U_g=\frac{9(1-\alpha)(1-\alpha+\alpha\beta)\alpha^2\beta^2k_{31}^2L^3}{2ABE_pwh^3}F^2 \tag{16}$$

3 Finite element analysis

3.1 The establishment of finite element model

In order to fit the low-frequency vibration environment, we establish the finite element model of the PZT unimorph beam with tip mass .The piezoelectric body is PZT-5H, which has high electro-mechanical coupling coefficient, high piezoelectric strain coefficient, non-aging and large time constant. It is suitable to be used as a transducer [13]. The metal substrate is made of phosphorus bronze, which has large modulus of elasticity and strong fatigue resistance, can bear larger deformation. The structure and material parameters of the piezoelectric vibrator with tip mass are as follows:

Table 1. Structure and material parameters of the PZT unimorph beam

Structure and material	PZT-	Metal
Length of the beam, (mm)	60	60
Width of the beam, (mm)	10	10
Thickness of the beam, (mm)	0.22	0.11
Density of the materials, (kg/m³)	7500	8920
Young's modulus, (Gpa)	5.6	106
Poisson's ratio		0.35

The electrical boundary condition of the PZT unimorph beam is open connection. The elastic stiffness coefficient matrix ($\vec{c}\times1010N/m^2$), piezoelectric stress constant matrix ($\vec{e}\times C/m^2$) and relative dielectric constant matrix ($\vec{\varepsilon}$) of PZT-5H are respectively [13]:

$$\left[\vec{c}\right]=\begin{bmatrix}12.6 & 7.95 & 8.41 & 0 & 0 & 0\\ 7.95 & 12.6 & 8.41 & 0 & 0 & 0\\ 8.41 & 8.41 & 11.7 & 0 & 0 & 0\\ 0 & 0 & 0 & 2.35 & 0 & 0\\ 0 & 0 & 0 & 0 & 2.30 & 0\\ 0 & 0 & 0 & 0 & 0 & 2.30\end{bmatrix}$$

$$\left[\vec{e}\right]=\begin{bmatrix}0 & 0 & -6.5\\ 0 & 0 & -6.5\\ 0 & 0 & 23.3\\ 0 & 0 & 0\\ 0 & 17.0 & 0\\ 17.0 & 0 & 0\end{bmatrix}\quad \left[\vec{\varepsilon}\right]=\begin{bmatrix}1700 & 0 & 0\\ 0 & 1700 & 0\\ 0 & 0 & 1470\end{bmatrix}$$

The finite element analysis of the PZT unimorph beam is a coupled field analysis. The coupled field analysis process usually consider the cross effect and interaction of two or more engineering physical fields. In this paper, direct coupling method is adopted. The PZT-5H adopts Solid5 piezoelectric coupling unit, the metal substrate and tip mass adopt Solid45 entity unit. The influence of the bonding layer is neglected and it is assumed that the force and displacement between PZT-5H layer and metal substrate are continuous [14]. The FEM is shown in figure 2:

Fig 2. FEM of the PZT unimorph beam with tip mass

3.2 Modal analysis

In order to study the factors that influence the power generating capability of the piezoelectric energy harvester, the impacts of different length, width and tip mass of the cantilever beam on the modal frequency of piezoelectric vibrator is studied respectively.

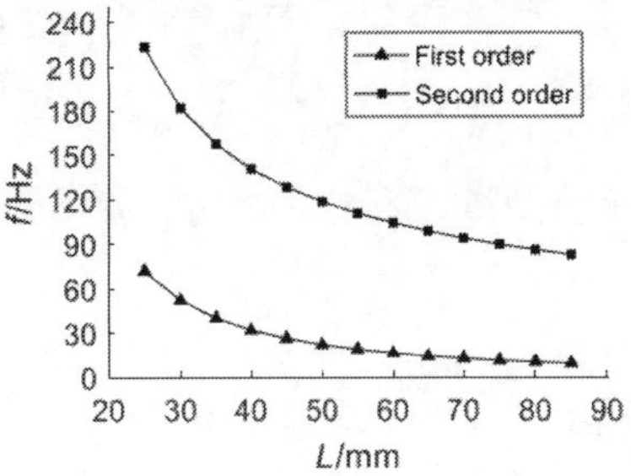

Fig 3. Variation of the PZT unimorph beam's first and second order modal frequencies with the length of the beam

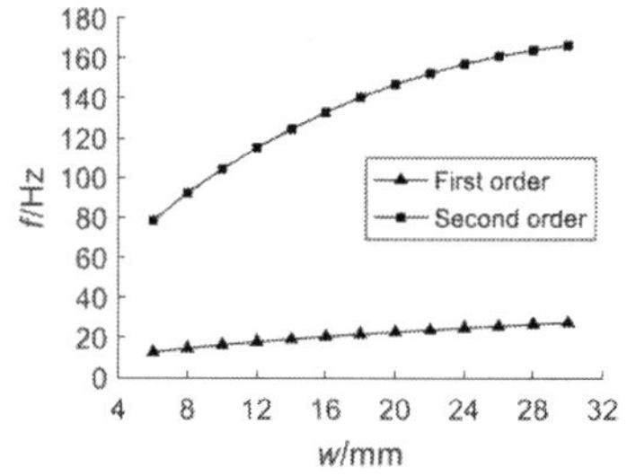

Fig 4. Variation of the PZT unimorph beam's first and second order modal frequencies with the width of the beam

The figure 3 shows that with the length of the cantilever beam increase, the first and second order modal frequencies are reduced. When the length of the beam is over 70mm, its effect gradually becomes weaker and weaker. Besides, we can see from the figure 4, with the width of the beam increasing, the first and second order modal frequencies are increased too.

What is shown in figure 5 is the trend of first and second order modal frequencies of the beam when the tip mass changed only. We can find that with the increase of tip mass, the resonant frequencies of first vibration mode are also gradually reduced.

According to the study, when the length is 60 mm ($L = 60mm$), the width is 10 mm ($w = 10mm$) and the tip mass at the end of the beam is 8.92g ($M = 8.92g$), the frequency at first vibration mode is 16.296 Hz.

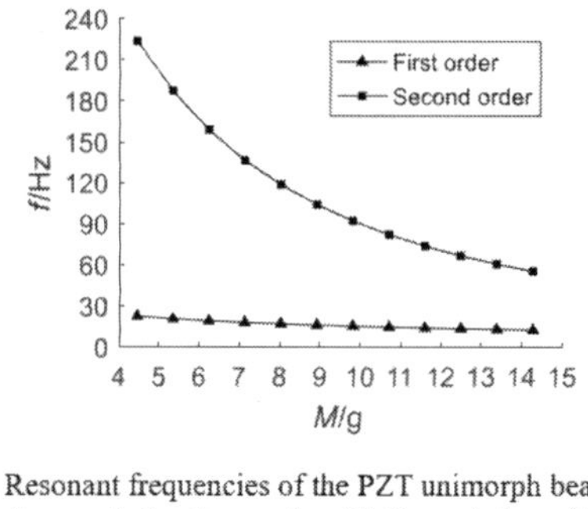

Fig 5. Resonant frequencies of the PZT unimorph beam under first and second vibration mode with the variation of the tip mass

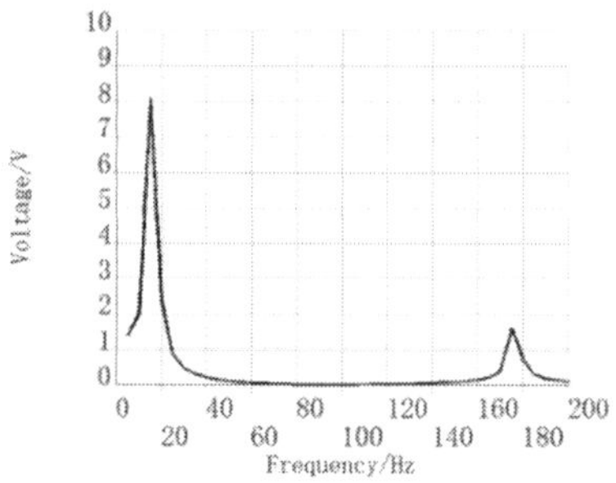

Fig 6. Relation between the voltage output and vibration frequencies

3.3 Harmonic response analysis

Harmonic response analysis was carried on when an external exciting force of 0.01N is applied in z direction on the tip mass. In ANSYS, the analysis frequency ranging from 0 to 200Hz, the voltage of lower PZT surface was set to 0, and the voltage on the upper surface of the PZT layer was coupled. Through the simulation, we can get the voltage output from the coupling point on top surface of the PZT beam. What is shown in figure 6 is the relation between voltage output and vibration frequencies.

We can see from figure 6, the peak voltage of first-modal frequency is about 8.08V, and because the second vibration mode of the PZT unimorph cantilever is distortion, the polarization of PZT layer along z axis is very small, which make the output voltage here is almost zero. Besides, the peak voltage output of the third-vibration mode is 1.65V.

4 Conclusion

In this paper, the energy harvester designed which based on the PZT unimorph beam can effectively convert the mechanical vibration energy into electrical energy in low-frequency vibration environment. The length, width and tip mass of the beam have different effect on the conversion ability. When the size of the PZT unimorph beam is 60x60x0.33 mm, the tip mass is 8.92g, the first-order modal frequency can be as low as 16.296 Hz. When an external exciting force of 0.01 N is applied in z axis at the tip mass at this time, the peak voltage output of the piezoelectric energy harvester at the first-modal is 8.08V, which can better meet the power requirement of small electronic components using in low-frequency vibration ambient environment. In practical applications, all the above parameters can also be optimized according to different ambient environment.

Acknowledge

The study is supported by Technology Division of China Railway (2017J004-H).

References

1. Shaikh F K and Zeadally S 2016 Energy harvesting in wireless sensor nesecondrks: A comprehensive review J. Renewable & Sustainable Energy Reviews. 55 1041-54
2. P Glynne-Jones, M J Tudor, S P Beeby and N M White 2004 An electromagnetic vibration-powered generator for intelligent sensor systems J. Sens. Actuator. A: Phys. 110 344-49
3. D Mallick, A Amann and S Roy 2015 A nonlinear stretching based electromagnetic energy harvester on FR4 for wideband operation J. Smart Mater. Struct. 24(1) 15013-026
4. Mitcheson P D, Miao P, Stark B H, Yeatman E M, Holmes A S and Green T C 2004 MEMS

electrostatic micropower generator for low frequency operation J. Sensors & Actuators A Physical. 115(2–3) 523-29

5. Sodano H A, Park G, Leo D J and Inman D J 2003 Use of piezoelectric energy harvesting devices for charging batteries C.Smart Structures and Materials 2003 Conference. 101-08
6. Wang S, Lam K H, Sun C L, Kwok K W, Chan H L and Guo M S 2007 Energy harvesting with piezoelectric drum transducer J. Applied Physics Letters. 90(11) 799
7. Cheng G M, Pang J Z, Tang K H Yang Z G, Ping Z and Kan J W 2007 Development of measuring system for electricity generating capacity of piezoelectric ceramics J. Journal of Jilin University
8. Roundy S, Wright P K and Rabaey J 2003 A study of low level vibrations as a power source for wireless sensor nodes J. Computer Communications 26(11) 1131-44
9. Jiang S, Li X, Guo S, Hu Y, Yang J and Jiang Q 2005 Performance of a piezoelectric bimorph for scavenging vibration energy J. Smart Materials & Structures. 14(4) 769
10. Jiang S N, Guo S H and Li X F 2012 Performance analysis for a unimorph cantilever piezoelectric harvester J. Journal of Vibration & Shock. 31(19) 90-94
11. Greve D W and Oppenheim I J 2003 Energy scavenging for sensor applications using structural strains J. Proceedings of SPIE - The International Society for Optical Engineering
12. Kan J W, Tang K. H, Wang S Y, Yang Z. G, Jia J and Zeng P 2008 Modeling and simulation of piezoelectric cantilever generators J Optics & Precision Engineering 16(1), 71-75
13. Luan G D, Zhang J and Wang R Q 2005 Piezoelectric Transducers and Arrays ed Wang Y and Sun Y (Peking University Press) chapter 3 pp 75-102
14. Peter Koh ke 1998 Ansys Coupled-Field Analysis Guide, Release5. Ansys Inc

Solar Power Plant Generation Short-Term Forecasting Model

Stanislav Eroshenko, *Elena* Kochneva, *Pavel* Kruchkov and *Aleksandra* Khalyasmaa

Automated Electrical Systems Department, Ural Federal University named after the first President of Russia B.N. Yeltsin, 19, Mira St., Yekaterinburg, Russia

Abstract. Recently, renewable generation plays an increasingly important role in the energy balance. Solar energy is developing at a rapid pace, while the solar power plants output depends on weather conditions. Solar power plant output short-term forecasting is an urgent issue. The future electricity generation qualitative forecasts allow electricity producers and network operators to actively manage the variable capacity of solar power plants, and thereby to optimally integrate the solar resources into the country's overall power system. The article presents one of the possible approaches to the solution of the short-term forecasting problem of a solar power plant output.

1 Introduction

The solar power plants in the power system leads to the emergence of an additional uncertainty source, in addition to the load behavior. This uncertainty implements to the calculating the balance of consumption and generation in the power system. The balance of the produced and consumed electric power is necessary both for maintaining the frequency in the power system and for ensuring the economical operation of energy system other generating units, as well as the possibility of trading in the electricity and capacity market. A high percentage of the solar power plants installed capacity in the energy system will inevitably lead to an increase in the influence of the uncertainty factor of the solar plant power output on the electric grid operation. The reserve created to cover this uncertainty should be as small as possible to maximize the benefits of the solar power plants deployment.

According to the Russian Federation Government Resolution №449 "About using the renewable energy sources stimulation mechanism in wholesale and retail energy markets" taken on 28 may 2013 there is competitive selection of the investment projects for the generating facilities based on the renewable energy sources using construction. These competitive selections are held 4 years ahead for solar power plants. The results of selections held in 2013-2014, 76 objects with total capacity 1081 MW have been selected. It is expected the selections in 2017-2018 will result in the solar power plants installed capacity increasing and covering the wide range of regions.

The solar energy market growth dynamics is characterized by the total volumes of installed capacities introduced and planned for commissioning. According to the list of renewable energy projects which were chosen during the competitive selection in 2017, 26 solar generating facilities are planned for commission in the period from 2020 to 2022 in unified energy system of Russia. The total installed capacity of these facilities is more than 500 MW [1-3]. The objects construction geography covers the whole southern part of the Russia territory. The single object installed capacity is in the range from 15 to 45 MW.

Thus, nowadays the solar power plants are independent participants of the wholesale electricity market and consequently can participate in competitive procedures. Thus, the function of solar power plant generation short-term forecast is required for trading on the day-ahead market. In turn, this function is a complex multicriteria task that is based on meteorological and climatic data, generating facility state data and the adjustment electrical network topological and mode data.

Taking into account the solar generation development intensity as well as the current market needs engineering company "ProSoft Systems" Ltd together with Ural Federal University named after the first President of Russia B.N. Yeltsin carried out the development and implementation of a short-term forecasting system for the electricity generation in photovoltaic plants in the framework of research and development work.

2 Initial data

Solar power plants generation short-term forecast is carried out on the "day ahead" interval for the purpose of ensuring the participation of photovoltaic generation in electricity markets commercial procedures and also on the 2-7-day perspective in order to solve medium-urgent object generation planning and power plant equipment maintenance and repair planning.

 https://doi.org/10.1051/matecconf/201820804004

Initial data for solar radiation energy flux density forecast at the ground surface falling on a horizontal surface forecast

Geographical coordinates		
Latitude, [degree]		
Longitude, [degree]		
Local time		
Daylight saving time, [h]		
Estimated time interval		
Date (number of day in the year), [units]		
Weather data		
Solar radiation energy flux density on a horizontal surface, [W/m^2]		(actual measurements)
Cloud cover, [%]		(actual measurements or weather provider data)
recommended retrospective depth is 14 days		
Constants	**1.**	
Solar constant, [W/m^2]		

Fig 1. Initial data for the forecast first stage

The short-term forecasting function is implemented into the software package Energosfera 8.0 produced by "ProSoft Systems" Ltd. "Energosfera 8.0" is designed for automated systems of fiscal and technical metering for various types of energy resources (electricity, heat, water, steam, gas, etc.) building [4]. Solar plant accounting and monitoring devices automated data collection, data storage, primary data processing and information presenting to the software product user in the reports, graphs, tables and logs form is provided.

The following main data sources are used in the automated meter reading system Energosfera 8.0 for the correct short-term forecasting module operation:

- Horizontally mounted pyranometer for measuring the total energy flux density of solar radiation, [kW/m^2];
- Online meteorological forecast data for solar power plant geographical coordinates [cloudiness, [%]; description of meteorological conditions and events);
- Data from local weather station installed on the solar plant territory (wind speed, [m/s]; wind direction, [degrees]; air temperature, [°C]; air humidity [%]).

3 Solar energy plant generation "a day ahead" forecast

The algorithm for calculating the short-term forecast is practically implemented in three stages [5-7]:

1. Calculation of the solar radiation energy flux density forecast at the ground surface falling on a horizontal surface, [kW/m^2].
2. Calculation of the energy solar radiation flux density at the ground surface incident on the solar panel inclined plane, [kW/m^2].
3. Calculation of the hourly solar power plant output based on the current data on the state of plant generating equipment, [kWh].

Measurement data obtained from the monitoring systems established in the solar plant territory, the weather forecast data for the present day and calculated values determining the sun\s movement nature are used for the solar radiation energy flux density forecast at a horizontal surface. Figure 1 presents the initial variables for the forecast first stage.

Solar radiation flux density change estimation when passing through the Earth's atmosphere is carried out using the normalized parameter called transparency coefficient. This coefficient takes into account the effects of such factors as the absorption of solar radiation by ozone, air molecules, water vapor, and cloud cover. The parameter "cloudness" in percentage terms is the most significant among the metrological forecast factors from the point of view of solar power plant generation forecast.

Meteorological condition short-term forecasting in the territory of Russian Federation is carried out in accordance with the current governing documents. For example, according to the national guidelines in the firld of meteorology the cloudness forecast should correspond to the 8-point scale (step 0,125 pu) [7]. However according to international standards the scale for the same variable should be divided into 10 points (step 0,100 pu). In a number of cases, a quantitative estimate of the cloudiness in % is not presented, and a verbal description of weather conditions and events such as "fog", "cloudy", "cloudy with clearings," etc. is provided. As part of the software package Energosfera 8.0, an algorithm of recalculating the weather conditions and events in the corresponding values of cloudiness in percent is implemented.

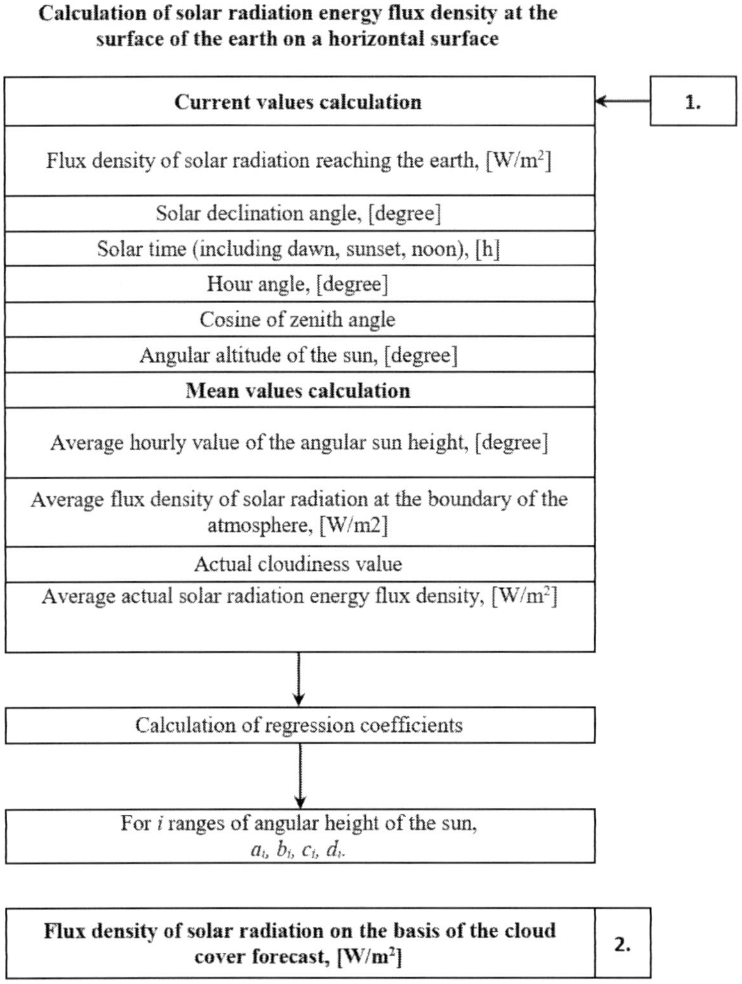

Figure 2. Block diagram of solar radiation flux density near the earth surface, falling on the horizontal plane calculation.

The transparency coefficient change is of a random nature, while the individual characteristics of the solar power plant location area have a great influence. Identification of the transparency coefficient statistical dependence on key characteristics is carried out using the multiple regression model. The block diagram of the program forecasting module "a day ahead" of the solar radiation flux density near the earth surface, falling on the horizontal plane, is presented in Figure 2.

After obtaining the required value of the solar radiation flux density according to the methodology presented above, a forecast is made for the solar power plant electricity generation. The procedure consists of three stages.

A. Determination of the photovoltaic panels direct current output power

The design power of one photovoltaic panel is determined by the equation [8-9]:

$$P_{PV} = Y_{PV} f_{PV} \left(\frac{\overline{G}_T}{\overline{G}_{T,STC}} \right) [1 + \alpha_p (\overline{T}_c - T_{c,STC})], \qquad (1)$$

where P_{PV} is power of one photovoltaic panel, [W]; Y_{PV} is rated photovoltaic panel power at standard conditions (nameplate value), [W]; f_{PV} is reduction coefficient of the overall panel efficiency as a result of degradation of photoelectric modules, [pu]; $\overline{G}_{T,STC}$ is the solar radiation energy flux density under standard conditions, is assumed to be 1000 W/m²; α_p is the temperature coefficient of power (passport value), [%/°C]; $\overline{T}_c$ is photovoltaic panel temperature, [° C]; $T_{c,STC}$ is photovoltaic panel temperature under standard conditions, is assumed to be 25 ° C.

The presented model of power calculation, produced by a single photovoltaic panel, is universal for all types of crystalline photocells and photocells used on the basis of amorphous silicon and allows to take into account external conditions affecting the efficiency of panel production, and also take into account the degradation of the panel in the process operation.

B. Determination of the inverters output power

Semiconductor devices that provide power to the AC network can be also highlighted as part of the main solar power plant equipment. The used inverter model allows us to take into account the losses that occur during the conversion of a direct current into a variable one [10]:

$$P_{ac} = \frac{P_{ac_0}(P_{dc} - P_{s_0})}{P_{dc_0} - P_{s_0}}, \qquad (2)$$

where P_{ac} is output power on the AC side of the inverter, [W]; P_{ac_0} is rated power on the AC side of the inverter (passport characteristic), [W]; P_{dc} is power generated at the DC side, [W]; P_{dc_0} is rated power on the DC side of the inverter (passport characteristic), [W]; P_{s_0} is power consumption of the inverter (passport characteristic), [W].

C. Definition of solar power plant electricity generation

Determination of solar power plant electricity generation hourly value will be presented in further papers. The main influence on the result at this calculation stage is rendered by the generating facility electrical connection scheme, as well as the composition of the main equipment i.e. the types and number of photovoltaic panels, the types and number of inverters/converters, the equipment of the substation distribution device.

4 Conclusion

The developed short-term forecasting system is based on the physical model of solar radiation propagation in the Earth's atmosphere and allows to fulfill a short-term forecast of solar power generation based on the weather forecast data provided by the meteorological provider and a retrospective measurement of solar radiation flux density obtained from meteorological sensors installed on the territory of solar power plant.

A technique for short-term forecasting of the solar power plant is proposed. Taking into account the volume of the forecast information, a physical model is used to determine the averaged solar radiation entering the unit area of the horizontal surface of the Earth. The subsequent accounting of the geometric location of solar panels at the station, the description of the model of solar panels, converters, inverters, as well as the passport data of the substation equipment, will allow to determine the generation of a solar power station for the forecasted weather conditions. Since the physical model uses the meteorological forecast with hourly averaging as its input, its accuracy is equivalent to the accuracy of the forecast itself.

References

1. List of renewable energy projects selected according to the results of competitive selection in accordance with the Registry of selection of investment projects for construction of generating facilities operating on the basis of renewable energy sources/ Association of noncommercial partnership "Market Council". URL: https://www.atsenergo.ru/vie/proresults
2. William Glassley, Jan Kleissl. Current state of the art in solar forecasting. California renewable energy collaborative final report. (2013)
3. Utility-Scale Solar Photovoltaic Power Plants^ A Project Developer's Guide. International finance company. (2015)
4. Software package Energosfera. ProSoft Systems Ltd. URL: http://www.prosoftsystems.ru/catalog/show/programmnyj-kompleks-jenergosfera1
5. Mohamed Abuella, Badrul Chowdhury. Solar Power Probabilistic nForecasting by Using Multiple Linear Regression Analysis. Department of Electrical and Computer Engineering, University of North Carolina, USA. (2015)
6. D.A. Snegirev, R.T. Valiev, S.A. Eroshenko, A.I. Khalyasmaa. Functional assessment system of solar power plant energy production. *Proceedings of the 8th International Conference on Energy and Environment (CEIM)*. Pp 397-353, (2017).
7. D.A. Snegirev, R.T. Valiev, S.A. Eroshenko, A.I. Khalyasmaa. Algorithmic realization of short-term solar power plant output forecasting. *Proceedings of the II* International *Conference on Control in tecgnical systems (CTS'2017)*. Pp 228-231. (2017).
8. Guidance Document 52.27.724-2009 «Manual on Short-Term Weather Forecasts of General Purpose» / URL: http://docs.cntd.ru/document/437152648
9. Chiou-Jye Huang, Mao-Ting Huang, and Chung-Cheng Chen. A Novel Power Output Model for Photovoltaic Systems. *International Journal of Smart Grid and Clean Energy*. – № 2. – Pp 139-147. (2013)
10. D.L. King, S. Gonzalez, G. M. Galbraith and W. E. Boyson. Performance Model for Grid-Connected Photovoltaic Inverters. Albuquerque, NM, Sandia National Laboratories. 47 p. (2007).

Acknowledgements

The work was supported by Act 211 Government of the Russian Federation, contract № 02.A03.21.0006.

Solar power generation short-term forecasting model's implementation experience

Elena Kochneva

Automated Electrical Systems Department, Ural Federal University named after the first President of Russia B.N. Yeltsin, 19, Mira St., Yekaterinburg, Russia

Abstract. Recently there is significant increase in the installed capacity of solar power plants in Russia. Thereby there are issues of solar power plants owners information support for participation in wholesale electricity market. The paper describes the experience of short term forecasting system practical implementation. The system is proposed for forecasting the solar power plant generation "a day ahead" as a part of the software for automatic meter reading systems "Energosfera". The short-term forecasting program modules structure, key parameters and characteristics used during the forecasting process description is presented.

1 Introduction

Energy consumption on the planet is constantly growing, and at the same time, humanity faces the task of reducing greenhouse gas emissions to prevent global climate change. Therefore, the development of renewable energy sources is becoming more and more important. Solar generation is one of the most promising and actively developing areas of renewable energy. Solar power plants are now being created in the territory of about 100 countries not by chance [1]. The pace of solar power plants development around the world in 2016 surpassed the pace of coal power station development for the first time. The International Energy Agency came to this conclusion in its report. Two-thirds of the cumulative increase in capacity of all power plants in the world fell to solar ones.

Today, in the world solar energy industry the solar energy production forecasting for short periods (day, several days, a week) does not have a tried and tested technology until the end and is often associated with large errors that may achieve 60-65%. At the same time, the longer the forecast period is the more accurately one can predict the result of solar power plant output. This rule is quite correct for the area, i.e. the point forecast is always less accurate. The accuracy of the forecast increases significantly during the geographical region area increasing of for which the forecast is made.

The electricity amount forecast produced by the solar station is the prediction of the solar radiation amount that will received by solar panels. It depends on many factors, the main ones are climatic and meteorological conditions, i.e. the position of the sun in the sky, the duration of daylight, cloudiness, precipitation, wind force, etc.

Short-term solar power plants output forecasting serves to

- decrease of uncertainty in planning modes;
- creation of a tool for the participation of station owners on the day-ahead market and the balancing market; creation of a tool for the financial planning of station owners.

2 Forecasting approaches

There are four main approaches to forecasting the solar power plants generation [2-4]:

Physical models describe physical relationships between weather conditions, solar radiation and station power generation. The input includes a numerical weather forecast, local meteorological measurements, terrain and surface type data, as well as data on the actual the power plant output. In addition, satellite systems can be used to track the direction and speed of the clouds, which allows predicting solar radiation in real time.

Statistical models describe the relationship between the solar radiation flux density, obtained from numerical weather prediction and the generation of solar energy directly by statistical analysis of time series by retrospective data without taking into account physical processes. This method is widely used for medium and long-term solar power generation forecasting.

Adaptive and trained models use artificial intelligence systems to derive a link between the forecasted weather conditions and output power, while "learning" the model is based on the selected retrospective data.

Initial data for calculating the solar power plant generation

Data on the photovoltaic panels position, on the terrain	
Angle of photovoltaic panels inclination, [degree]	
Azimuth of the photovoltaic panel, [degree]	
Surface albedo, [units]	
Meteorological data	
Relative humidity, [units]	forecast
Air temperature, [°C]	forecast
Wind speed, [m/s]	forecast
Solar power plant reference data	
Station wiring diagram	
Number of consecutively connected panels, [pieces]	
Number of parallel panels, [pieces]	
Passport data of photovoltaic panels	
Rated power in standard conditions, [kW]	
Efficiency, [units]	
Temperature coefficient of power, [units/%]	
Degradation of panels, [units]	
Passport data of the inverter	
Number of inverters, [units]	
Inverter rated power at alrternative current in standard conditions, [kW]	
Inverter rated power at constant current in standard conditions, [kW]	
Temperature coefficient of power, [units/%]	
Maximum inverter efficiency, [%]	

Initial data for calculating the power of a solar power plant

Calculation of the solar radiation flux density on an inclined plane
Scattered radiation, [W/m^2]
Direct radiation, [W/m^2]
Anisotropy coefficient, [units]
The solar radiation flux density on the inclined panel, [W/m2]
Calculation of panel operation mode
calculation of the photocells temperature, [°C]
Average panel power for 1 hour, [kW]
Calculation of inverters operating mode
Power consumption of inverters, [kW]
Output power of inverters, [kW]

↓

Solar power plant output, [kW]

Fig 1. Solar power plant output forecast block diagram.

Unlike the statistical approach, the learning approach uses algorithms that can implicitly describe complex nonlinear relationships between weather conditions and the power generated by the power plant, which can not always be taken into account in statistical analysis. For both statistical and learning models, the determining value for the accuracy of the forecast renders a sample of the input data used in constructing the model.

Combined forecasting models are often a combination of physical and statistical models. The physical approach is not always justified because of the complexity or impossibility of taking into account certain factors, which requires the introduction of a statistical approach for their determination.

The statistical approach can have considerable accuracy in calculating the averaged values of solar radiation power over a long period of time (day, month, year), but it has much less accuracy in a shorter time interval (minutes, hours) when physical processes, for example, cloudiness, can not be averaged over a given interval. Optimum accuracy of combined models is achieved by adjusting statistical models to constantly changing conditions described by physical models.

Taking into account the specifics of the task to be solved such as the creation of a solar power plant operational and short-term forecasting system, as well as the initial information limited and uncertainty in the absence of retrospective data, it is most appropriate to use physical models that make possible to obtain a forecast for the station based on the minimum composition of the forecast meteorological information. In order to refine the estimates obtained, statistical models are additionally implemented. The solar power plant generation forecast is based on the calculation model of solar radiation intensity at the power plant territory.

 https://doi.org/10.1051/matecconf/201820804005

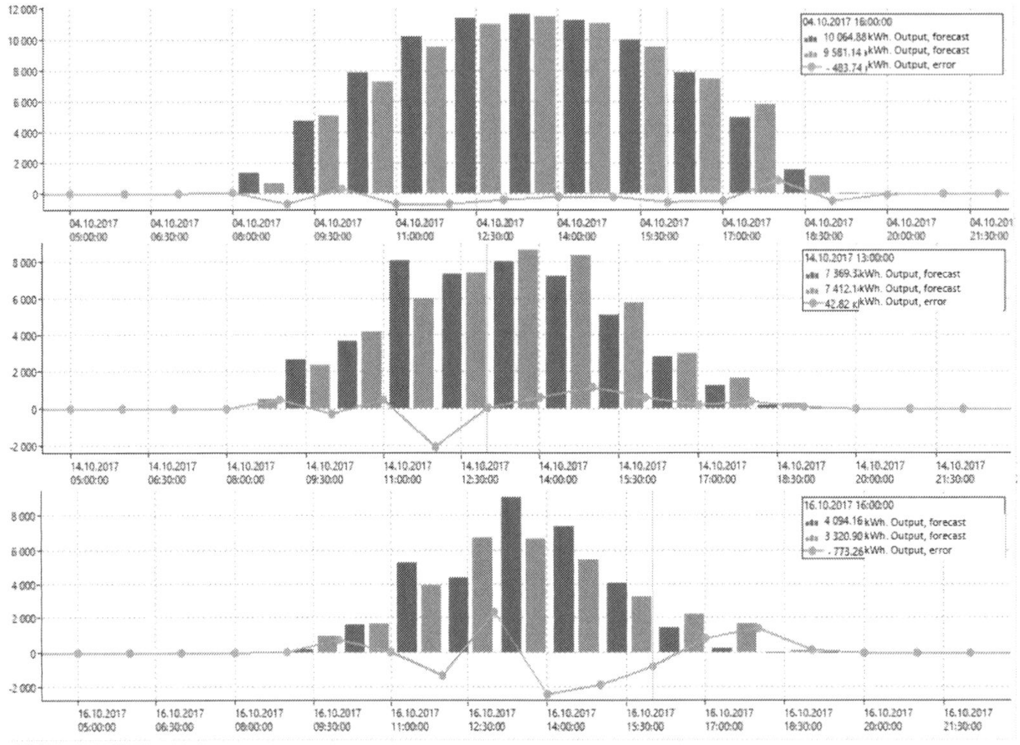

Fig 2. Power plant generation short-term forecast examples

Table 1. Short-term forecast error parameters

Date	Average cloudiness, p.u.	Average error for the period, MAPE, %	Standard deviation, %
04.10.2017	0,00	1,426	2,818
14.10.2017	0,76	1,088	5,294
16.10.2017	0,90	0,491	9,126
01.10.2017 - 31.10.2017	0,55	-2,740	20,731
01.11.2017 - 30.11.2017	0,62	-4,753	17,253
01.12.2017 - 31.12.2017	0,63	-6,318	19,058

3 Solar power plant output forecast

The solar power plant generation forecast is based on the calculation model of solar radiation intensity at the power plant territory [5]. Calculation of the solar radiation incident on the photovoltaic panel surface includes the determination of three values, which are solar radiation on the boundary of the earth's atmosphere, solar radiation on the horizontal surface of the earth and solar radiation on the inclined surface [6-7].

The value of solar radiation on the boundary of the earth's atmosphere is calculated from the geometric model of the Sun's motion. The value of the solar constant which is 1367 W/m2, the angle of solar declination for a given day in a year, as well as the dependence of the solar radiation on the geographical position (longitude and latitude) of the calculated point are taken into account.

To estimate the change in the solar radiation during its passing through the Earth's atmosphere, the transparency coefficient is used. The coefficient of transparency takes into account the absorption and reflection of solar radiation into outer space by different layers of atmosphere. The calculation of this coefficient is performed on the basis of the regression model, including the actual and forecast cloud data, as well as the mean values of the angular height of the sun within the specified time step [8].

The cumulative energy flux density (intensity) of solar radiation at the earth's surface incident on the inclined surface is the sum of direct radiation and scattered radiation. Direct radiation is determined from the calculation of solar radiation on the boundary of the earth's atmosphere, transparency coefficient, the angle of inclination of the panels, and the scattered radiation depends on the reflectivity (albedo) of the earth's surface and the humidity of the air.

After receiving the values of the solar radiation falling on the photovoltaic panels, a forecast is made for the generation of electricity from the solar power plant. The nominal power of the photoconductor, the reduction in the overall efficiency of the panel, the calculated value of intensity) of solar radiation at the earth's surface incident on the inclined surface, the panel efficiency, the panel's temperature coefficient, ambient temperature and wind speed are taken into account when calculating the generation power of a photoconductor.

To calculate the solar plant total power, the total number of photovoltaic panels participating in the generation of electricity for a given time interval is used. In addition, losses in AC and DC installations, i.e. inverters, transformers and conductors of the power station electrical network are taken into account [9-10]. Determination of solar power plant electricity generation hourly values is carried out in accordance with the block diagram shown in figure 1.

4 Forecast error evaluation

The average absolute error in percent of MAPE is used to analyze transparency coefficient short-term forecasting models. During the foreign experience analysis the impossibility of MAPE error evaluation criteria in a traditional form using for the statistical evaluation of the significance and comparative analysis of the solar energy forecasting model efficiency was established. The solar power plant installed power is assumed as a value, in relation to which the generation forecast MAPE error is calculated [7].

$$MAPE = \frac{1}{N}\sum_{i=1}^{N}\left|\frac{\overline{P}_i^{fact} - \overline{P}_i^{forecast}}{P_{PV}}\right|\cdot 100\% \quad (1)$$

where MAPE is the average absolute error in percent, [%]; N is the quantity of observations; i is the number of the observation; $\overline{P}_i^{fact}$ is the solar power plant active power generation actual value for i-th observation, [kW]; $\overline{P}_i^{forecast}$ is predicted value of solar power station active power generation for i-th observation, [kW]; P_{PV} is the installed capacity of solar power station, [kW].

5 Model practical realization

At the moment the short-term forecasting system is introduced into trial operation as part of the software package "Energosfera 8.0" at Zavodskaya solar power plant.

As part of the solar power plant main equipment, there are solar panels with a nominal power of 270 watts (the total number is 56,600 units). The total number of installed inverters is 16.

The solar panels have a south orientation and an inclination angle of 30 degrees, there is no solar tracker.

Meteorological information comes with a discreteness of 1 hour from an external weather provider and includes: forecast cloudiness, air temperature, humidity, wind direction and speed.

Examples of calculating the solar power plant generation forecast "a the day ahead" are presented in figure 2 for the following types of weather conditions: clear, cloudy with clearings, overcast respectively. The forecasting error summary parameters in percent relative to the installed capacity of the power plant for October-December 2017 are presented in Table 1.

6 Conclusion

As a result of the system pilot integration, the necessary statistical data was collected at the operating solar power plant. These data are necessary for modernization and improvement of the forecasting system. The forecast error established to the solar power plant for the period from October 1, 2017 to December 31, 2017 was 4.6% which is comparable to the accuracy of global practice forecasts. The error standard deviation for the period under consideration is estimated at 19%. Deviation occurs due to the random nature of weather conditions, as well as the provider weather forecast error.

The presented system of solar power plant generation short-term forecasting "a day ahead" is universal and can be used at any operating solar power plant facility as part of the software package "Energosfera 8.0" produced by ProSoft Systems Ltd.

To date, ProSoft Systems Ltd together with with Ural Federal University is developing a short-term forecast prompt correction system based on solar radiation energy flux density current measurements. This correction will ensure the possibility of solar power plant owners participation in wholesale market intradaily trading procedures.

References

1. List of renewable energy projects selected according to the results of competitive selection in accordance with the Registry of selection of investment projects for construction of generating facilities operating on the basis of renewable energy sources/ Association of noncommercial partnership "Market Council". URL: https://www.atsenergo.ru/vie/proresults
2. Heinemann, D., Lorenz E., Girodo M. Forecasting of solar radiation. Oldenburg University, Institute of Physics, Energy Meteorology Group.
3. Bacher, P., Madsen, H., HA Nielsen Online short-term solar power forecasting. Solar Energy. Vol 83, Issue 10, October 2009: 1772-1783. (2009)
4. C. Chen, S. Duan, T. Cai, and B. Liu, "Online 24-h solar power forecasting based on weather type classification using artificial neural network," Solar Energy, vol. 85, no. 11, pp. 2856-2870. (2011)
5. Software package Energosfera / ProSoft Systems Ltd. URL: http://www.prosoftsystems.ru/catalog/show/programmnyj-kompleks-jenergosfera1
6. D.A. Snegirev, R.T. Valiev, S.A. Eroshenko, A.I. Khalyasmaa. Functional assessment system of solar power plant energy production. *Proceedings of the* 8th International *Conference on Energy and Environment (CEIM)*. Pp 397-353. (2017).

7. D.A. Snegirev, R.T. Valiev, S.A. Eroshenko, A.I. Khalyasmaa. Algorithmic realization of short-term solar power plant output forecasting. *Proceedings of the II International Conference on Control in tecgnical systems (CTS'2017).* Pp 228-231. (2017).
8. Guidance Document 52.27.724-2009 «Manual on Short-Term Weather Forecasts of General Purpose» / URL: http://docs.cntd.ru/document/437152648
9. Chiou-Jye Huang, Mao-Ting Huang, and Chung-Cheng Chen. A Novel Power Output Model for Photovoltaic Systems. *International Journal of Smart Grid and Clean Energy.* № 2. – Pp 139-147. (2013)
10. D.L. King, S. Gonzalez, G. M. Galbraith and W. E. Boyson. Performance Model for Grid-Connected Photovoltaic Inverters. Albuquerque, NM, Sandia National Laboratories. 47 p. (2007)

Acknowledgements

The work was supported by Act 211 Government of the Russian Federation, contract № 02.A03.21.0006.

A Cognitive Framework to Secure Smart Cities

Shahab Tayeb[1], Neha Raste[1], *Matin* Pirouz[2], and *Shahram* Latifi[1]

[1]Department of Electrical and Computer Engineering, University of Nevada, Las Vegas, USA
[2]Department of Computer Science, University of Nevada, Las Vegas, USA

Abstract. The advancement in technology has transformed Cyber Physical Systems and their interface with IoT into a more sophisticated and challenging paradigm. As a result, vulnerabilities and potential attacks manifest themselves considerably more than before, forcing researchers to rethink the conventional strategies that are currently in place to secure such physical systems. This manuscript studies the complex interweaving of sensor networks and physical systems and suggests a foundational innovation in the field. In sharp contrast with the existing IDS and IPS solutions, in this paper, a preventive and proactive method is employed to stay ahead of attacks by constantly monitoring network data patterns and identifying threats that are imminent. Here, by capitalizing on the significant progress in processing power (e.g. petascale computing) and storage capacity of computer systems, we propose a deep learning approach to predict and identify various security breaches that are about to occur. The learning process takes place by collecting a large number of files of different types and running tests on them to classify them as benign or malicious. The prediction model obtained as such can then be used to identify attacks. Our project articulates a new framework for interactions between physical systems and sensor networks, where malicious packets are repeatedly learned over time while the system continually operates with respect to imperfect security mechanisms.

1 Introduction

The world is at the brink of a new digital revolution and Cyber Physical Systems (CPS) based on the Internet of Things (IoT) networks mark the next frontier. IoT allows companies to increase productivity, city services to converge, vehicles to become autonomous, and homes to become smarter. There has been much research on the design, evaluation, testing, and verification of CPS and its associated IoT. Nonetheless, research on the development of security models and frameworks for IoT networks is very limited. A key challenge is that security solutions for IoT should not hinder the openness of the network, nor should they introduce additional latency or overhead to communications across the network. These requirements are achieved by incorporating security into the design of IoT infrastructures. This project is focused on two main principles: "adaptive security architecture" and "IoT-based CPS or ICPS" both of which are listed on Gartner's 2016 top 10 strategic technology trends.

Dozens of hardware platforms of embedded systems are gaining popularity as prototypes of IoT [1-2]. Smart objects and embedded sensors are currently secured based on the same best practices of traditional networks without considering the limitations imposed by the proliferation of smart nodes in terms of processing power and memory. This is mainly due to limited research in this field. Encapsulation of protocol stack layers is done on a single hardware processor and thus, leaving the lower layers unprotected has detrimental effects. With so many new forms of data, new forms of threats will come to existence targeting them.

Firewalls, Intrusion Detection Systems (IDS), and Intrusion Prevention Systems (IPS) can be found as standalone platforms, or as modules integrated into other hardware, or even as software applications with the two categories of IDS being Network-based and Host-based IDS. New generations of devices bring along newer and more sophisticated generation of threat agents and attacks. This concern is addressed by integrating security in design and thus, preventing the problem from happening. ICPS lack a secure design for implementation. Because IoT systems utilize diverse protocols and technologies encompassing a wide array of technology concepts such as Application Programming Interfaces (APIs), sensor-equipped edge devices, and messaging protocols, they are prone to different attacks. Additionally, lack of standardization to support IoT increases heterogeneity of these networks and introduces inoperable components which will create vulnerabilities. Because of utilizing a wide array of heterogeneous and often unreliable smart objects, there is a need for a reliable design model capable of supporting bandwidth-intensive applications.

The design objectives of this framework are twofold: first, to address security concerns; and second, to provide on-demand security guidelines for the next generation of CPS. The research questions are: a) What are the security vulnerabilities and challenges presented by the emerging technologies (e.g. 802.11.5, ZigBee, GPRS,

LTE) in providing IoT connectivity? b) Can Deep Learning (DL) be as successful on IoT security as it has been in computer vision and speech recognition? c) Can security by design guideline and frameworks outperform the existing security patches and protocols? and d) How different are the security gaps for smart city sensors and gateways from those of traditional networks.

2 Vulnerability analyses

2.1. IoT Security

The IoT is composed of many layers of technologies, each with its own set of security challenges. Smart devices are more capable of gathering and curating sensed data which makes them more susceptible to being targeted by a variety of attack types from single target impersonation, rogue nodes, and privileged access to batched ones such as botnets and DDoS. It has been reported by FCC's Technological Advisory Council (TAC) that hackers have the lead in breaching the IoT security. The reasons are threefold: i) Conventional network security wisdom is not applicable to the IoT realm. IoT is an ecosystem driven by business gaps, rather than just a myriad of devices; ii) IoT manufacturers don't prioritize security and lack a security culture. IoT vendors compromise security to gain functionality and openness for a broader target market. IoT manufacturers follow agile manifesto for their development process which opens up many security gaps; and iii) There are inherent vulnerabilities in individual IoT nodes: a) For many types of IoT devices, physical access cannot be restricted, thus devices that expose critical information on internal nodes can be compromised; b) Although chip manufacturing innovations have led to the emergence of embedded chips with hardware-based security (e.g. ARM TrustZone) and hardware with cryptography support (e.g. ARMv8), the inclusion of such chips in every device is cost prohibitive. Thus, it makes sense to look for network security solutions that do not require modification of existing and emerging IoT devices; and c) IoT nodes generally don't support advanced networking capabilities and in particular security protocols. The proposal aims to advance insight to IoT and identify its vulnerabilities, while attempting to develop methodologies to guard against cyber-attacks that can penetrate the IoT layers through a wide range of heterogeneous devices. Securing systems from a network design perspective defines security zones and layers based on data requirements of each network segment, independently of device type and location. This is different from encrypted IoT chips and restricted physical access to IoT nodes, and enhances protection against zero-day attacks and well-known threats.

2.2 Smart City

Cities are rapidly converging toward digital technologies to provide advanced information services, efficient management, and resource utilization that will positively impact all aspects of our life and economy. This has led to the proliferation of ubiquitous connectivity to critical infrastructures (electrical grid, utility networks, health care, finance, etc.) that are used to deliver advanced information services to homes, businesses, and government. On the other hand, such smart systems are more complex, dynamic, heterogeneous, and have many vulnerabilities that can be exploited by cyberattacks. Protecting and securing the resources and services of smart cities become critically important due to the disruptive or even potentially life-threatening nature of a failure or attack on smart cities' infrastructures.

A resilient architecture that protects smart cities' communications, controls, and computations based on autonomic computing and Moving Target Defense (MTD) techniques was proposed in [3]. The key idea was to make it extremely difficult for the attackers to figure out the current active execution environments used to run smart city services by randomizing the use of these resources at runtime.

An important part of Smart City is wireless communication networks which are pervading the IoT realm due to their fast, easy, and inexpensive deployment. Pervasive wireless technologies have higher security requirements. Even though the existing security protocols for wireless communications address the privacy and confidentiality issues, various unaddressed vulnerabilities exist. Such vulnerabilities target cyber and physical availability of the systems, spoof data link and network layer addresses protocols, or even upper layer session hijacking.

Table 1. Mapping known attacks to smart city

Attacks	Main Characteristics	Mapping to Smart City
DoS	Rendering a device unusable through exhaustion of target's resources	Smart City sensors/loggers have more limited resources (e.g. CPU & RAM)
DDoS	A type of DoS where the source are thousands of zombies	50 billion devices targeted to become zombies and the same 50 billion are potential victims
IP Spoofing	An unauthorized use of someone else's logical address	More valid addresses increase susceptibility of spoofing attacks
Physical Attack	Someone getting physically close to network components	More connected devices equal easier physical access to them
Eaves-dropping	Type of reconnaissance	More data leads to a higher probability of reconnaissance gaps
Sybil	Subversion of reputation systems by forged identities in peer-to-peer networks	Wireless Sensor Networks are the main target for Sybil attacks
Black hole	Packet drop on intermediary devices	Limited resources on Smart City sensors and nodes are easy targets

2.3 Smart city data analytics

Smart city can be illustrated as a complex network with different types of relationships. These relationships can be as simple as a one direction data connection to as complicated as a weighted prioritized two-way connection between a gateway and a data logger. Smart nodes are placed in communities of similar purpose devices. Based on the Confidentiality, Integrity, Security (CIA) mechanisms and addressing such vertices using Authentication, Authorization, Accounting (AAA), smart city security is adding different layers of access for every user of the network. Finding communities of similar devices with similar purposes is possible through evaluating similar relationship between devices which are known as nodes in the networks. Finding these communities can help level out and separate different levels of domains for various type of relationships and access.

In today's world, networks are as big as billions of nodes and smart cities are no different. To secure them, we need to put them in partitions and secure each partition both separately and as a group. To find these partition, also known as communities, there are big data community detection algorithms that could be used. Also ranking the partition could facilitate finding out which partitions can achieve a higher level of security. Security can be better handled if appropriate set of partitions are identified within the networks. With sub-partitioning, systems such as Hadoop can make the parallel data handling possible [4].

2.4 Existing methods

Alipour et al. [5] analyzed intrusion detection systems for Media Access Control (MAC) and Physical Layer (PHY) specifications using an anomaly-based behavioral analysis to detect abnormal behaviors, which are likely to be triggered by threat agents. They did this by monitoring the n-consecutive transitions of the protocol state machine. Then, sequential Machine Learning techniques were applied to model the n-transition patterns in the protocol. The probabilities of these transitions were normalized, reaching a low false positive of less than 0.1%. Spoofers impersonate legitimate users to exploit the user services and privileges. The Semi-Global Alignment algorithm (SGA) is an efficient technique to detect spoof attacks. The limitation of SGA is that it cannot be applied to large scale, multiuser systems due to high false positive rates. Kholidy et al. [6] proposed the Data-Driven SGA which improves the scoring systems using distinct alignment parameters per user. It also adapts to changes in the user behavior by updating the signature of a user according to his/her current behavior. The main objective of this proposal is to design a secure architectural framework for implementation of IoT-based, small to large-scale CPS in Smart City. This is important because of the inevitable migration to IoT networks and the unsafe and insecure nature of the underlying sensor-embedded smart objects, which interact with the physical world. Traditional security solutions might address security needs of IoT in part but there are challenges such as platform security limitations, ubiquitous mobility, mass quantities, and cloud-based operations that are not addressed.

3 The proposed approach

This research proposes a tunable underlying framework for IoT networks of different sizes which will, in turn, open many new research opportunities in IoT security. In addition, this research will facilitate and expedite adoption of small to large-scale IoT-based. But in the CPS context, security takes new forms and some of the previously used solutions such as Host-based IDS are not practical due to limited hardware resources on endpoint sensors. Adding to the issue is the fast-growing number of such sensors and their faster adoption by the public resulting in their widespread use without taking into account the many security gaps. Together with scientific advances in sensing and communications technologies, many consumers are using body sensors, connecting their generated data to their online profiles, or storing them on their smartphones or laptops. This project employs four technologies or methods as discussed below. The logical relation among these pieces is presented in Figure 1.

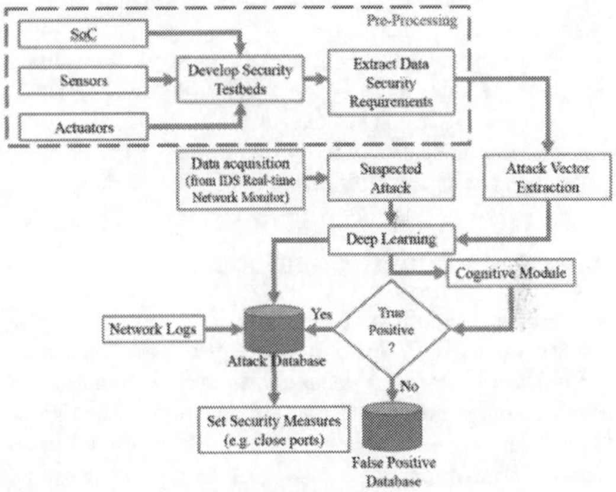

Fig. 1. Framework-Development Process

Anomaly-based, also called behavior-based, methods assume that attackers behave differently than normal users. The advantage of this method is the ability to spot a threat without first knowing its signature. Historically, this advantage has been offset by high false positive rates, the difficulty of training a system in a highly dynamic environment, and computational expense [7]. Some instances of the targeted vulnerabilities are presented in Table 2. It should be noted that most of the new attacks are typically minor mutations of the known ones, which leads us to believe that the DL approach can be successful on detecting imminent attacks. DL methods are successfully incorporated in various domains because DL relies on local proximity (typically spatial and/or temporal) among patterns to find and construct higher order patterns (Figure 2).

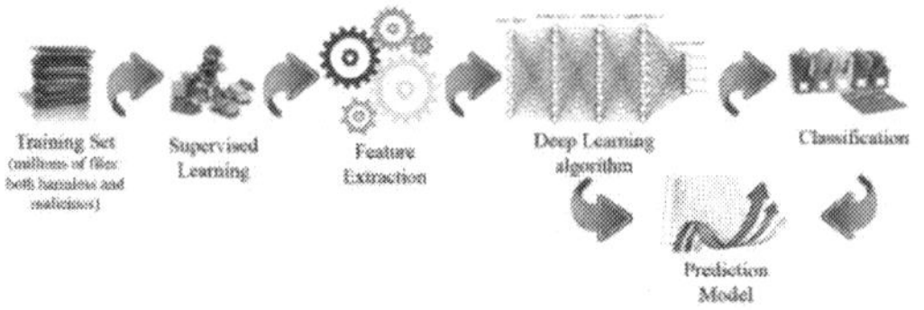

Fig. 2. Overarching Scheme of Threat Prediction

The factors moving Machine Learning tools and techniques from the research lab to the operational domain include both the phenomenal growth in inexpensive compute power and bandwidth and the overwhelming amount of data generated and dumped into Security Information and Event Management (SIEM) tools daily. Although Machine Learning tools can be very effective, they produce very different results depending on the source and quality of data being analyzed. Specific domain knowledge related to security—as opposed to clinical research or finance, for example—is needed to design a threat detection system using appropriate Machine Learning mathematical and statistical algorithms. A data scientist must apply security domain knowledge to identify primary and secondary sources of data, determine how to clean and transform acquired data and select the best Machine Learning analytical method or algorithm for the problem at hand. Primary sources for the security domain include network packets, Machine Learning -based analysis of which reveals otherwise invisible communication patterns from an attacker inside the network. Secondary sources are logs routinely collected from other devices, which may provide additional depth to the analysis but not direct evidence of activity due to the nature of logs' role in providing security defences [7].

4 Results and discussions

DL [13] is a field that encompasses machine learning so it can be used to learn intricate patterns from large volumes of data. It is generally an architecture formed out of neural network activation functions. Supervised and Unsupervised learning refer to labeled and non-labeled data respectively [14]. For supervised learning, techniques such as Recurrent Neural Networks (fast and efficient), Convolutional Neural Networks (Time consuming, but suitable for special data, such as images), Long Short Term Memory (which can be used for vanishing gradient problem [15] which occurs nearing the end of training, where gradient is supposed to be really less) apart from the traditional neural networks such as Deep Boltzmann Machines or Deep Belief Networks as well as fully connected, slow-to-train Multi-Layer Perceptron [14]. Each layer of the deep models shown below can be consisting of linear or complex activation functions depending on the overall complexity of the problem. For instance, for malware detection problem, we stacked two layers with linear activations with two layers of Rectified Linear Units in between them. This was implemented to get the best accuracy of prediction for the given data [16]. In this study, CSIC 2010 HTTP Dataset was used to detect web attacks using session IDs and indices. This data set has been widely used for abnormal behavior detection. Regarding web traffic, some of the problems of this data set are that it is out of date and also that it does not include enough actual attacks and hence, it is criticized by security researchers.

This is a proven benchmark initially set for researchers to compare different methods of detection and classification of network attacks. It was built as an improvement over the earlier KDD Cup 1999 dataset in the form of a reduction in redundant records, proportionate number of records in each difficulty level group [16]. Experts believe that new attacks can be mostly identified by the signature of the known attacks. According to this principle, we train the data on the features given in this dataset, and some derived from them. They include, but are not limited to the duration, protocol type, destination network service, source and destination lengths in bytes, flags, number of wrong fragments, the number of high QoS packets, etc. The results show the logistic regression classification, where the Dependent Variable is categorical, can perform anomaly detection efficiently. The ROC curve characterizing the preliminary results is outlined in Figure 3. As illustrated in Figure 3, a simple logistic regression classification with two parameters can achieve a performance of 64%. Utilizing DL techniques with a multitude of features results in higher accuracy. Logistic regression classifies data into two categories, and the Receiver operating characteristics (ROC) curve indicates the area under the curve which signifies the percentage of accurate classification. According to the results on the dataset, accuracy percentage is 86%. A standard metric to evaluate logistic regression classification is accuracy, which is calculated by dividing true Positives over the sum of false positives and true positives.

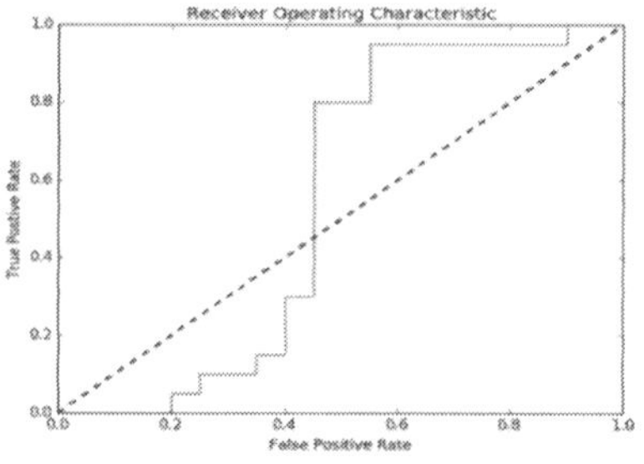

Fig. 3. ROC Curve of Logistic Regression

The learning model consists of an input vector X. Logits are outputs of linear functions – that are continuous and differentiable. Logits need to be converted into a scale of probabilities [0,1]. The weight and bias parameters need training. This linear block can be cascaded with multiple different linear blocks that sum up to learn different features of the input. However, to increase the complexity to define finer features of the input, we need a combination of non-linear elements that can do so. This can be achieved by combining rectified linear units that scale inputs. Once Softmax function converts logits into probabilities we can use these values to be given to series of Rectified Linear Units.

Rectified Linear Units (ReLUs) can be shown as:

$$f(x) = \sum_{i=1}^{inf} \sigma(x - i + 0.5) \approx \log(1 + e^{x})$$

where, x = input vector, f(x) = output of a rectified linear unit. A series of ReLUs [17] cascaded together can form a computationally expensive, although fairly efficient non-linear differentiable model to model complexities of a function. ReLUs can be replaced by functions such as Sigmoid, Tanh, etc. Figure 4 presents the comparison diagram of such non-linear functions for the preliminary results. Sigmoid functions are used for logistic regression functions. ReLUs outperform sigmoid and give better classification accuracy only by a slight margin. Thus, both are widely used and give comparable results. They introduce non-linearity while pooling up layers of one convolutional layer on others. Components of the learning model are depicted in Figure 5.

In general, the linear blocks or layers can be stacked upon each other, with non-linear interface as shown below. DL Model (forward propagation) [18] is the very basis of learning in which features from the first layer is carried forward to the next layer. For training, a widely popular algorithm is Back-Propagation [19], in which gradients or relative difference between iterations of calculating weight functions are minimized by a backward-looking architecture as shown below. Back propagation is a mean-squared-error function which is differentiable.

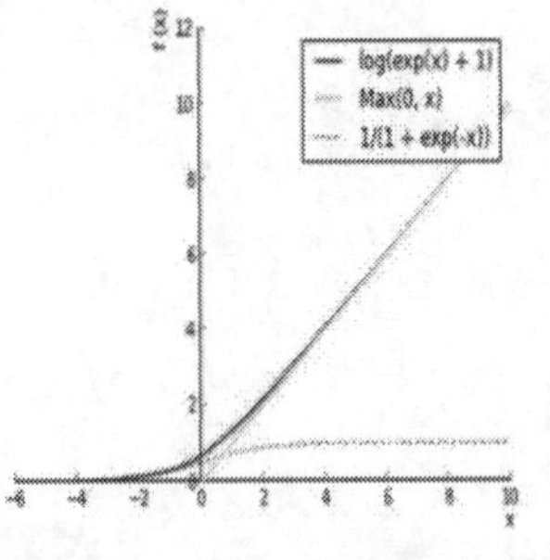

Fig. 4. ReLUs and Max-Pooling/Sigmoid Functions

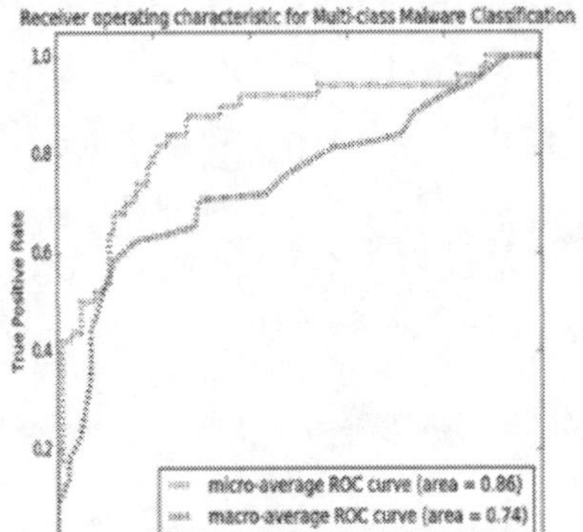

Fig. 4. Classification Results

This work is supported in part by the Doctoral Graduate Research Assistantship from UNLV Graduate College and in part by the NSF award #EPS-IIA-1301726 (EPSCoR NEXUS).

References

1. Sadeghi, A.-R.; Wachsmann, C.; Waidner, M., "Security and privacy challenges in industrial Internet of Things," in Design Automation Conference (DAC), 2015 52nd ACM/EDAC/IEEE, pp.1-6, 8-12, June 2015.
2. S. Tayeb, S. Latifi and Y. Kim, "A survey on IoT communication and computation frameworks: An industrial perspective," 2017 IEEE 7th Annual Computing and Communication Workshop and Conference (CCWC), Las Vegas, NV, 2017, pp. 1-6. doi: 10.1109/CCWC.2017.7868354
3. J. Pacheco, C. Tunc and S. Hariri, "Design and evaluation of resilient infrastructures systems for smart cities," 2016 IEEE International Smart Cities Conference (ISC2), Trento, 2016, pp. 1-6. doi: 10.1109/ISC2.2016.7580756
4. Pirouz, Matin, Justin Zhan, and Shahab Tayeb. "An optimized approach for community detection and ranking." Journal of Big Data 3.1 (2016): 22.
5. H. Alipour, Y. B. Al-Nashif, P. Satam and S. Hariri, "Wireless Anomaly Detection Based on IEEE 802.11 Behavior Analysis," in IEEE Transactions on Information Forensics and Security, vol. 10, no. 10, pp. 2158-2170, Oct. 2015. doi: 10.1109/TIFS.2015.2433898
6. H. A. Kholidy, F. Baiardi and S. Hariri, "DDSGA: A Data-Driven Semi-Global Alignment Approach for Detecting Masquerade Attacks," in IEEE Transactions on Dependable and Secure Computing, vol. 12, no. 2, pp. 164-178, March-April 1 2015. doi: 10.1109/TDSC.2014.2327966
7. Barbara Filkins, "the Expanding Role of Data Analytics in Threat Detection", SANS Whitepaper, October 2015.
8. Farooq MU, Waseem M, Khairi A, Mazhar S. A critical analysis on the security concerns of internet of things (IoT). International Journal of Computer Applications. 2015 Jan 1;111(7).
9. Kolias C, Stavrou A, Voas J, Bojanova I, Kuhn R. Learning Internet-of-Things Security" Hands-On". IEEE Security & Privacy. 2016 Jan;14(1):37-46.
10. S. Ray, S. Bhunia, Y. Jin and M. Tehranipoor, "Security validation in IoT space," 2016 IEEE 34th VLSI Test Symposium (VTS), Las Vegas, NV, 2016, pp. 1-1. doi: 10.1109/VTS.2016.7477288
11. R. Mahmoud, T. Yousuf, F. Aloul and I. Zualkernan, "Internet of things (IoT) security: Current status, challenges and prospective measures," 2015 10th International Conference for Internet Technology and Secured Transactions (ICITST), London, 2015, pp. 336-341. doi: 10.1109/ICITST.2015.7412116
12. S. A. Zonouz, R. Berthier, H. Khurana, W. H. Sanders and T. Yardley, "Seclius: An Information

Flow-Based, Consequence-Centric Security Metric," in IEEE Transactions on Parallel and Distributed Systems, vol. 26, no. 2, pp. 562-573, Feb. 2015. doi: 10.1109/TPDS.2013.162

13. G. E. Dahl, J. W. Stokes, L. Deng and D. Yu, "Large-scale malware classification using random projections and neural networks," 2013 IEEE International Conference on Acoustics, Speech and Signal Processing, Vancouver, BC, 2013, pp. 3422-3426
14. Salakhutdinov, Ruslan, and Geoffrey E. Hinton. "Deep Boltzmann Machines." AISTATS. vol. 1. 2009.
15. S. Yuan, X. Wu and Y. Xiang, "Incorporating Pre-Training in Long Short-Term Memory Networks for Tweets Classification," 2016 IEEE 16th International Conference on Data Mining (ICDM), Barcelona, Spain, 2016, pp. 1329-1334
16. Dataset for KDD Cup and Workshop, ACM SIGKDD, San Jose, California, Aug 12, 2007. James Bennett, Charles Elkan, Bing Liu, Padhraic Smyth, and Domonkos Tikk. 2007. KDD Cup and workshop 2007. SIGKDD Explor. Newsl. 9, 2 (December 2007), 51-52. doi: 10.1145/1345448.1345459
17. Nair, Vinod, and Geoffrey E. Hinton. "Rectified linear units improve restricted boltzmann machines." Proceedings of the 27th international conference on machine learning (ICML-10). 2010.
18. R. Sánchez, A. Arpi and L. Minchala, "Fault Identification and Classification of Spur Gearbox with Feed Forward Back Propagation Artificial Neural Network," 2012 VI Andean Region International Conference, Cuenca, 2012, pp. 215-215.
19. Y. Liu, W. Jing and L. Xu, "Cascading model based back propagation neural network in enabling precise classification," 2016 12th International Conference on Natural Computation, Fuzzy Systems and Knowledge Discovery (ICNC-FSKD), Changsha, 2016, pp. 7-11

An Early Performance Comparison of CUDA and OpenACC

Xuechao Li[1] and *Po-Chou* Shih[2]

[1]Department of Computer Science, Concordia University Chicago, River Forest IL, USA.
[2]Institute of Industrial and Business Management, National Taipei University of Technology, Taipei, Taiwan.

Abstract. This paper presents a performance comparison between CUDA and OpenACC. The performance analysis focuses on programming models and underlying compilers. In addition, we proposed a Performance Ratio of Data Sensitivity (PRoDS) metric to objectively compare traditional subjective performances: how sensitive OpenACC and CUDA implementations are to change in data size. The results show that in terms of kernel running time, the OpenACC performance is lower than the CUDA performance because PGI compiler needs to translate OpenACC kernels into object code while CUDA codes can be directly run. Besides, OpenACC programs are more sensitive to data changes than the equivalent CUDA programs with optimizations, but CUDA is more sensitive to data changes than OpenACC if there are no optimizations. Overall we found that OpenACC is a reliable programming model and a good alternative to CUDA for accelerator devices.

1. Introduction

High-performance computing has been applied into more and more scientific research applications such as chemical reaction process, information retrieval system [1], explosion process and fluid dynamic process. Currently the most popular high-performance computing device is the NVIDIA accelerator. However, the CUDA programming framework [13] requires programmers to fully grasp solid knowledge of software and hardware, and it is an error-prone and time-consuming work if legacy code needs to be rewritten by CUDA. To overcome the above limitations of CUDA and reduce developers' coding workload, the OpenACC programming model [14] has been released. It provides simple directives to achieve similar functions in CUDA and the OpenACC code can be run under different platforms. Hence, it is important to compare OpenACC and CUDA especially for parallel programming beginners because OpenACC can extremely reduce programmers' workload if they have the similar performance.

In terms of directive-based GPU programming models, some research work has been done on OpenMP accelerator [17], hiCUDA [18] and R-stream [19]. But with OpenACC motivation of simplifying CUDA programming with similar functions, in this paper we investigated the performance comparison of CUDA and OpenACC with the following two factors: (1) CUDA is one of the most popular parallel programming models and OpenACC is an easily learned and simplified high-level parallel language especially for programming beginners; and (2) One motivation of OpenACC development aims to simplify low-level parallel language such as CUDA. Finally, in this paper we make the following key contributions:

(1) We used 19 kernels in 10 benchmarks covering real-world applications and synthetic applications to present a detailed performance comparison between OpenACC and CUDA. To our best knowledge, this is the first paper of the performance comparisons between OpenACC and CUDA with a tremendous amount of benchmarks and performance metrics proposed.

(2) We made a quantitative comparison and a qualitative conclusion of CUDA and OpenACC with regard to programming models, optimization technologies, underlying compilers and data sensitivity.

The remainder of this paper is organized as follows: Section 2 discusses related work on performance comparison of diverse parallel programming models; Section 3 shows our configuration, testbeds and the selected benchmarks; Section 4 presents a methodology used in this paper; Section 5 discribes an overall performance analysis result, identifies and explains the main differences between CUDA and OpenACC; Section 6 presents the background of CUDA and OpenACC; Section 7 concludes this paper and presents our future work.

2. Related work

NVIDIA GPU devices are currently one of the most widely used accelerators so the CUDA programming language is used by most programmers, which requires professional software and hardware knowledge. To alleviate or avoid this limitation especially for beginners, OpenACC was developed by CAPS Enterprise, The Portland Group (PGI), Cray Inc. and NVIDIA. A good performance comparison between OpenACC and CUDA can effectively assist programmers to make a decision on which language is better for their specific research work. Suejb *et al* [20] investigated OpenCL, OpenACC, CUDA and OpenMP, the threat that CUDA is not able to be run on Intel GPUs made conclusions less convinced. Also, compared to other papers [2-4] where only 1–2 benchmarks were used, we use 19 kernels in 10 benchmarks to fully evaluate OpenACC and CUDA programming models and use the PRoDS to objectively compare the data sensitivity. Although the results from the paper [2] showed that in general OpenACC performance is slower than CUDA's, we found that all results were based on only two micro benchmarks and one application. The use of few benchmarks cannot fully evaluate their performances so based on results from 19 kernels in 10 benchmarks in this paper, we find that CUDA codes still outperforms OpenACC codes in terms of kernel running time. In this section, a number of performance comparisons among different parallel programming models are presented.

Christgau *et al.* [4] presented a very interesting application--Tsunami Simulation EasyWave—for comparing CUDA and OpenACC using two different GPU generations (Nvidia Tesla and Fermi). Through listing runtime values under different hardware-specific optimizations—memory alignment, call by value and shared memory—three key points were revealed that (1) even the most tuned code on Tesla does not reach the performance of the unoptimized code on the Fermi GPU; (2) the platform independent approach does not reach the speed of the native CUDA code; (3) memory access patterns have a critical impact on the compute kernel's performance.

Tetsuya *et al.* [2] used two micro benchmarks and one real-world application to compare CUDA and OpenACC. The performance was compared among four different compilers: PGI [15], Cray, HMPP [16] and CUDA with different optimization technologies: thread mapping and shared memory in Matrix Multiplication micro benchmark and branch hoisting and register blocking were applied into 7-Point stencil micro benchmark. Finally, a performance relative to the CPU performance was compared in computational fluid dynamics application with kernel specification, loop fusion and fine-grained parallelization in the matrix free Gauss-Seidel method optimization. The results showed that OpenACC performance is 98% of CUDA's with careful manual optimization but in general it is slower than CUDA.

In [9] authors compared kernel execution times and data transfer times between host and device and end-to-end application execution times for both CUDA and OpenCL. The only application tested was Adiabatic Quantum Algorthms (AQUA) [10] which was implemented by complex, near-identical CUDA and OpenCL kernels. All run times and data transfer times were listed with different input sizes-Qubits. The conclusion was made that CUDA performed better when transferring data and CUDA's kernel execution was also consistently faster than OpenCL's.

An overview and comparison of OpenCL and CUDA was also presented in [11]. Five application benchmarks—Sobel filter, Gaussian filter, Median filter, Motion Estimation, Disparity Estimation—were tested under NVIDIA Quadro 4000. This paper compared and analyzed C, OpenCL and CUDA and the two kinds of APIs—CUDA runtime API and CUDA driver API. Detailed data comparison generated by different benchmarks generally showed CUDA had a better performance compared with OpenCL.

3. Experiment

3.1 Testbeds

All experimental results were obtained by nvprof (NVIDIA 2015) from the NVIDIA profiler toolkit and generated on a server which consists of the host CPU and the attached GPU. The host processor was a 10-core Intel Xeon E5-2650 at 2.3 GHz with 16 GB DDR4 RAM and 25 MB cache; the device was an NVIDIA Tesla K40c at 0.75 GHz with 11,520 MB global memory. We used the NVIDIA CUDA compiler v7.0.17 and the PGI Accelerator compilers (for OpenACC) v15.4, running on CentOS 7.1.1503.

3.2 Benchmarks selection

Benchmarks were selected from the Rodinia suite [5] and a code sample from NVIDIA. Also we developed a particular scientific computation application with an emphasis on computing speed. To get a "balanced" comparison data, we followed a similar benchmark categories standard in the paper [6] to select benchmarks from synthetic applications to real-world applications and following the guideline of the Dwarfs [7], details of categories, Dwarfs, domains and descriptions were described in Table 1 where Syn denotes a synthetic application and R-W denotes a real-world application. Finally, more details about each benchmark in the Rodinia suite are described in
the papers [6]. Following a similar explanation in the paper [6], we presented a different meaning of Real-world applications and Synthetic applications in this paper.

- Real-world applications. To solve realistic problems in other scientific research areas, such applications usually use well-known and classic algorithms to do iteration calculations.
- Synthetic applications. For the purpose of pure theoretical measurement, sometimes researchers need highly-intensive computation to explore the biggest machine performance. Here, MatMul in both CUDA and OpenACC, Jacobi (CUDA) are developed by ourselves (denoted by SELF).

4. Methodology

4.1 Data sensitivity

An interesting phenomenon was found when time was measured with different input sizes in some benchmarks under different programming models. Time change does not follow the pace of input size change. For those well-known or classic algorithms, the time complexity can clearly tell the relationship of time and input size such as an O(n log n) in the Fast Fourier Transform(FFT) algorithm, but it cannot be perfectly applied to our benchmarks. For example, in CFD the input size is mainly showed as a file size and in some benchmarks, there are no classic algorithms. Also, in this paper we focus on the impact of different programming models on run time, rather than the algorithms' impact. We call this kind of impact Performance Ratio of Data Sensitivity (PRoDS) – how sensitive both OpenACC and CUDA programming models are to data changes in this paper specifically.

In order to objectively and quantitatively measure PRoDS, we propose the following formula to calculate the PRoDS. It is described:

$$PRoDS = \frac{|Difference_{time\_cuda}|}{|Difference_{time\_OpenACC}|}$$

If PRoDS < 1, the performance of OpenACC is more sensitive to data than CUDA; if PRoDS > 1, the performance of CUDA is more sensitive to data than OpenACC. If |1-PRoDS| < 0.1, we assume that both OpenACC and CUDA have the same data sensitivity as the definition in the paper (Fang, Varbanescu, and Sips 2011).

For the PRoDS formula, $Difference_{time\_cuda}$ means a difference of CUDA running time with different input sizes. For example, in Matrix Multiplication, if 10 seconds were spent when input size is 1024x1024 and 6 seconds were spent when input size is 512x512, $Difference_{time\_cuda}$ is 4 seconds. A similar explanation is in $Difference_{time\_OpenACC}$.

5. Data evaluation

5.1 Comparision on data sensitivity

To investigate which programming model is more stable when input size changes over a wide range, we proposed a metric of the Performance Ratio of Data Sensitivity (PRoDS) to convert a subjective measurement to an objective one. And an input size from benchmarks was listed in Table 2. Compared to the formula in [6], one obvious difference in the PRoDS formula is to add a condition to judge whether or not both programming models have the same data sensitivity, which means that PRoDS just denotes a ratio of performance changing extent. To obtain a "native" comparison of the data sensitivity, the performance in the thread mapping version of OpenACC was selected to compare to the CUDA version. Because many different input sizes were used in measuring the performance of each benchmark, we selected the "worst" gradient values as comparison samples, which have biggest values. Finally, to make a fair comparison of data sensitivity,

we divided all benchmarks into two categories: one without optimization and one with optimizations, and the PRoDS caused by different optimization technologies is ignored.

In non-optimization category, the CUDA implementation was more sensitive to data than the OpenACC implementation. This is particularly true for the first kernel in CFD (*cfd_k1*) and the second kernel in BFS (*bfs_k2*). As Figure 2 shows, in cfd_k1, the PRoDS of CUDA is 1.4 times more sensitive than one of OpenACC while in bfs_k2 CUDA is 13% more sensitive to data than OpenACC. Because the thread mapping version was used in this experiment, we explicitly added specific values for gang and vector parameter in kernel construct of OpenACC. In this way, the PGI compiler's flexibility of compiling parallelizable loops in OpenACC implementation was significantly limited. In contrast, CUDA not only allowed programmers to set up Gid/Block dimensions, but also gave programmers permission on operations of index such as *blockDim.x, threadIdx.x*.

In the optimization category, the OpenACC implementation was more sensitive to data than the CUDA implementation. This is particularly true for Hotspot and Lud with memory-coalescing optimization. As Figure 1 presented, in Hotspot the OpenACC implementation was 60% more sensitive to data than the CUDA one. In Lud the OpenACC implementation was 51% more sensitive to data

Table 1: Benchmarks selection

Benchmark	Suite	Category	Dwarf	Domain	Description
Jacobi	NVIDIA	Syn	Dense Linear Algebra	Image Processing	A Jacobi iteration on Laplace2D
MatMul	SELF	Syn	Dense Linear Algebra	Linear Algebra	Matrix multiplication
BFS	Rodinia	R-W	Graph Traversal	Graph Algorithms	Graph breadth first search
SRAD	Rodinia	R-W	Structured Grid	Image Processing	Removing speckles in an image
Gaussian	Rodinia	R-W	Dense Linear Algebra	Linear Algebra	Solving for variables in a linear system
Pathfinder	Rodinia	R-W	Dynamic Programming	Grid Traversal	Find a path from bottom row to top row
LUD	Rodinia	R-W	Dense Linear Algebra	Linear Algebra	Calculate solutions of linear equations
NN	Rodinia	R-W	Dense Linear Algebra	Data Mining	Find the k-nearest neighbors

Hotspot	Rodinia	R-W	Structured Grid	Physics Simulation	Thermal simulation
CFD	Rodinia	R-W	Unstructured Grid	Fluid Dynamics	A solver for Euler equations

than the CUDA implementation. The OpenACC programming style itself affected the data sensitivity. For memory coalescing, whether or not neighboring elements can be accessed mainly depends on loop order in OpenACC implementation, but the lack of architecture-specific directives in the OpenACC programming style limits programmers' ability in defining data attribution such as *shared* memory. In contrast, CUDA programmers can explicitly declare all data used as a shared type so that all results measured with different input sizes were less sensitive than the results in OpenACC implementation.

Through analyzing the PRoDS above, we concluded that the CUDA implementation was more sensitive to data than the OpenACC implementation without optimizations while the OpenACC implementation was more sensitive to data than the CUDA implementation with optimizations.

Table 2. Input size for selected benchmarks

Benchmark	Input Size
MatMul	Matrix: 1024x1024
Jacobi	Matrix: 1024x1024
BFS	4096x4096
Gaussian	Matrix: 1024x1024
Pathfinder	Row: 10000, Col:100, Iteration:100
NN	Number of records: 15690
CFD	File size: 0.2M

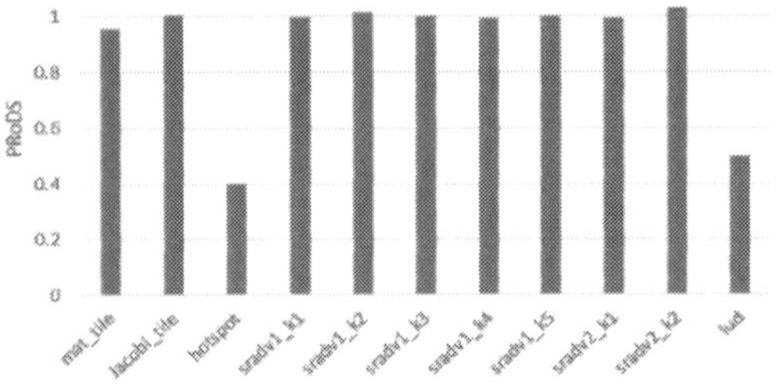

Figure 1. Performance comparison of data sensitivity with optimization

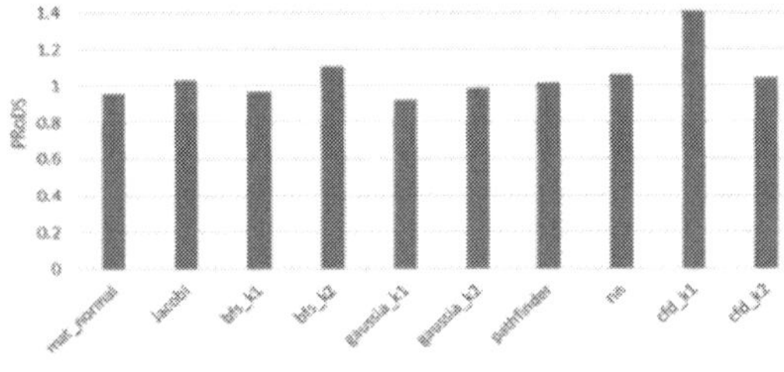

Figure 2. Performance comparison of data sensitivity without optimization

5.2 Comparision on programming model

Gang and *vector* parameters in *kernels* construct or *num_gangs* and *vector_length* parameters in parallel construct can assist OpenACC to achieve a similar dimension setup of Grid and Block in CUDA so that OpenACC and CUDA can implement and achieve similar acceleration in parallel loops. But as far as device memory hierarchy is concerned, CUDA can explicitly support specific kinds of memory in devices while OpenACC fails to do it because it is designed to be portable across platforms from multiple vendors [2]. For example, the use of architecture-specific device memory shared is impossible in OpenACC while this kind of memory can be explicitly declared and used in CUDA. Although cache directive can, to some extent, compensate performance loss caused by the lack of shared memory, the lack of barrier synchronization extremely limits effects of the use of this directive to read-only data [2].

6. Background

The goal of high-performance parallel languages is to accelerate run time and to reduce the unnecessary waiting time of computing operations and resources schedule by utilizing optimization technologies such as kernel parameters tuning, the use of local memory, data layout in memory and avoiding CPU-GPU data transfer. In this section, we simply introduce languages used in this paper: CUDA and OpenACC.

CUDA is a parallel computing programming model that can be run only under NVIDIA's GPUs. It fully utilizes hardware architecture and software algorithms to accelerate high-performance computation with architecture-specific directives such as shared, or global. But this utilization requires programmers to fully understand each detail of hardware and software. For example, in order to utilize memory-coalescing or tiling technology, the software-managed on-chip and off-chip memory knowledge should be gained by users before programming, which is a big challenge, especially for programmers who have little knowledge of software and hardware. Also, in order to successfully launch CUDA kernels, all configurations, such as Grid/Block dimensions, computing behaviors of each thread, and synchronization problems need to carefully be tuned.

OpenACC is a cross-platform programming model and a new specification for compiler directives that allow for annotation of the compute region and data region that are offloaded to accelerators [2] such as GPUs. The OpenACC Application Programming Interface [8] offers a directive-based approach for describing how to manage data and execute sections of codes on the device, where parallelism often appears in regular repetition constructs such as Fortran "DO" loops and C "FOR" loops. In OpenACC, porting of legacy CPU-base code only requires to add several lines of annotations before the sections where they need to be accelerated, without changing code structures [2]. But over-simplification parallel programming model also brings to users limitations which may prevent full use of the available architectural resources, potentially resulting in greatly inferior performance when compared to highly manually tuned CUDA code [2]. For example, programmers can use synchronization to manage all threads of one block in CUDA while they fail in OpenACC. Some architecture-specific directives such as global, shared and private have been provided by CUDA while OpenACC does not provide them.

7. Conclusions and future work

The analysis of performance gaps has been shown above in 19 kernels of 10 benchmarks. We used kernel execution time and data sensitivity as main standards referenced when conclusions were made. A comparison of data sensitivity is a new index to explore an easily-ignored problem that how both programming models are sensitive to changes of data sizes. The appearance of the PRoDS formula brings us an objective comparison rather than a subjective comparison. Through the analysis of Figure 1 and Figure 2, the OpenACC programming model is more sensitive to data than the CUDA with optimizations while CUDA is more sensitive than OpenACC without optimizations. Overall, the OpenACC performance is similar to CUDA under a fair comparison and OpenACC can be a good alternative to CUDA especially for beginners in high-level parallel programming.

In the future work, we will continue to explore the influence of optimization techniques on CUDA and OpenACC models. Also, we will also use the autotuner [12] to optimize parameters in OpenACC so that the best performance can be used in the comparisons.

References

1. X. Li, C. H, Li. and Y. Xie. 2011. "A Retrieval System of Vehicles Based on Recognition of License Plates". Proceedings of 2011 International Conference on Machine Learning and Cybernetics (ICMLC), IEEE, Guilin, pp.1453-1459.
2. Hoshino, T., Maruyama, N., Matsuoka, S. and Takaki, R. 2013. "CUDA vs OpenACC: Performance Case Studies with Kernel Benchmarks and a memory-bound CFD Application". 13th IEEE/ACM International Symposium on Cluster, Cloud and Grid Computing (CCGrid), pp. 136-143.
3. Herdman, J.A. Gaudin, W.P. Mclntosh-Smith, S. and Boulton, M. 2012. "Accelerating Hydrocodes with OpenACC, OpenCL and CUDA". 2012 SC Companion: High Performance Computing, Networking, Storage and Analysis (SCC). pp. 465-471.
4. Christgau, S., Spazier, J., Schnor, B., Hammitzsch, M., Babeyko, A. and Waechter, J. 2014. "A comparison of CUDA and OpenACC: Accelerating the Tsunami Simulation EasyWave". 27th International Conference on Architecture of Computing Systems (ARCS). pp. 1-5.
5. Che, S., Jeremy, W., Sheaffer, Michael, B., Lukasz G. S., Liang. W., and Kevin, S. 2010. "A characterization of the rodinia benchmark suite with comparison to contemporary cmp workloads", in Proceedings of the IEEE International Symposium on Workload Characterization (IISWC'10), pp. 1-11
6. Fang, J., Varbanescu, A and Sips, H. 2011. "A Comprehensive Performance Comparison of CUDA and OpenCL". International Conference of Parallel Processing (ICPP), pp. 216-225.
7. Krste, A. Ras, B. Bryan, C. Joseph, G. Parry, H. Kurt, K. David, P. William, P. John, S. Samuel, W. and Katherine, Y. 2006. "The landscape of parallel computing research: a view from Berkeley". Tech. Rep. UCB/EECS-2006-183, Electrical Engineering and Computer Science, University of California at Berkeley
8. The OpenACC Application Programming Interface, Version1.0, November 2011.
9. Kamran, K. Neil, D. and Firas, H. A Performance Comparison of CUDA and OpenCL.
10. http://aqua.dwavesys.com
11. Po-Yu Chen ; Chun-Chieh Lan ; Long-Sheng Huang and Kuo-Hsuan Wu. Overview and Comparison of OpenCL and CUDA Technology for GPGPU. IEEE Asia Pacific Conference on Circuits and Systems (APCCAS). pp. 448 – 451. Dec. 2012
12. Calvin,M, Jeffrey, O. and Xuechao, L. Autotuning OpenACC Work Distribution via Direct Search.To appear at Extreme Science and Engineering Discovery Enviroment (XSEDE15), Jul. 2015
13. CUDA, "NVIDIA CUDA [online]. available: http://developer.nvidia.com/category/zone/cuda-zone," 2012.
14. OpenACC. https://www.openacc.org/
15. PGI Accelerator, "The Portland Group, PGI Fortran and C Accelarator Programming Model [Online]. Available: http://www.pgroup.com/resources/accel.htm," 2009
16. HMPP, "HMPP Workbench, a directive-based compiler for hybrid computing [Online]. Available: www.caps-entreprise.com/hmpp.html," 2009.
17. J. C. Beyer, E. J. Stotzer, A. Hart, and B. R. de Supinski, "OpenMP for Accelerators." in IWOMP'11, 2011, pp. 108–121
18. T. D. Han and T. S. Abdelrahman, "hicuda: High-level gpgpu programming," IEEE Transactions on Parallel and Distributed Systems, vol. **22**, no. 1, pp. 78–90, 2011
19. A. Leung, N. Vasilache, B. Meister, M. Baskaran, D. Wohlford, C. Bastoul, and R. Lethin, "A mapping path for multi-GPGPU accelerated computers from a portable high level programming abstraction," in Proceedings of the 3rd Workshop on General-

Purpose Computation on Graphics Processing Units, ser. GPGPU '10. New York, NY, USA: ACM, 2010, pp. 51–61

20. J. Dongarra, P. Beckman, T. Moore, P. Aerts, G. Aloisio, J.-. Andre, D. Barkai, J.-. Berthou, T. Boku, B. Braunschweig, F. Cappello, B. Chapman, X. Chi, A. Choudhary, S. Dosanjh, T. Dunning, S. Fiore, A. Geist, B. Gropp, RobertHarrison, M. Hereld, M. Heroux, A. Hoisie, K. Hotta, Y. Ishikawa, Z. Jin, F. Johnson, S. Kale, R. Kenway, D. Keyes, B. Kramer, J. Labarta, A. Lichnewsky, T. Lippert, B. Lucas, B. Maccabe, S. Matsuoka, P. Messina, P. Michielse, B. Mohr, M. Mueller, W. Nagel, H. Nakashima, M. E. Papka, D. Reed, M. Sato, E. Seidel, J. Shalf, D. Skinner, M. Snir, T. Sterling, R. Stevens, F. Streitz, B. Sugar, S. Sumimoto, W. Tang, J. Taylor, R. Thakur, A. Trefethen, M. Valero, A. van der Steen, J. Vetter, P. Williams, R. Wisniewski, and K. Yelick, "The International Exascale Software Project RoadMap," Journal of High Performance Computer Applications, vol. **25**, no. 1, 2011

A Dynamic Hierarchical Evaluating Network for Real-Time Strategy Games

Weilong Yang[1], *Qi* Zhang[1], *Yong* Peng[1]

[1]Department of Modeling and Simulation, National University of Defense Technology, Changsha, China

Abstract. Researches of AI planning in Real-Time Strategy (RTS) games have been widely applied to human behavior modeling and combat simulation. State evaluation is an important research area for AI planning, which ensures the decision accuracy. Since complex interactions exist among different game aspects, the weighted average model usually cannot be well used to compute the evaluation of game state, which results in misleading player's generation strategy. In this paper, we take dynamic changes and player's preference into consideration, analyze player's preference and units' relationships base on game theory and propose a dynamic hierarchical evaluating network, denoted as DHEN. Experiments show that the modified evaluating algorithm can effectively improve the accuracy of task planning algorithm for RTS games.

1 Introduction

Real-Time Strategy (RTS) games are popular real-time combat simulation games in which players instruct units to gather resources, build structures, destroy opponent's buildings to win the game. As typical agent-based game, RTS games pose a huge challenge for AI researchers due to the large state space, limited decision time and dynamic adversarial environment involved. AI planning becomes an important research area for real-time adversarial planning and non-determination decision [1]. State evaluation is an important part of AI planning. By calculating correlative factors and player's preference, the evaluation of game state is obtained, which can be used to judge whether the state is advantageous for player or not. As basic algorithm for player's planning and decision, evaluation method can influence the decision process, and improve the performance of AI planning in games.

Current evaluating algorithms take related factors into consideration to obtain a state score. This method is applicable to simple and fixed game scene. However, in RTS games, the evaluating factors are constantly changing, for which fixed evaluating algorithm cannot accurately describe the game state at different game phases. For example, the current evaluating function does not consider the spatial relationship between units, resulting in that units at different positions still are calculated by carried resources and hit-point value.

This paper focuses on the evaluating process in RTS games and constructs a hierarchical network similar to HTN planning method. By decomposing evaluating factors, the relationships between different factors are analysed. Considering the changing weights, a dynamic hierarchical evaluating network is constructed and a dynamic weights calculating algorithm is proposed. In the remainder of this paper, after discussing related work, we first present the definition of hierarchical factor network. Then we propose dynamic weight calculating method, followed by extensions to apply it to μRTS games, a minimalistic RTS game used for planning algorithm evaluation.

2 Related Works

2.1. RTS games

RTS games are regarded as a simplification of combat simulation and could therefore serve as a test bed for investigating activities such as real-time adversarial planning and decision making under uncertainty [2]. Compared with conventional board games, RTS games have the following primary differences [3]:

1. Players can pursue actions simultaneously with the actions of other players, and need not take turns.

2. Player actions can be conducted over very short decision times, allowing for rapid sequences of actions. While Player actions are durative, in that an action requires numerous time steps to be executed.

3. The state space and branch factors are typically very large. For example, a typical 128 × 128 map in "StarCraft" generally includes about 400 player controllable units. Considering only the location of each unit, the number of possible states is about 101685, whereas the state space of chess is typically estimated to be around 1050.

4. The environment of an RTS game is dynamic and non-determinability. The opponent's action is not

controllable and the executing task may be broken down by the uncertain environment.

Research has been conducted to modify game tree search methods to address these differences. Chung investigated the applicability of game tree Monte Carlo simulations in RTS games [4]. Balla and Fern applied the upper confidence bound for trees (UCT) algorithm in an RTS game to address the complications associated with durative actions [5]. Churchill addressed the complications associated with simultaneous and durative actions by extending the alpha–beta search process [6]. Methods such as combinatorial multi-armed bandits have attempted to address the challenge of large branching factors [7-8]. Ontañón and Buro combined the hierarchical task network (HTN) planning approach with game tree search to develop what was denoted as adversarial HTN [9]. Among these planning algorithms, evaluating algorithm is needed to calculate the current and play-out game state.

2.2 State Evaluation in RTS games

Alexander [10] first described evaluating process in RTS game state, proposing an evaluating function by calculating the hit-point and attack ability of units. LTD (Life-Time Damage) is a modified evaluating algorithm, which based on the lifetime damage each unit can inflict. And by weaken the contribution of hit-point, LTD2 algorithm is proposed [9]. Tung [11] combined the LTD method and the unit portfolio evaluation to evaluate game state, but did not consider the relationship between two algorithms.

Taking nonterminal position into consideration, Stanescu [12] modified the state evaluating algorithm by taking other properties into account, such as resources, visibility, security, and goals. By putting forward the concept of visibility and detection, a formula is given to calculate the efficiency of detection. Graham Erickson [13] proposed a global evaluating function model for predicting which side would win after a period of game play-out. Taking military features, economic and player skill into consideration, the logistic regression is used to learn the model weights. Alberto [14] used a contrastive evaluating algorithm to obtain the score by calculating the difference between player and its opponent. Bakkes [15] generated evaluating function by using a central data store of samples of gameplay experiences.

Besides traditional evaluating function method with expert knowledge, learning method is also used to evaluate game state. Li [16] used a GA-based reinforcement learning method named ELM to calculate the optimal unit combination strategy, and building strategy. Marius [17] proposed the use of neural network method for evaluation, the spatial relationships between units were also considered in the category. Learning method can achieve good performance but need pre-learning and large dataset, and if the game parameters changes, it will be useless unless the dataset for changed parameters is provided.

Previous researches mostly use fixed factors and weights to evaluate the state, the relationships between factors are not considered. In this paper, we use hierarchical evaluating network to express factors and relations. By taking game state and player's evaluating principle into consideration, a dynamic weight generation algorithm is proposed, which can help evaluating game state more accurately.

3 Hierarchical Evaluating Networks

Evaluating algorithm is the basis of decision-making method, which can lead the planning process. There are two important aspects in evaluating game state: evaluating factors and factors' weight. This section employs a hierarchical network to manage evaluating factors and use regression algorithm to calculate the dynamic weights.

3.1. Hierarchical factor network

The evaluation of game state reflects the players' judgment of game state by taking different aspects factors into consideration. For different level issues, the evaluating factors are also different. In simple games such as Super Mario, the winning gold is the only factor which reflects state value. For RTS games, considering the complexity and the diversity of game process, the more factors are considered, the more comprehensive the judgment can be.

HTN is an automatic planning method using hierarchical approach to decompose complicated task into sub-tasks until all the sub-tasks can be executed [18]. Since the planning process is similar to human decision process, HTN algorithm achieves satisfying results in solving complex problem and is more sufficient to handle large state space than other planning algorithms [19]. The domain knowledge of HTN planning includes compound actions, atomic actions, and patterns.

Inspiring by HTN process of solving problems, we propose a hierarchical evaluating network for RTS games. Similar to HTN, HEN is a tree structure network, in which nodes can be classified as compound nodes and primitive nodes. Compound nodes represent tasks and missions, which are composed of primitive nodes and compound nodes. Primitive nodes stand for state attributes that can be directly calculated. By analysing game state from different levels and perspectives, the evaluating aspects are systematically considered. For RTS games, HTN networks are mostly divided into two or three layers [20]. Two-layer includes macro layer and micro layer, macro layer represents for player goal and global strategy and micro layer represents for teams' or units' control and actions. Three-layer includes strategy layer, tactical layer and reaction layer. Each layer corresponds to different planning level.

We use the three-layer network to structure the hierarchical evaluating network since it can describe the game state more comprehensive. The strategy layer stands for player's top strategy and winning goal. The tactical layer focuses on the player's tactical strategy, such as opponent modelling, task assignment and development strategy selection. The reaction layer is indicators of tactical task which can be calculated directly, such units' position for tactical position selection, and map's occupied information for path planning. By decomposing factors from strategy layer to reaction layer, a factor tree is constructed. In the tree, upper layer is decomposed into the lower layer.

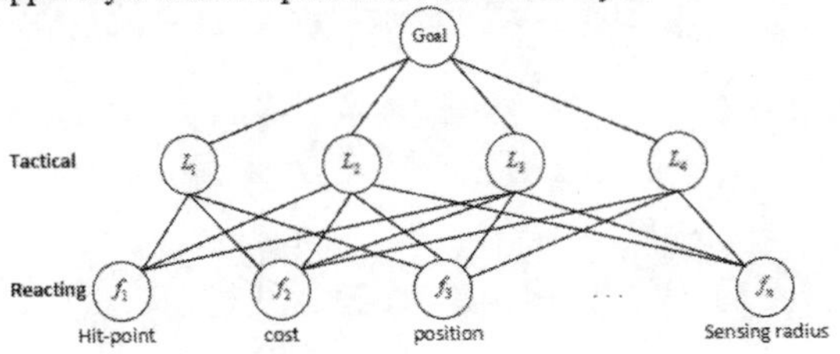

Fig 1. Illustration of a network generated by the HEN algorithm for depth 3.

As shown in Figure 1, in the HEN network, aspects are integrated and all layers can be decomposed to factors. The connection lines between indicators stand for the relationships among the indicators at each level. By constructing network between indicators, it can be found that one sub-indicator can affect multiple upper-level indicators. The evaluation of game state and factor relationships can be expressed as follow:

$$\begin{cases} E(s) = \sum_{i=1}^{n}\sum_{j=1}^{m} E_i(f_j) \\ E(f) = (factor_{father}, factor_{sub}, \text{Attribute}, Weight) \end{cases} \tag{1}$$

The evaluation of game state $E(s)$ is the summation of all aspects' scores. $E_i(f_j)$ stands for the evaluation of each factors. $factor_{father}$ stands for the father factor, $factor_{sub}$ stands for the sub factor. *Attribute* stands for the attributes of current factor. *Weight* is the current factor's weight of its father task.

Using hierarchical network to calculate the state situation, a comprehensive attention can be given to both overall and detail factors. In Table 1, the domain knowledge used in DHEN to construct the hierarchical network is listed. The tactical layer includes 4 different aspects, and the reacting layer includes 10 indicators which can be calculated.

Table 1. Domain Knowledge of DHEN.

Tactical	Reacting
Economic Military Building Sensor radius	Resources in units Resource in base Total resource Unit number Work number Light rush number Heavy rush number Building cost Building hit-point Unit position

3.2. Dynamic weight

The essence of RTS game is a two-player zero-sum game. Traditional planning algorithms usually use pre-given factor's weights to evaluate game state, which means the evaluation is fixed from game beginning to the end. But in fact, along with the game evolving, player's preference will change, which means the evaluating principle should change. In this section, we employ three dynamic indexes to describe game dynamic changes: player's strategy S, game time T, and game state *Situation*.

Player's strategy S refers to the initial strategy adopted by each player for gaming. For RTS games, it refers to the game AI used in different players, which designed by hardcode or auto-planning technology. For example, for aggressive decision makers, in the early stages of game, they prefer to create initial forces to attack; for technology decision makers, they prefer to upgrade technology. $G = (\min, \max, p)$ stands for player's goal. Different game principles can be sorted as three types: radical, conservative, balanced. For example, for radical players, they try to destroy his enemy leaving out consideration about their own loss, and their principles can be expressed as $G = \min(b)$. In Table 2 shows the principles of typical RTS AI methods:

Table 2. Typical AI methods.

AI	Strategy
Random	A random strategy AI which executes actions randomly.
Worker Rush	A hardcoded rush strategy that constantly produces workers and sends them to attack.
Light Rush	A rush AI which builds a barracks, and then constantly produces light military units to attack the nearest target (it uses one worker to mine resources).
Heavy Rush	Identical to LightRush, except for producing slower but stronger heavy units.
Monte Carlo	A standard Monte Carlo search algorithm: for each legal player action, it runs as many simulations as possible to estimate their expected reward.
AHTN	An Adversarial Hierarchical Task Network, which combines minimax game tree search with HTN planning.

Game time T affects the evaluating principle in two aspects: one is that since the total game time is given, players' strategies must consider the game time comparing with the final game time. On the other hand, even the same situation may have different effects at different times. Using an exponential decline method, the player's preference for is described as the game time increases.

The game situation *Situation* refers to the comparison of two forces. There are two levels of comparison, the relationship between your own forces and your own basic decision-making strategies. When the strength of one's own side does not meet the basic needs of resource collection, it is better to send units first to collect resources rather than attack. The second level is the comparison of the strength of the enemy and the enemy. When the strength of the enemy is far greater than the strength of one's own, it should be rather defensive rather than offensive.

Based on the above three aspects, we dynamically adjust the calculation of weights $f\_e(p,t,evaluation(s))$ as follow:

$$f\_e(p,t,evaluation(s)) = g(G(\max,\min,p) \cup \frac{t}{\max\_time}) \cdot evaluation(s) \quad (2)$$

$g(x)$ stands for the union relation between $G(\max,\min,p)$ and time proportion, $evaluation(s)$ stands for the contrast evaluation in state s.

The process of dynamic weight generating algorithm is as follow: Line 1-3 show that, player obtain the game time t and player principle p under the evaluated state. Line 4-10 show that player get the evaluation of its own evaluation(s, max) and its opponent's evaluation(s, min). Line 11-13 show the generation of evaluating weights under p, t and evaluation(s).

Algorithm 1 Dynamic Weight Generation(s)

```
1.  While state s need to be evaluated
2.      Get the game time denoted as t
3.      Get player principle denoted as p
4.      For unit in unit(s)
5.          If unit belongs to max then
6.              add unit to evaluation(s, max)
7.          Else If unit belongs to min then
8.              add unit to evaluation(s, min)
9.          End If
10.     End For
11.     dynamic_weight= f_e(p, t, evaluation(s))
12. End While
13. Return dynamic_weight
```

4 Experiments and Result

In order to verify the proposed algorithm, an empirical study based on μRTS game is carried out, and the performance of DHEN is compared to the performances of other state-of-the-art evaluating algorithms developed for RTS games.

4.1. Experiment Environment and Setting

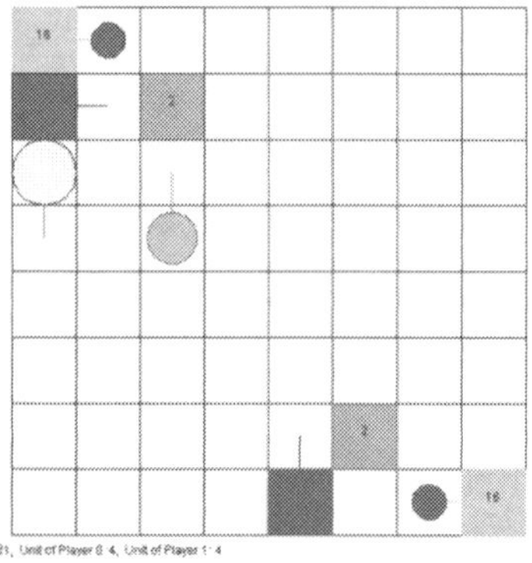

Fig 2. A screenshot of the μRTS game environment. The two players are distinguished according to the blue and red outline colors. The green squares are resources. The white squares are bases. The gray circles are workers, the yellow circles are heavy attackers, and the orange circles are light attackers.

We evaluated the performance of DHEN algorithm using the free-software μRTS (https://githubs.com/ santiontanon/ microrts), which has been used by several researchers to validate new algorithms for RTS games. Figure 2 shows a screenshot of a μRTS game, in which two players compete to destroy opponent's units.

The maps used in our experiments are three: M1 (8 × 8 tiles), M2 (12 × 12 tiles), and M3 (16 × 16 tiles). Maximum game time of M1 is limited to 3000 cycles, of M2 is 3000 cycles, and of M3 it is to 10000 cycles.

To evaluate each algorithm, we conduct a round-robin tournament, in which each algorithm plays 50 games against all other algorithms in each of 3 different maps. The method used to compute the score of each algorithm is: the winner of each game is awarded 1 point, and both algorithms are awarded 0.5 points in the event of a tie. Each of the two AI players in all competitions begins with a single base, an equivalent resource value, and a single worker.

4.2. Experimental Results and Analysis

In this section, we compare the performance of DHEN with other evaluating algorithms in terms of the average score in three maps. We compared DHEN algorithm with simple LTD evaluating function, optimal LTD evaluating function and lanchester evaluating function by applying them in AHTN planning algorithm and IDABCD planning algorithm.

Figure 3 and Figure 4 presents the comparison of the average scores obtained for using different evaluating algorithm in AHTN and IDABCD algorithm. According to the results, applied in both planning algorithm, the DHEN evaluating function can achieve better results. Since the AHTN algorithm has more approaches and depending heavier on evaluation function, the preference is better. In addition, the performance of DHEN varies little with the increasing scale of the maps. The performances of both players deteriorate in map M1 because the map is relatively smaller, so the difference between optimal action and normal action is not obvious for results. With respect to other algorithms, we note that Lanchester evaluation function and optimal evaluation function do not achieve an anticipant performance as it performed in MCTS algorithm. But both algorithms perform better in AHTN algorithm, reveals that the optimal parameters is changing when applied in different planning algorithms.

Table 3. Average Scores of each HTN algorithm with different domain knowledge.

Planning Algorithm	Simple eval.	Optimized eval.	Lanche ster	DHE N
IDABCD	81.6	80.3	73.3	84.3
AHTN	78.3	73	79.3	89

Table 3 shows the average scores of each evaluating algorithm applied in IDABCD and AHTN. We can find that the DHEN algorithm applied in IDABCD improves win rate about 3% compared with simple evaluation function, which performs best among the three benchmark algorithms. The performance of DHEN applied in AHTN brings an improvement of winning rate about 12% compared with the lanchester evaluating function. We can find that, compared with planning process, the evaluation time cost little time.

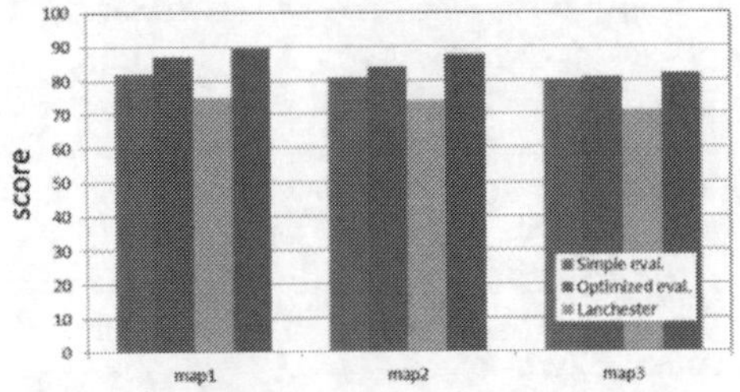

Fig 3. The average score of each evaluation used in AHTN algorithm for map M1, M2, M3 with respect to the CPU time 100 ms.

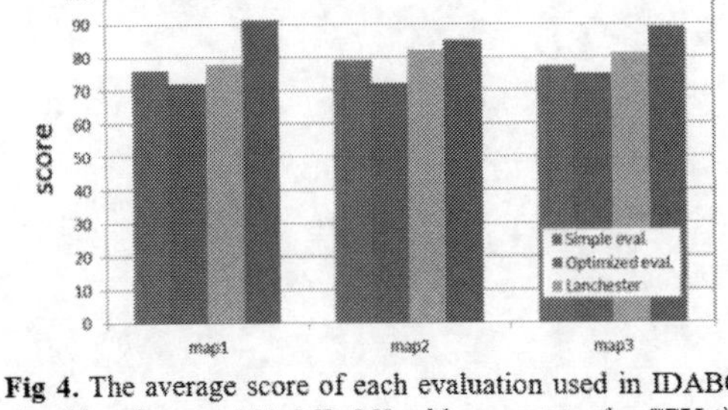

Fig 4. The average score of each evaluation used in IDABCD algorithm for map M1, M2, M3 with respect to the CPU time 100 ms.

Figure 5 shows the average decision times of four evaluating functions. Since the AHTN algorithm is flexible than IDABCD, it needs more decision time. We find that DHEN cost almost the same time as simple LTD evaluating function and optimal LTD evaluating function, because they are calculated according to the same evaluating structure. Since lanchester evaluating function is an iterative process, it needs more time. The DHEN algorithm increase the decision time about 10% compared with simple/optimal LTD algorithm. Therefore, the modification in DHEN brings little time increment compared to LTD evaluation algorithm, which is within an acceptable range.

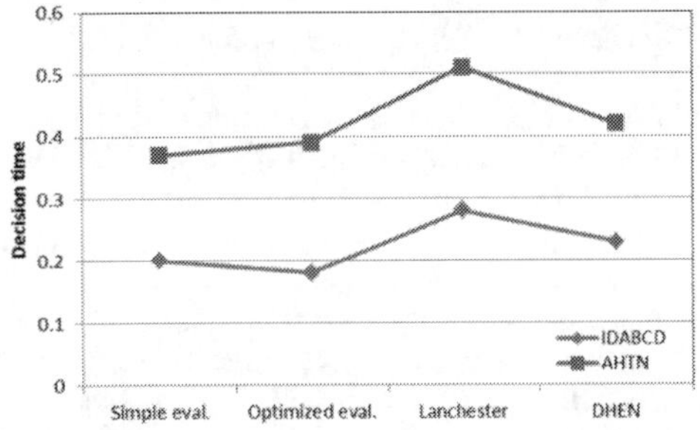

Fig 5. The average decision times of DHEN applied in IDABCD and AHTN.

5 Conclusions

In summary, we have performed both experimental and theoretical study of evaluating RTS game state. A dynamic hierarchical evaluation network is proposed to handle the evaluation problem accompany with the planning process. The experimental results in μRTS game successfully verify the algorithm's validation.

In future work, the DHEN algorithm can be extended in multiple directions. Our experiments demonstrate that domain knowledge of the HEN has a significant influence on the performance of DHEN. However, encoding perfect evaluation network is a difficult and time-consuming process. The automatic extraction of HTN domain knowledge from thousands of RTS game replays may provide an efficient approach.

In addition, we note that using deep learning algorithm to calculate the weights of factors may improve the veracity of planning. Therefore, a modified DHEN algorithm using learning method may enhance its performance in huger game environment.

Acknowledgment

The work described in this paper is sponsored by the National Natural Science Foundation of China under Grant No. 61473300.

References

1. Buro, M. Call for AI research in RTS games. In Proceedings of the AAAI-04 Workshop on Challenges in Game AI (2004), pp.139-142.
2. Ontañón, S., Synnaeve, G., Uriarte, A., Richoux, F., Churchill, D., & Preuss, M. (2013). A survey of real-time strategy game ai research and competition

in starcraft. IEEE Transactions on Computational Intelligence & Ai in Games, 5(4), pp. 293-311.

3. Ontañón, S. Experiments with game tree search in real-time strategy games. *Artif. Intell.* 2012, arXiv:1208.1940.
4. Chung, M., Buro, M., & Schaeffer, J. (2005). Monte Carlo Planning in RTS Games. IEEE Symposium on Computational Intelligence and Games (pp.117-124). DBLP.
5. Balla, R.K.; Fern, A. UCT for tactical assault planning in real-time strategy games. In *Proceedings of the 21st International Jont Conference on Artifical Intelligence*; AAAI Press Palo Alto,, CA, USA, 2009; pp. 40–45.
6. Churchill, D.; Saffidine, A.; Buro, M. Fast Heuristic Search for RTS Game Combat Scenarios. In Proceedings *of the Artificial Intelligence and Interactive Digital Entertainment (AIIDE)*; AAAI Press: Palo Alto, CA, USA, 2012, pp. 112-117
7. Ontañón, S. The combinatorial multi-armed bandit problem and its application to real-time strategy games. In *Proceedings of 9th Artificial Intelligence and Interactive Digital Entertainment Conference (AIIDE)*; AAAI Press: Palo Alto, CA, USA, 2013.
8. Shleyfman A, Komenda A, Domshlak C. On Combinatorial Actions and CMABs with Linear Side Information[J]. Frontiers in Artificial Intelligence & Applications, 2014, 263:825-830.
9. Buro, M. (2015). Adversarial hierarchical-task network planning for complex real-time games. International Conference on Artificial Intelligence (Vol.8, pp.1652-1658). AAAI Press.
10. Kovarsky, A., & Buro, M. (2005). Heuristic search applied to abstract combat games. Lecture Notes in Computer Science, 3501, pp. 66-78.
11. TungDucNguyen, KienQuangNguyen, & Ruck, T. (2015). Heuristic search exploiting non-additive and unit properties for rts-game unit micromanagement. Journal of Information Processing, 23(1), pp. 2-8.
12. Stanescu, M., Hernandez, S. P., Erickson, G., Greiner, R., & Buro, M. (2013). Predicting army combat outcomes in StarCraft. AAAI Conference on Artificial Intelligence and Interactive Digital Entertainment (pp.86-92). AAAI Press.
13. Erickson, G., & Buro, M. (2014). Global state evaluation in StarCraft. Tenth AAAI Conference on Artificial Intelligence and Interactive Digital Entertainment (pp.112-118). AAAI Press..
14. Uriarte, A., & Ontañón, S. (2014). Game-tree search over high-level game states in RTS games. AIIDE. pp.73-79.
15. Bakkes, S., Spronck, P., & Herik, J. V. D. (2013). Phase-dependent evaluation in rts games.
16. Li, Y. J., Ng, P. H. F., & Shiu, S. C. K. (2015). A fast evaluation method for rts game strategy using fuzzy extreme learning machine. Natural Computing(3), pp. 1-13.
17. Stanescu, M., Barriga, N. A., Hess, A., & Buro, M. (2017). Evaluating real-time strategy game states using convolutional neural networks. Computational Intelligence and Games. IEEE.
18. Malik Ghallab, Dana Nau, Paolo Traverso. Automated Planning: Theory and Practice. 2008. Elsevier(Singapore) Pte.Ltd.
19. Sacerdoti, E. D. (1975). The nonlinear nature of plans. International Joint Conference on Artificial Intelligence (pp.206-214). Morgan Kaufmann Publishers Inc.
20. David Churchill. [D]. Heuristic Search Techniques for Real-Time Strategy Games. University of Alberta. 2016.

MATEC Web of Conferences **208**, 05005 (2018) https://doi.org/10.1051/matecconf/201820805005
ICMIE 2018

Trust Evaluation Model for P2P Networks based on Time and Interaction

KE Xuemeng[1], *ZHOU* Guofu[1, 2], *DU* Zhoumin[2]

[1]State Key Laboratory of Software Engineering, Wuhan University, China
[2]Computer School, Wuhan University, China

Abstract. The traditional centralized trusting mechanism does not meet the requirements of the modern P2P network, so it is necessary to establish a distributed trusting mechanism to strengthen the system's reliability. Factors including time attenuation factor, interactive frequency factor, interactive size factor and average online time factor markedly influence the trust of the nodes in a P2P network. This paper defines the precise effects of these factors on trust and proposes a comprehensive trust model and global trust model based on direct trust and indirect trust.

1 Background

P2P networks have become the focus of the Internet due to their application in file distribution, streaming media transmission, distributed computing, and more [1]. The sharp increase in the number of P2P network users and the openness of P2P networks themselves has created an increasing complexity of trust between network nodes. In 1994, Marsh first extended the "trust relationship" concept from the social network to the computer network and research on trust management in computer networks began in earnest [2]. The traditional centralized trusting mechanism does not meet P2P network requirements, so it is necessary to establish a new distributed trusting mechanism to build trust relationships between nodes, as the basis of the interaction among them.

The interactive success rate of nodes that select interactive objects with higher trust is higher. The block chain[3][4], a relatively new P2P network application, is a mode of distributed data storage [5], a point-to-point transmission consensus mechanism, and an encryption algorithm [6], The consensus mechanism is a method of building trust between nodes and accessing rights in a block chain system [7].

2 Related Work

Many P2P-based trust models have been proposed in recent years. EigenTrust [8][9], for example, is one of the first models of dynamic computing the global trust value of nodes. In EigenTrust, the local trust value of a node is calculated after its interactions; the results are fed back to calculate the global trust value of the node. When applied to a large-scale network, EigenTrust suffers poor scalability, inappropriate convergence, and high computational complexity. FCTrust [10] also uses the feedback concept, under which the trust of nodes is determined by interaction frequency. However, it equally assigns weight to all the transactions instead of giving more importance to recent ones in calculating local trust, resulting in an inaccurate evaluation of the target node.

Based on the above literature review, we examined the factors which significantly influence the trust among nodes including time attenuation [11], interactive frequency, interactive size and average online time. We established a novel model of trust evaluation based on time and interaction accordingly, discussed in detail below.

3 The Trust Factor

3.1 Time Attenuation Factor

Definition 1 (Time Attenuation Factor): The attenuation of the influence of nodes' interaction history on its trust over time is represented by h and calculated as follows:

$$\rho^h = \min\left(\rho_e^{|t_h - t_n|}, 1\right) \quad (1)$$

where tn is for the current moment; th is for the historical moment of interaction. The closer to the current moment the moment of historical interaction this, more influence the interaction has on the trust evaluation values. Conversely, there is less impact to the trust evaluation farther from the current moment[12][13]. The EigenTrust 89 model ignores the impact of time attenuation on trust calculation and thus is not sufficiently objective. In our model, the time attenuation factor is introduced into the calculation of each trust factor.

ρ_e is the base of the exponential attenuation function. ρ_e takes the value e, i.e., ρ_e=e. The time attenuation is always less than or equal to 1. The time attenuation factor is greater, approaching to 1 when the moment of historical interaction is closer to the current time. When the moment of historical interaction occurred a long time ago, the time attenuation factor value approaches infinitesimal. Time attenuation factor h is represented by ρ^h_{sat} and ρ^h_{unsat} according to a successful or failing result, respectively.

ρ^h_{sat}: The time attenuation factor in the case of a successful interaction. In this case, t_h represents the historical moment of a successful interaction.

ρ^h_{unsat}: The time attenuation factor in the case of a failing interaction. In this case, th represents the historical moment of a failing interaction.

The moment of historical interaction has a great influence on trust value. The effect of historical interaction between nodes on the trust evaluation values attenuates gradually as time goes by, and this attenuation is not linear. When the historical interaction occurred a long time ago, the influence on trust shrinks towards zero[14]. The difference between the moments of historical interaction and the current moment is rounded down unlike the calculation method of time attenuation factor in previous studies[14]. Each time attenuation factor is matched with a corresponding time period. The time attenuation factor is h when the interaction occurs within the period from $\lfloor t_h - t_n \rfloor$ to $(\lfloor t_h - t_n \rfloor + 1)$.

3.2 Interactive Frequency Factor

Definition 2 (Interactive Frequency Factor): The interactive evaluation factor C_{ij} measures the effect of successful interaction frequency and failing interaction frequency on trust degree. It is calculated as follows:

$$s_{ij} = sat(i,j) - unsat(i,j), \tag{2}$$

$$C_{ij} = \frac{\max(s_{ij}, 0)}{\sum_j \max(s_{ij}, 0)}. \tag{3}$$

Based on the time attenuation factor, the definition of $sat(i,j)$ and $unsat(i,j)$ is :

$$sat(i,j) = \sum_{k=1}^{n} vs^k_{ij} \cdot \rho^h_{sat}, \tag{4}$$

$$unsat(i,j) = \sum_{k=1}^{n} vf^k_{ij} \cdot \rho^h_{unsat}. \tag{5}$$

sat(i,j) represents the number of successful interactions between node i and node j having taken time attenuation factor into consideration. Node i is a requesting node and node j is a requested node. In sat(j,i), node j is a requesting node and node i is a requested node. Therefore, sat(i,j) is different from sat(j,i), and their values may differ.

unsat(I,j) represents the number of failing interactions between node i and node j with time attenuation factor taken into consideration. Node i is a requesting node and node j is a requested node. In the same way, unsat(i,j) is different from unsat(j,i).

k represents a period from $\lfloor t_h - t_n \rfloor$ to $(\lfloor t_h - t_n \rfloor + 1)$ approaching the current moment; k has the same meaning in the post.

vs^k_{ij} represents the number of successful interactions between node i and node j within the period k. vf^k_{ij} represents the number of failing interactions between node i and node j within the period of k.

In order to avoid unfairness to new nodes and to distinguish new nodes from bad nodes, node i is considered to be a new node if:

$$\sum_j (sat(i,j) + unsat(i,j)) = 0. \tag{6}$$

Because the new nodes have no interaction with any other node, the trust evaluation of new nodes to other nodes can only be evaluated by the indirect trust degree and the attributes of the nodes. No node has interacted with a new node in the network, so each new node is given an initial global trust $GTrust_{new}$ which is equal to the global trust value of its invitation node, that is to say, the higher the global trust of its invitation node, the higher the initial global trust that the new node is given. If

$$\sum_j (sat(i,j) + unsat(i,j)) \neq 0, \tag{7}$$

Node i is not a new node. The interactive evaluation factor for node i and node j is calculated according to (2) and (3).

3.3 Interactive Size Factor

Definition 3 (Interactive Size Factor): The interactive size factor measures the impact of interactive data volume on trust degree between node. It is denoted as D_{ij} and calculated as follows:

$$data_{ij} = data_{sat}(i,j) - data_{unsat}(i,j), \tag{8}$$

$$D_{ij} = \frac{\max(data(i,j), 0)}{\sum_j \max(data(i,j), 0)}. \tag{9}$$

where data(i,j) represents the interactive data volume between node i and node j. $data_{sat}(i,j)$ represents the data volume of successful interaction between node i and node j. $data_{unsat}(i,j)$ represents the data volume of failing interaction between node i and node j.

The time attenuation factor is also calculated during calculation of the interactive data volume. The definitions of $data_{sat}(i,j)$ and $data_{unsat}(i,j)$ are:

$$data_{sat}(i,j) = \sum_{k=1}^{n} satdata_{ij}^{k} \cdot \rho_{sat}^{h} \quad (10)$$

$$data_{unsat}(i,j) = \sum_{k=1}^{n} unsatdata_{ij}^{k} \cdot \rho_{unsat}^{h} \quad (11)$$

$satdata_{ij}^{k}$ represents the data volume of successful interaction between node i and node j within period k. $unsatdata_{ij}^{k}$ represents the data volume of failing interaction between node i and node j within period k.

The interactive size factor can prevent malicious nodes from performing well in smaller-sized interactions, but cheating in larger interactions while the trust is still good. The EigenTrust89 model ignores the impact of interactive size on trust calculation.

3.4 Average Online Time Factor

The node is more likely to be online in the future, and considered to be generally more reliable, when it has a longer average online time. The average online time factor reflects the influence of average online time on trust. Again, the EigenTrust89 model ignores the impact of time factor on trust calculation.

Definition 4 (Average Online Time Factor): Average online time factor measures the impact of the node's average online time on its trust degree. It is represented by Tj and calculated as follows:

$$T_j = \frac{t_j}{N_j} \quad (12)$$

t_j represents the total online time of node j. Nj represents the number of times that node j has gone online.

4 Trust Model

A P2P network can be modeled as a directed graph G={V,E}. The vertex set is represented by V={P1,P2,⋯,Pn} where n represents the number of nodes in the network. The edge set is represented by E={(Pi,Pj)|i,j=1,2⋯,n}, where $E \subseteq V \times V$. e_{ij} is the weight of edge (Pi,Pj) calculated by the number of interactions between Pi and Pj having taken the time attenuation into consideration. Pi is a requesting node and Pj is a requested node. It is defined as follows:

$$e_{ij} = sat(i,j) + unsat(i,j) \quad (13)$$

We also introduced the concept of a set Qj into the trust model, where Qj represents a collection of nodes that have interactions with P_j.The definition of set Qj is:

$$Q_j = \{P_k \in V | e_{kj} > 0\} \quad (14)$$

4.1 Direcct Trust

Definition 5 (Direct Trust): Direct trust[15], also referred to as local trust, is represented by DTrust(i,j). It is the trust degree that P_i has established over the history of direct interaction with P_j.

The following penalty function is introduced into the direct trust model:

$$pen(i,j) = -\frac{\sum_i DTrust(i,j)}{|Q_j|} \times \frac{unsat(i,j)}{sat(i,j)} \quad (15)$$

Pen(i,j) is the penalty function of node i to node j. unsat(i,j) is the number of failing interactions between node i and node j, and sat(i,j) is the number of successful interactions between node i and node j. The more the node fails to interact and the less it succeeds, the greater the punishment.

When calculating direct trust, EigenTrust89 adopts an iterative approach, which requires excess computation resources. FCTrust10 equally assigns weight to all the factors, which obscures the importance of each factor.

The greater the interactive frequency factor, the greater the interactive size factor, the greater the average online time factor, and the greater the direct trust value of this node, here, we built a direct trust model by multiplying these factors. The mathematical definition of direct trust of Pi to Pj is :

$$DTrust(i,j) = C_{ij} \times D_{ij} \times T_j + pen(i,j)\frac{1}{1+\rho_e^{-n}} \quad (16)$$

$\frac{1}{1+\rho_e^{-n}}$ represents the accelerating factor and $\rho_e = e$. When the node's behavior becomes malicious, the acceleration factor reduces the trust value rapidly. When the node's behavior becomes normal, the trust value slowly increases.

4.2 Indirect Trust

Definition 6 (Indirect Trust): Indirect trust[16], also referred to as recommendation trust, is represented by ITrust(i,j); it is the trust degree that nodes build through indirect interactive history.

When Pi has no direct interaction or very few direct interactions with P_j, the trust degree of Pi to P_j .depends on the recommendations from other nodes that have direct interaction with P_j.

For any $P_k \in Q_j$, the direct trust degree of P_k to P_j can be calculated according to (16). The definition of indirect trust ITrust(i,j) of Pi to Pj:

$$ITrust(i,j) = \frac{\sum_{P_k \in Q_j} DTrust(k,j)}{|Q_j|} \quad (17)$$

4.3 Comprehensive Trust

Definition 7 (Comprehensive Trust): Comprehensive trust is the trust established by direct trust and indirect trust. The comprehensive trust of P_i to P_j is represented by STrust(i,j) and calculated as follows:

$$q_1 = \frac{e_{ij}}{\sum_i e_{ij}}, \quad (18)$$

$$q_2 = 1 - \frac{e_{ij}}{\sum_i e_{ij}}, \quad (19)$$

$$STrust(i,j) = q_1 \cdot DTrust(i,j) + q_2 \cdot ITrust(i,j) \quad (20)$$

q_i, the weight of STrust(i,j) is a dynamic value. q_i is related to the frequency of interaction between Pi and Pj.

The more frequently Pi interacts with Pj, the greater q_1 is, the smaller q_2 is and the more important the direct trust is. The less frequently Pi interacts with Pj, the greater q_2 is, the smaller q_1 is and the more important the indirect trust is.

$e_{ij} = 0$ indicates that Pi does not interact with P_j. Therefore $q_1 = 0$ and $q_2 = 1$. The trust value to P_j is measured by the indirect trust.

When Pi only interacts with Pj and has no interaction with other nodes, then $q_1 = 1$ and $q_2 = 0$. The trust value is measured by the direct trust.

4.4 Global Trust

Definition 8 (Global Trust): To measure the trust of each node based on the entire network, the global trust of Pi is calculated according to the comprehensive trust of other nodes in the network to P_i. The global trust degree of P_i is represented by GTrust(i).

Definition 9 (Approve): When the comprehensive trust degree of Pi to Pj exceeds a certain value, Pj is approved by Pi and denoted by R(i,j)=true. Otherwise, Pj is not approved by Pi and denoted by R(i,j)=false.

When STrust(i,j)> λ , P_j is approved by Pi and denoted by R(i,j)=true. When STrust(i,j)< λ ,P_j is not approved by Pi and denoted by R(i,j)=false.

The λ variable mentioned above is an approbation threshold which distinguishes good nodes from bad nodes. If the threshold is too high, the global trust of nodes in the network is generally low. Conversely, if the threshold is too low, the global trust is generally high and good nodes may not be sufficiently distinguishable from bad nodes. We set λ =0.2 in our experiment.

The greater the proportion of nodes that recognize this node in the network, the higher the global trust degree the node has; therefore the global trust degree of P_i can be calculated by the percentage of the nodes approve P_i in the entire network. The global trust degree has global uniqueness.

A node set W_i is defined to represent the set of nodes that approve P_i in the network. Definition of W_i is shown in

$$W_i = \{P_k \in V | R(k,i) = true\} . \quad (21)$$

The global trust of P_i is calculated by the percentage of the approve nodes with all nodes in the network. The global trust degree is defined as follows:

$$GTrust(i) = \frac{|W_i|}{|V|} . \quad (22)$$

5 Experiment And Analysis

We designed several nodes to simulate the application scenarios of a P2P network in order to verify the proposed trust evaluation model. The historical interaction records of nodes served as inputs to calculate the trust degree.

In the number of initial nodes is too large, the calculation is may be overly complex and the result may not be conducive to observation and comparison. If the number of nodes is too small, the advantages of the model are not effectively reflected. We set the number of nodes to 10 in our experiment. Node 5 and node 10 are bad nodes which provide one bad service for every two good services. Other nodes are normal nodes which consistently provide good service. In the P2P network, all shared files are randomly distributed evenly. The nodes in the network randomly download or upload files from other nodes. During the experiment, online status of nodes, number of interactions, requesting node, requested node, volume of interactive data and results of interaction were recorded. These records were then used as input data to calculate the direct trust, indirect trust, comprehensive trust and global trust of each node in the network. For the sake of comparison, we choose a normal node and a bad node for separate recordings of changes in global trust over defined time period. We also selected a node to record the attenuation of the node global trust degree as time progressed.

Figure 1 shows the global trust between nodes after a period of interaction as-calculated using the proposed model and EigenTrust, respectively. Node 5 and node 10 showed the lowest global trust degree, as each provided bad service once time. We found it much easier to distinguish the good nodes from the bad nodes using the proposed model compared to EigenTrust.

Figure 2 shows the changes in global trust degree of nodes 3 and 5. One period is defined as the end of the previous service to the completion of the present service. Node 3 is a normal node and always provides good service; node 5 provides two good services followed by a bad one. Node 3's global trust degree continually increased over the observation period, while the global trust of node 5 fluctuated. When providing a good service, node 5's global trust rebounded to some extent before decreasing when it subsequently provided bad service. The amplitude of its decline was always greater than the amplitude of increase due to the penalty function. When the node is behaving badly, the node is punished and the trust degree quickly drops. When node's behavior gets better, its trust degree increases, but slowly.

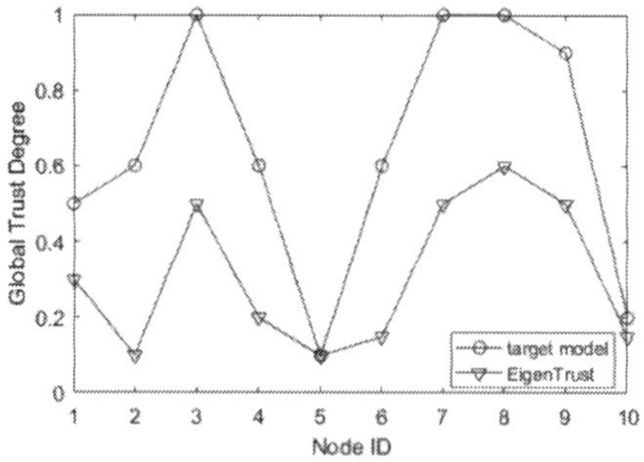

Fig 1. Global trust degree.

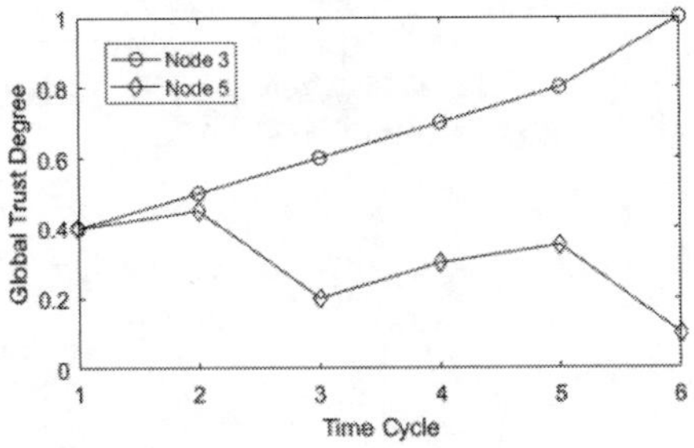

Fig 2. Changes in global trust after interaction.

Figure 3 shows the change curve of the global trust degree of node 3. The closer to the current moment, the greater impact the interaction of nodes on trust degree. The farther from the current moment, the smaller the impact the interaction has on the trust degree. Node 3 appears to provide good service in the previous interaction and has a high level of trust. Once node 3 has not provided any service for a long time, its trust degree drops off.

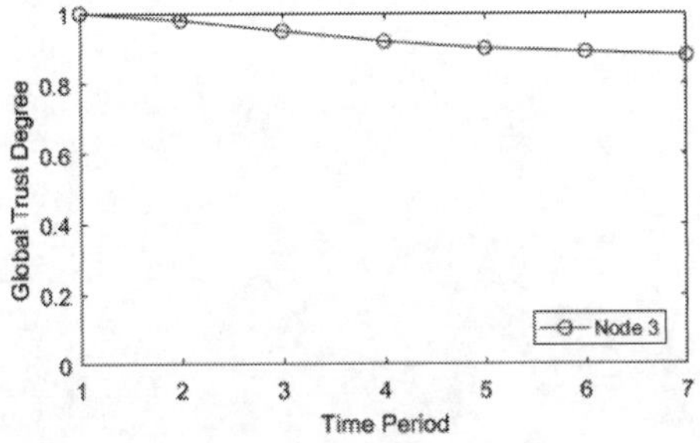

Fig 3. Changes in global trust over time.

6 Conclusion

There is no third-party authentication in the P2P network, so it is necessary to judge and select appropriates node with which to interact to ensure system reliability. This paper proposed a trust evaluation model based on time and interaction which resolves trust problems in P2P networks. Multiple factors including the time attenuation factor, interactive frequency factor, interactive size factor and average online time factor were introduced to calculate the direct trust degree and indirect trust degree respectively. The comprehensive trust degree and global trust degree can be obtained according to the direct trust degree and indirect trust degree.

In the future, we plan to apply the trust evaluation model in actual P2P networks, including the block chain. We will choose appropriate trust nodes for effective and safe interactions to download and upload files, and thus realize distributed storage and access for files in the P2P network.

References

1. HuiGuo Dong. The application and implementation of search technique based on p2p networks. Computer Knowledge and Technology, 2010.
2. Stephen Paul Marsh.Formalising trust as a computational concept. University of Stirling, 1994.
3. Aaron Wright and Primavera De Filippi. Decentralized blockchain technology and the rise of lex cryptographia. Social Science Electronic Publishing, 2015.
4. Xianmin Yang, L. I. Xin, W. U. Huanqing, and Keyun Zhao. The application model and challenges of blockchain technology in education. Modern Distance Education Research, 2017.
5. Hoang Giang Do and Wee Keong Ng. Blockchain-based system for secure data storage with private keyword search. In IEEE World Congress on Services, pages 90–93, 2017.
6. Yuqin Xu, Qingzhong Li, Xingpin Min, Lizhen Cui, Zongshui Xiao, and Lanju Kong. E-commerce blockchain consensus mechanism for supporting high-throughput and real-time transaction. 2016.
7. Yong Yuan and Feiyue Wang. The development status and prospect of blockchain technology. Automated journal, 42(4):481–494, 2016.
8. Sepandar D. Kamvar, Mario T. Schlosser, and Hector Garcia-Molina. The eigentrust algorithm for reputation management in p2p networks. In International Conference on World Wide Web, pages 640–651, 2003.
9. Kenli Li, Yan He, Xiaoling Liu, and Ying Wang. Security-driven scheduling algorithms based on eigentrust in grid. In International Conference on Parallel and Distributed Computing, Applications and Technologies, pages 1068–1072, 2005.
10. Jianli Hu, Quanyuan Wu, and Bin Zhou. Fctrust: A robust and efficient feedback credibility-based distributed p2p trust model. pages 1963–1968, 2008.
11. Blackburn and Joseph. The time factor. National Productivity Review, 9(4):395–408, 1991.
12. Limin Tao. Research on Subjective Trust Management Based on Uncertainty Theory in Open Network Environment. PhD thesis, Zhejiang University of Technology, 2013.
13. Yu Bao, Guosun Zeng, Liansun Zeng, Bo Chen, and Wei Wang. A measure of trust in p2p networks to prevent deception. Journal of communication, 29(10):215–222, 2008.
14. Yanxia Cui. Study and Improvement of EigenTrust Trust Model Based on P2P Network. PhD thesis, Tianjin university of technology, 2016.
15. Kiefhaber R, Jahr R, Msadek N, et al. Ranking of Direct Trust, Confidence, and Reputation in an Abstract System with Unreliable Components[C]// Ubiquitous Intelligence and Computing, 2013 IEEE, International Conference on and, International

 https://doi.org/10.1051/matecconf/201820805005

Conference on Autonomic and Trusted Computing. IEEE, 2014:388-395.

16. Barsoum A, Hasan A. Enabling Dynamic Data and Indirect Mutual Trust for Cloud Computing Storage Systems[J]. IEEE Transactions on Parallel & Distributed Systems, 2013,24(12):2375-2385.

AUTHOR INDEX